Thermal Management in Lithium-Ion Batteries: Latest Advances and Prospects

Thermal Management in Lithium-Ion Batteries: Latest Advances and Prospects

Guest Editors

Xianglin Li
Chuanbo Yang
Prahit Dubey

MDPI

Basel • Beijing • Wuhan • Barcelona • Belgrade • Novi Sad • Cluj • Manchester

Guest Editors

Xianglin Li
Department of Mechanical Engineering and Materials Science
Washington University in St. Louis
St. Louis
USA

Chuanbo Yang
Energy Conversion and Storage Systems Center
National Renewable Energy Laboratory
Golden
USA

Prahit Dubey
Thermal Systems and Software
General Motors
Milford
USA

Editorial Office
MDPI AG
Grosspeteranlage 5
4052 Basel, Switzerland

This is a reprint of the Special Issue, published open access by the journal *Batteries* (ISSN 2313-0105), freely accessible at: https://www.mdpi.com/journal/batteries/special_issues/36WM4V34VD.

For citation purposes, cite each article independently as indicated on the article page online and as indicated below:

Lastname, A.A.; Lastname, B.B. Article Title. *Journal Name* **Year**, *Volume Number*, Page Range.

ISBN 978-3-7258-5693-0 (Hbk)
ISBN 978-3-7258-5694-7 (PDF)
https://doi.org/10.3390/books978-3-7258-5694-7

Contents

Preface

This Special Issue explores the latest advancements in thermal management strategies for batteries. It includes a broad range of studies that perform the following:

- Review cutting-edge technologies and challenges in understanding the thermal behaviors of lithium-ion batteries (LIBs) and battery packs.
- Investigate battery aging due to temperature variations and analyze the thermal responses of batteries under various stress conditions.
- Develop innovative system designs and detection techniques for battery thermal management (BTM) systems.
- Highlight the integration of phase change materials (PCMs) with heat pipes, showcasing enhanced effectiveness in BTM systems.

This compilation of work provides comprehensive insights and foster innovation in the ongoing development of BTM solutions.

Xianglin Li, Chuanbo Yang, and Prahit Dubey
Guest Editors

batteries

MDPI

Editorial

Thermal Management in Lithium-Ion Batteries: Latest Advances and Prospects

Xianglin Li [1,*], Chuanbo Yang [2] and Prahit Dubey [3]

1 Department of Mechanical Engineering and Materials Science, Washington University in St. Louis, St. Louis, MO 63130, USA
2 National Renewable Energy Laboratory, Golden, CO 80401, USA; chuanbo.yang@nrel.gov
3 General Motors, Detroit, MI 48243, USA; prahit.dubey@gm.com
* Correspondence: lxianglin@wustl.edu

We are excited to present a Special Issue (SI) for *Batteries* on battery thermal management systems (BTMS). This SI aims to address the evolving demands of the field and foster insights into effective thermal management strategies. This SI includes 10 papers that review state-of-the-art technologies, characterize the thermal behaviors of lithium-ion batteries (LIB) and battery packs, and design new BTMS. Several papers have reviewed state-of-the-art technologies, challenges, and perspectives.

- Ahmadian-Elmi and Zhao [1] evaluated thermal management strategies for cylindrical Li-ion battery packs. They assessed the performance, efficiency, cost, and applicability of air cooling, liquid cooling, phase change material (PCM)-based management, and hybrid thermal management, discussing both strengths and limitations.
- Tai and Lee [2] focused on BTM systems for electric vehicles (EVs), emphasizing thermal runaway prevention and suppression. They reviewed publications from 2020 to 2025, covering indirect liquid cooling, water mist cooling, immersion cooling, PCM cooling, and hybrid cooling techniques. The review analyzed mechanisms, effectiveness, and practicality, offering future design directions for next-generation EV BTMS.
- Ortiz et al. [3] discussed single- and multi-phase cooling technologies, advanced materials, structures, sensors, models, and numerical simulations. They highlighted the use of machine learning (ML) for detecting and predicting battery thermal issues and analyzed the challenges of BTM under ultra-fast charging and low-temperature operations.

Several papers characterized the thermal behaviors of lithium-ion batteries (LIB) and battery packs, our understanding of battery aging due to temperature gradient, and thermal responses of batteries under different abuse conditions.

- Saxon et al. [4] studied commercial LIB using calorimetry, examining the impact of temperature, C-rates, and formation cycles, and noted additional heat generation through module interconnects, leading to about 20% more heat per cell compared to standalone cells at application currents of around 8C. Their testing of a 5 kWh battery pack revealed high temperature non-uniformity (12 °C) due to insufficient cooling. The study proposes a bottom-up approach to integrate battery thermal characterization with multi-domain models to avoid costly module prototype testing.
- Yao et al. [5] experimentally compared the thermal runaway (TR) of LIB triggered by overheating versus overcharging. They found different TR mechanisms, suggesting varied venting gas composition and voltage behavior. The study recommended safety

https://doi.org/10.3390/batteries11090335

strategies incorporating material-specific modifications and system-level controls to prevent TR due to these factors.

- Iriyama et al. [6] investigated the degradation of liquid-cooled LIB packs caused by non-uniform temperature distributions. They observed slower degradation in cells at the center compared to those near the cooling plates, attributing faster degradation to larger temperature gradients at the sides. Post-mortem analyses identified anode degradation as the main contributor. Interestingly, cells in liquid-cooled packs showed capacity recovery after low C-rate performance tests and extended rest, likely aiding lithium diffusion and redistribution in the anode.

Several studies focus on system designs and detection in BTM systems. Experimental and modeling studies integrate PCMs with heat pipes to improve the effectiveness of BTM systems.

- Sharifi et al. [7] developed a hybrid BTMS prototype integrating PCM with copper-water heat pipes. This novel design eliminates direct contact between PCM and the battery to reduce design complexity and improve BTMS effectiveness. The performance of the BTMS was assessed experimentally and numerically simulated using a 3D ANSYS-FLUENT 2023 model.
- Jia et al. [8] integrated Extended Kalman Filter with a 1D electro-thermal aging coupled model to estimate the real-time core temperature, state of charge (SOC), and capacity of a cylinder cell using surface temperature and voltage measurements. This model framework showed improved accuracy over existing electro-thermal models when batteries experienced high-rate cycling or aging-induced capacity degradation.
- Rahmani et al. [9] created a CFD model to study the effectiveness of hybrid BTMS on a 20-cell battery pack under 5C discharge and charge cycles. The BTMS contained a porous media (aluminum foam, copper foam, silicon carbide, aluminum oxide, or graphite) surrounding the batteries, allowing airflow to dissipate heat. Graphite demonstrated the best performance for maintaining low maximum cell temperature and temperature difference.
- Oyewola et al. [10] applied CFD models to compare air-cooled BTMS flow channel shapes with one or two outlets. Step-like plenum designs reduced the maximum temperature by 3.5K compared to the original Z-shape design. Plenum designs with two outlets showed a smaller pressure drop (resulting in a pressure drop reduction ranging from 3.66 to 5.91 Pa at an airflow rate of 3 m/s) than the Z-shape design.

Continued innovation, comprehensive research and development, and demonstrations of cost-effective prototypes are essential for advancing BTMS technologies. The discoveries and insights presented in these 10 papers help pave the way for safer and more efficient energy storage solutions. The necessity of preventing thermal runways and ensuring optimal battery operation highlights the critical need for ongoing development in this field.

We extend our sincere thanks to all the authors for their significant contributions, the editors for their diligent manuscript handling, and the anonymous reviewers for their valuable feedback. This SI could not have been possible without all your contributions. We encourage readers to engage with these publications, benefit from the studies, and propel the field forward together.

Conflicts of Interest: The authors declare no conflict of interest.

References

1. Ahmadian-Elmi, M.; Zhao, P. Review of Thermal Management Strategies for Cylindrical Lithium-Ion Battery Packs. *Batteries* **2024**, *10*, 50. [CrossRef]

2. Tai, L.D.; Lee, M.-Y. Advances in the Battery Thermal Management Systems of Electric Vehicles for Thermal Runaway Prevention and Suppression. *Batteries* **2025**, *11*, 216. [CrossRef]
3. Ortiz, Y.; Arévalo, P.; Peña, D.; Jurado, F. Recent Advances in Thermal Management Strategies for Lithium-Ion Batteries: A Comprehensive Review. *Batteries* **2024**, *10*, 83. [CrossRef]
4. Saxon, A.; Yang, C.; Santhanagopalan, S.; Keyser, M.; Colclasure, A. Li-Ion Battery Thermal Characterization for Thermal Management Design. *Batteries* **2024**, *10*, 136. [CrossRef]
5. Yao, Y.; Liu, L.; Gu, J.; Xing, H.; Liu, H.; Cheng, Y.; Wang, Y.; Yue, S.; Qiu, Y.; Zhang, Z. Characteristic Differences of Thermal Runaway Triggered by Overheating and Overcharging in Lithium-Ion Batteries and Multi-Dimensional Safety Protection Strategies. *Batteries* **2025**, *11*, 242. [CrossRef]
6. Iriyama, T.; Carter, M.; Cavalheiro, G.M.; Poudel, P.; Nelson, G.J.; Zhang, G. Effects of Non-Uniform Temperature Distribution on the Degradation of Liquid-Cooled Parallel-Connected Lithium-Ion Cells. *Batteries* **2024**, *10*, 274. [CrossRef]
7. Sharifi, N.; Shabgard, H.; Millard, C.; Etufugh, U. Hybrid Heat Pipe-PCM-Assisted Thermal Management for Lithium-Ion Batteries. *Batteries* **2025**, *11*, 64. [CrossRef]
8. Jia, Y.; Brancato, L.; Giglio, M.; Cadini, F. Online Core Temperature Estimation for Lithium-Ion Batteries via an Aging-Integrated ECM-1D Coupled Model-Based Algorithm. *Batteries* **2025**, *11*, 160. [CrossRef]
9. Rahmani, A.; Dibaj, M.; Akrami, M. Enhancing Battery Pack Cooling Efficiency Through Graphite-Integrated Hybrid-Battery Thermal Management Systems. *Batteries* **2025**, *11*, 113. [CrossRef]
10. Oyewola, O.M.; Idowu, E.T.; Labiran, M.J.; Hatfield, M.C.; Drabo, M.L. Air-Outlet and Step-Number Effects on a Step-like Plenum Battery's Thermal Management System. *Batteries* **2025**, *11*, 87. [CrossRef]

batteries

Review

Review of Thermal Management Strategies for Cylindrical Lithium-Ion Battery Packs

Mohammad Ahmadian-Elmi and Peng Zhao *

Department of Mechanical, Aerospace & Biomedical Engineering, UT Space Institute, University of Tennessee, Knoxville, TN 37388, USA; mahmadi4@utk.edu
* Correspondence: pzhao12@utk.edu

Abstract: This paper presents a comprehensive review of the thermal management strategies employed in cylindrical lithium-ion battery packs, with a focus on enhancing performance, safety, and lifespan. Effective thermal management is critical to retain battery cycle life and mitigate safety issues such as thermal runaway. This review covers four major thermal management techniques: air cooling, liquid cooling, phase-change materials (PCM), and hybrid methods. Air-cooling strategies are analyzed for their simplicity and cost-effectiveness, while liquid-cooling systems are explored for their superior heat dissipation capabilities. Phase-change materials, with their latent heat absorption and release properties, are evaluated as potential passive cooling solutions. Additionally, hybrid methods, such as combining two or more strategies, are discussed for their synergistic effects in achieving optimal thermal management. Each strategy is assessed in terms of its thermal performance, energy efficiency, cost implications, and applicability to cylindrical lithium-ion battery packs. The paper provides valuable insights into the strengths and limitations of each technique, offering a comprehensive guide for researchers, engineers, and policymakers in the field of energy storage. The findings contribute to the ongoing efforts to develop efficient and sustainable thermal management solutions for cylindrical lithium-ion battery packs in various applications.

Keywords: lithium-ion batteries; cylindrical cells; thermal management; air cooling; liquid cooling; phase-change material; hybrid cooling

1. Introduction

Global warming is an urgent concern driven by the emission of greenhouse gases [1]. To combat this existential threat, there is an urgent need to shift our energy paradigm toward cleaner alternatives, particularly renewable energy sources like solar, wind, and hydropower. Unlike fossil fuels, these sources produce virtually no greenhouse gas emissions and are sustainable in the long run [2]. One of the key challenges with renewables, such as solar and wind, is their inherent variability and intermittency. Energy storage addresses this issue by capturing excess energy during periods of high production and releasing it at high demand or low supply conditions [3].

Energy storage systems can be categorized based on the type of energy stored into electrochemical, mechanical, electrical, chemical, and thermal energies [4]. Electrochemical energy storage refers to the process of storing electrical energy in the form of chemical energy and converting it back into electrical energy when needed. There are a few representative kinds of electrochemical energy storage systems: rechargeable batteries [5], electrochemical capacitors [6], and fuel and electrolysis cells [7]. Among these options, lithium-ion batteries (LIB) have become ubiquitous in modern electronics and electric vehicles because of their good energy density, relatively lightweight structure, and excellent rechargeability [8]. These batteries operate by the transfer of lithium ions between the electrodes, creating a current [9]. Various types of Li-ion cells, including cylindrical [10], prismatic [11], and pouch [12] designs, are accessible in the market for a wide range of applications.

Cylindrical lithium-ion batteries are a prevalent and versatile type of rechargeable power source with a distinctive tubular form. These batteries are widely utilized across numerous applications, including electronics, electric vehicles, and portable devices. Cylindrical lithium-ion battery cells comprise a rolled assembly, known as a jelly roll, which includes a cathode, an anode, a separator, and two current collectors for a unit layer. Common sizes, such as 18650 [13] and 21700 [14], have become industry standards, reflecting the popularity and reliability of cylindrical lithium-ion battery technology.

Similar to other chemical devices, battery performance is intrinsically influenced by temperature. As such, a suitable temperature range has to be maintained to guarantee performance and safety. Research suggests that to achieve optimal performance, it is advisable to limit the operation of LIBs to a specific temperature range, typically between 15 and 35 °C [15]. Temperature impacts batteries primarily through undesired chemical reactions and material degradations. Under low-temperature conditions, battery performance suffers from slow charge transfer and electrochemistry, as manifested by a substantial increase in internal resistance and a reduction in output power. High temperatures during the application of LIBs accelerate aging, negatively affecting performance and reducing overall lifespan. At even higher temperatures, thermal runaway may occur, where exothermic reactions are triggered among the battery materials [16], leading to subsequent fires and explosions.

In addition to temperature control, ensuring uniform temperature distribution within a battery has been a significant challenge for battery thermal management [17,18]. Inherent reversible entropic heat and irreversible polarization heat generations are associated with the normal operation of batteries. The escalating volumetric heat generation rate, associated with higher energy density and charge rates, necessitates an increased demand for cooling power to dissipate the accumulating battery heat. The situation is further complicated by the large number of cells included in a battery, many of which are subject to different thermal boundary conditions. The difference in thermal boundary conditions eventually leads to a temperature gradient within the battery pack, which further amplifies cell-to-cell variations among the pack in that cell; therefore, degradation strongly depends on temperature [19].

In fact, monitoring the thermal state of a battery itself is essential for maintaining it within safety limits and detecting potential thermal faults early on, preventing hazardous incidents like thermal runaways. Substantial efforts have been dedicated to implementing efficient onboard monitoring of battery thermal states. The primary method involves measuring battery temperature using devices like thermistors and thermocouples [20,21], through placing these temperature sensors on the battery surface to monitor its temperature during operations. Because of the restricted quantity of onboard temperature sensors and their incapacity to measure the internal temperature of the battery, estimating temperature becomes pivotal in overseeing the thermal condition of lithium-ion batteries (LIBs). Various metrics, including surface temperature, core temperature, bulk temperature, and temperature distribution, are employed as diverse measures for handling battery systems, each offering unique performance characteristics. Following that, existing techniques for estimating temperature can be classified into three categories: impedance/resistance-based estimation, thermal-model-based estimation, and data-driven estimation. Despite advancements at the cell level, challenges stemming from the intricate thermal dynamics of battery cells and packs, varying operating conditions, inadequate sensing of the pack, and limitations in the online applicability of current methodologies continue to hinder the effective monitoring of the thermal state at the pack or system level. Overcoming these challenges necessitates interdisciplinary efforts, including the application of advanced sensing technologies in the battery system, the derivation of high-fidelity control-oriented thermal models based on mathematical techniques, the utilization of state-of-the-art artificial intelligence (AI) technologies for precise and robust estimation, and the implementation of image processing methods to reconstruct temperature information [22].

The thermal conductivity of a cell plays a critical role in thermal management. The anisotropic layered structure of LIBs leads to a highly anisotropic thermal conductivity. In-plane heat conduction is primarily governed by the high thermal conductivity (k) of current collectors. The effective in-plane thermal conductivity falls within the range of 20 to 35 W/(m·K) [23,24]. Conversely, the effective cross-plane thermal conductivity is approximately 0.15–1.40 W/(m·K) [23]. This lower cross-plane k is a result of both the electrodes having low thermal conductivity and the presence of high thermal contact resistance between the layers [25]. In the particular case of cylindrical cells, the radial thermal conductivity is at least one order of magnitude lower compared to values in the longitudinal and azimuthal directions. A smart thermal management system should take advantage of the much faster in-plane heat conduction to remove heat more effectively.

To tackle these issues, lithium-ion batteries can be fitted with a battery management system (BMS) that oversees the regular functioning of the battery and optimizes its operation. Ensuring the safe functioning and extending the lifespan of a battery necessitates the presence of an efficient thermal management system. This paper presents a comprehensive review of the thermal management strategies employed in cylindrical lithium-ion battery packs. The review covers four major thermal management techniques: air cooling, liquid cooling, phase-change materials (PCM), and hybrid methods. Air-cooling strategies are analyzed for their simplicity and cost-effectiveness, while liquid-cooling systems are explored for their superior heat dissipation capabilities. Phase-change materials, with their latent heat absorption and release properties, are evaluated as potential passive cooling solutions. Additionally, hybrid methods, combining two or more strategies, are discussed for their synergistic effects in achieving optimal thermal management.

2. Battery Thermal Management Systems (BTMS)

Thermal management is a critical aspect of lithium-ion (Li-ion) battery packs to ensure their safe and efficient operation. The design of thermal management systems for cylindrical lithium-ion battery packs involves specific criteria to optimize performance and safety. First and foremost is the need for effective temperature control to maintain the battery within its optimal operating range, preventing overheating and potential safety hazards. Heat dissipation capability is equally vital, requiring the incorporation of materials with high thermal conductivity or other cooling mechanisms to transfer heat away from the battery cells efficiently. A well-designed thermal management system should also address temperature uniformity, ensuring that all cells within the battery pack experience consistent thermal conditions to avoid imbalances in performance and degradation. Additionally, the system should consider aspects such as thermal insulation to mitigate cold temperature effects and the prevention of thermal runaway events, emphasizing the importance of a comprehensive and multifaceted approach in managing the thermal challenges of lithium-ion batteries. Furthermore, achieving these criteria in a cost-effective manner is essential for the widespread adoption of lithium-ion batteries in various applications, including cost considerations of energy, material, and space. Therefore, an effective thermal management system for cylindrical lithium-ion battery packs must address these specific considerations to ensure both optimal performance and safety.

In addition to all the criteria that should be considered for the BTMS, two important features specific to cylindrical batteries should also be considered. The thermal conductivity of cylindrical cells differs in different directions, so the axial thermal conductivity is much more than the radial thermal conductivity. This feature makes the heat dissipation from the cell to the surrounding environment easier through an axial form of heat transfer. The second issue is the geometry of cylindrical packs. Due to the geometry of the cylindrical cells, their placement in the modules inevitably creates interstitial space among cells, which not only affects the volumetric energy density in the pack level but also adds extra constraints for heat removal and cooling system design.

BTMS for cylindrical lithium-ion battery packs can be broadly categorized into active and passive systems, each offering distinct approaches to address heat dissipation and temperature control. Active systems incorporate mechanisms that actively remove heat from the battery pack, such as liquid cooling or forced air convection. Liquid-cooling systems use coolants to absorb and transfer heat away from the cylindrical cells, while air-cooling systems rely on fans or other methods to facilitate heat exchange. These systems are effective in regulating temperature and preventing overheating during high-demand scenarios. On the other hand, passive systems leverage natural heat dissipation methods, often incorporating materials with high thermal conductivity, like heat sinks or phase-change materials, to passively dissipate heat without the need for external mechanisms. The choice between active and passive systems depends on factors such as application, space constraints, and specific thermal management requirements, highlighting the need for a tailored approach to optimize the performance and safety of cylindrical lithium-ion battery packs. Commercial electric vehicles commonly employ air-cooling and liquid-cooling systems for the thermal management of lithium-ion batteries. For instance, vehicles like the Nissan Leaf and Mahindra e2o Plus utilize air cooling. On the other hand, models such as Tesla, Chevrolet Volt, Hyundai Kona, Ford Focus, and Jaguar I-Pace opt for liquid-cooling methods [26]. Recently, newer methods such as heat pipe [27–31], immersion cooling [32,33], and thermoelectric [34] have been used for the thermal management of lithium batteries. Figure 1 categorizes the different thermal management methods of cylindrical lithium-ion batteries.

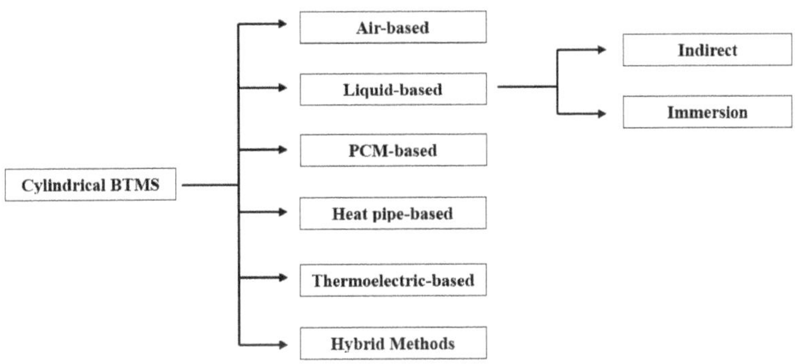

Figure 1. Classification of battery thermal management systems.

2.1. *Air-Cooled BTMS*

Air cooling is one of the most popular ways of heat dissipation in battery thermal storage systems (BTMS), due to its extensive advantages, such as its simple structural design, low cost, lightweight nature, ease of maintenance, long lifespan, simplicity in replacing a single cell, and little power consumption at a low battery discharge rate. In air-cooling systems, the battery is often equipped with cooling fins or other heat-dissipating structures that increase the effective surface area exposed to ambient air. As air flows over these structures, it carries away the heat, preventing the battery from overheating. Additional components, such as fans, may be incorporated to actively enhance air circulation and cooling efficiency.

Table A1 (presented in the Appendix A) summarizes the existing studies related to air-cooled cylindrical BTMS. This table classified the previous studies based on their method, battery type, number of cells, load, pack arrangement, airflow regime, airflow inlet, their main focus, and their achievements. Different battery types, including 18650, 21700, 26650, 32700, and 42110 were studied at various C-rates.

Researchers tried to study the air-cooling systems for different numbers of LIBs, from single cells to packs. Yang et al. [35] proposed a novel radiator featuring a surface structure inspired by bionics for each battery cell (Figure 2). The battery module comprises cells, fans, radiators, walls, and insulated frames. The cells are arranged in a parallel structure, with three rows along the battery axis. Insulated frames support and insulate the cells, positioned between each row. Cooling air, propelled by the fan, enters each cell row and exits the battery module. Four radiators are placed in the gaps between four cells, with curved contact surfaces touching the outer cell surfaces. The curvature radius matches that of a cylindrical battery. Additionally, bionic surface structures with equilateral triangular cross-sections are affixed to the convex curved surface. In the air-cooled BTM system, cell-generated heat is conducted to the radiator through solid heat conduction. The cooling air then dissipates the heat, effectively controlling the battery module's temperature. To enhance cooling performance and minimize power consumption, they endeavored to optimize the radiator parameters, such as square plate thickness, and the height, shape, and length of the bionic surface structure. Their findings indicate that the novel radiator thickness and bionic surface structure height significantly reduce the maximum temperature values and temperature differences among the cells.

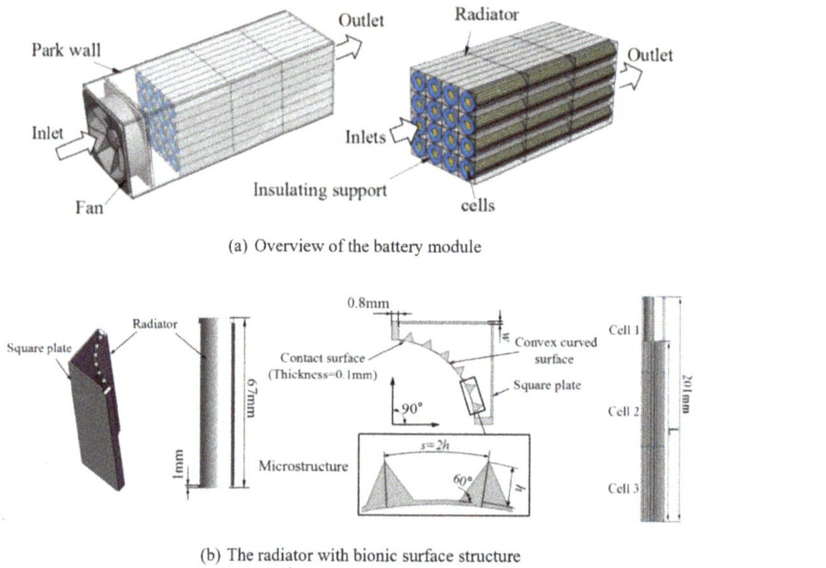

Figure 2. Scheme of the air-cooled battery management module with a bionic surface structure [35].

Designing a geometric structure for better air distribution has been one of the challenges of air-cooling systems. Yang et al. [10] explored the thermal characteristics of an air-cooled battery thermal management system specifically created for a cylindrical lithium-ion battery module featuring a honeycomb structure. They developed an air distribution plate (ADP) (refer to Figure 3) to improve temperature consistency within the battery module. Through the use of computational fluid dynamics (CFD), they conducted a numerical analysis of the cooling efficiency and airflow distribution. Modifying the width of the ADP channel was found to enhance the evenness of air distribution in the battery module spaces. Additionally, the utilization of a bionic heat sink demonstrates superior cooling effectiveness, significantly improving temperature consistency across the battery pack. Research has been conducted trying to provide geometric changes to improve air distribution. In another study, Zhou et al. [36] introduced an innovative cooling approach utilizing air distribution channels for cylindrical lithium-ion battery modules and conducted a numerical investigation (Figure 4). They examined the impact of orifice diameter, orifice rows, and inlet pressure. Their findings indicate that enlarging the orifice diameter and increasing the number of orifice rows contribute to lowering the maximum temperature and enhancing temperature uniformity. Importantly, this optimization comes with minimal additional power consumption.

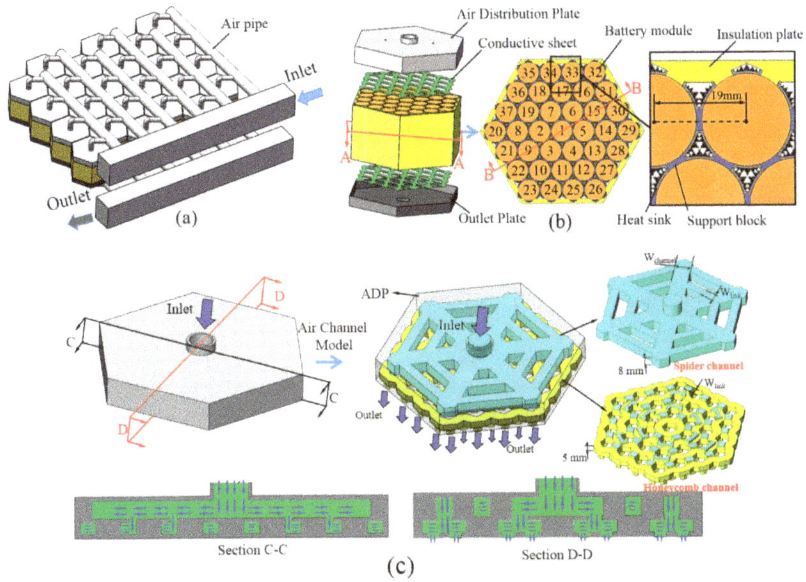

Figure 3. Structures and components of air-cooled BTMS: (**a**) battery pack, (**b**) battery module, and (**c**) air distribution plate [10].

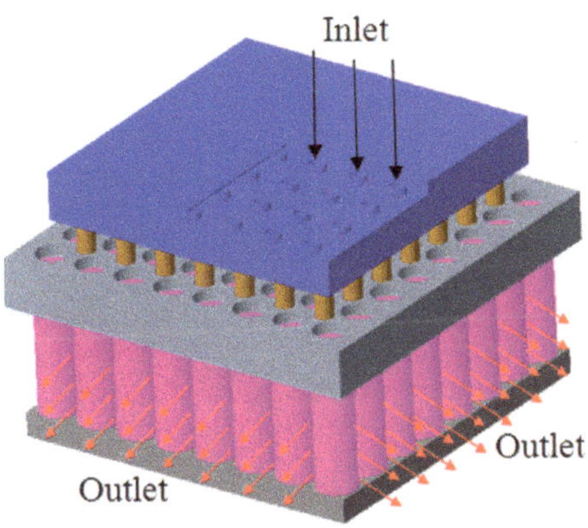

Figure 4. Schematic of the battery module using the air distribution pipe [36].

Changing the battery module structure can also be useful. Wang et al. [37] examined the heat efficiency of a battery module employing diverse cell arrangement structures such as rectangular, hexagonal, and circular configurations. The investigation also delved into optimizing air-cooling approaches by situating fans at different positions inside the battery module to enhance temperature uniformity. The research delved into factors influencing cooling capability based on simulations conducted using three-dimensional computational fluid dynamics (CFD). Heat generation rates were determined through thermal insulation experiments, and a lumped model was applied for individual cells. The findings revealed that placing the fan at the upper part of the battery module resulted in the most effective cooling performance, regardless of the structure. Among the structures tested, the 5 by 5 cubic arrangement demonstrated the best balance between cooling capability and cost. Meanwhile, the hexagonal structure with 19 cells was identified as the most favorable option for maximizing space utilization and ensuring effective cooling.

To enhance the cooling system's performance via geometric alterations, researchers designed jacket fins for each cell. Kummitha [38] introduced an innovative approach to achieve consistent temperature distribution within a battery module and reduce the maximum temperature reached by the battery cells. The novel module design incorporates cell holders at varying heights, serving as trays to support the battery cells and functioning as fins to enhance uniform temperature distribution. Comparative analyses were conducted on the base and new battery modules under different discharge conditions while maintaining consistent boundary conditions and physical parameters. In their design, jacket fins were implemented to enhance heat dissipation from the battery surface, ensuring a uniform thermal distribution. The heat removal plate, constituting the cell holder's jacket, was constructed using lightweight aluminum. Both models were evaluated under various discharge conditions using plane contours and temperature profiles for each cell within the battery module.

Air mass flow rate and channel design have a great impact on the performance of the air-cooling system. Lu et al. [39] developed a three-dimensional numerical model and an analytical thermal resistance model for a battery pack arranged in a staggered fashion. This was done to investigate how cooling channel design and air supply strategy affect the cooling efficiency of the battery pack. Their findings indicated that a larger cooling channel size can reduce the highest temperature and enhance the cooling energy efficiency of the stagger-arranged battery pack, although at the expense of space utilization, which subsequently influences the power density of the battery pack. Considering factors such as peak temperature, space utilization, and energy effectiveness, the optimal cooling channel size was determined to be 1 mm for the 18650 lithium-ion battery. Ji et al. [40] devised a lithium-ion cylindrical battery module incorporating distinct cell distribution strategies to enhance uniformity. They created a three-dimensional model to examine and enhance the configuration of the battery module. Introducing variations in the separation among cells, particularly through an arithmetic ratio arrangement, proves to be an effective method for improving homogeneity in the battery module. Furthermore, the highest temperature variation within the battery module decreased with an increase in the initial spacing. Comparing the arithmetic ratio and geometric ratio arrangements, it was determined that the arithmetic ratio arrangement yields superior temperature and discharge uniformities for the battery module. While the location of cells reaching the maximum temperature in the module changes over discharge time, they appear in a consistent position regardless of the configuration and initial gap. The process of changing positions was observed to be slower in arrangements with larger initial spacing and arithmetic ratios.

Using high thermal conductivity materials for components like thermal interface materials improves heat transfer and dissipation. Kausthubharam et al. [41] introduced a battery thermal system incorporating a commercially available thermal interface material, and they conducted a comparative analysis with a conventional battery pack across various discharge rates. Their approach demonstrates a noteworthy 25% decrease in peak temperature compared to the standard configuration. Lee et al. [42] created 21700 lithium-ion battery modules equipped with thermal interface materials (TIM). Their research revealed that the parallel thermal conductivity of the TIM significantly influences the temperature uniformity across the battery module. Meanwhile, the vertical thermal conductivity is more closely linked to the maximum battery temperature and State of Health (SoH).

One recent technology to increase the performance of cooling systems is the use of reciprocating flow, which is particularly useful in reducing cell-to-cell variations subjected to dissimilar thermal boundary conditions. Mahamud et al. [43] examined a novel battery thermal management technique employing reciprocating airflow for cylindrical Li-ion cells (Figure 5). Their analysis was conducted through numerical simulations utilizing a two-dimensional computational fluid dynamics (CFD) model, a lumped-capacitance thermal model for battery cells, and a flow network model. The results demonstrated that a shorter reciprocating period correlated with lower cell temperature differences and reduced maximum cell temperatures, both in instantaneous and time-averaged scenarios.

The enhancements achieved through reciprocating flow were primarily attributed to the redistribution of heat and the disruption of boundary layers formed on the cells due to the periodic reversal of airflow.

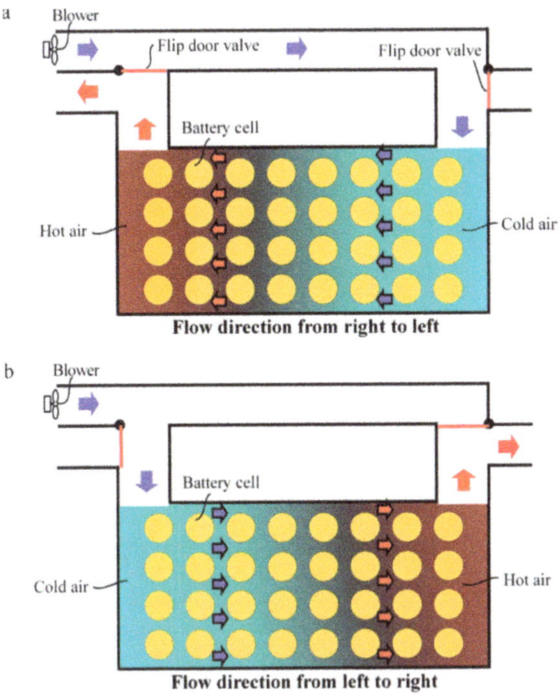

Figure 5. Schematic view of a battery system using a reciprocating airflow for battery thermal management. The flow directions are (**a**) from right to left and (**b**) from left to right [43].

Through advanced engineering and design strategies, researchers and engineers aim to optimize the airflow, cooling mechanisms, and overall thermal architecture of air-cooled lithium-ion battery systems. This includes the exploration of novel cooling materials, improved heat exchanger designs, and the development of smart thermal control systems. The goal is to achieve uniform temperature distribution within the battery pack, minimizing thermal gradients and preventing localized overheating. Wang et al. [44] introduced a multi-objective optimization approach using a BPNN-PSO method for an air-based BTMS. Initially, a database was established through CFD simulations before optimizing the design parameters. The optimization focused on channel spaces in the X-direction and air inlet velocity to enhance thermal performance and minimize energy consumption. Following the optimization of these parameters, further refinement of the BTMS was achieved by tuning the channel spaces in the Y-direction. The effectiveness of the BPNN-PSO method was validated against CFD simulation results under optimal conditions. The study revealed that the power consumption of the system, under optimal conditions, was reduced to 41.19% compared to the original system. Simulation analysis demonstrated the efficacy of the proposed multi-objective optimization method in significantly lowering the power consumption of the BTMS.

Precisely predicting temperature is a crucial challenge for enhancing battery performance and averting thermal runaway. The intricate nonlinear characteristics of heat generation and dissipation in lithium-ion batteries, coupled with susceptibility to external factors, make accurately forecasting battery temperature challenging. In recent years, the use of artificial neural networks (ANNs) has become widespread in various aspects of lithium-ion batteries. Their distinct advantages in addressing highly nonlinear problems, including battery modeling, state of charge (SOC) estimation, residual useful life (RUL) prediction, and battery temperature prediction, have led to their extensive application in this field [45]. Li et al. [46] achieved the optimal design of battery thermal management systems through the implementation of a three-dimensional electro-thermal model combined with an artificial neural network (ANN) model. The three-dimensional electro-thermal model was utilized to compute the temperature distribution and was validated using previously gathered experimental data. Numerical case studies were conducted to train the proposed ANN model, revealing the correlation between geometric parameters, operating conditions, and cooling efficiency. Wang et al. [47] constructed a thermal model for lithium-ion batteries using an artificial neural network. Specifically, BP-NN, RBF-NN, and Elman-NN models were chosen to analyze the impact of network structures and modeling techniques on temperature prediction performance. The dataset for network training was acquired through a numerical model, and all three neural network models were configured with the same training parameters. The findings reveal that, under sample working condition testing, the RBF-NN model exhibits superior prediction accuracy and greater applicability, whereas the Elman-NN model performs less effectively.

According to the studies conducted in the air-cooling systems of LIBs, existing efforts to improve the performance of these systems can be classified into three aspects: (1) study and modify the geometry of these systems, such as channel dimensions, outlet and inlet positions, and changes in the number of air channels; (2) study of the airflow including change of flow direction, unidirectional and reciprocating flow, as well as inlet mass flow rate and velocity, type of flow regime (turbulent or laminar); (3) study battery working conditions such as different battery loadings and ambient air temperature.

In summary, air-cooled lithium-ion battery systems, despite their widespread use, present several drawbacks and limitations. One significant challenge is their limited cooling capacity, particularly in high-power applications where heat dissipation becomes crucial. These systems can be sensitive to ambient temperatures, making them less effective in extreme climates. The reliance on air cooling can lead to inconsistent temperature distribution within the battery pack, potentially resulting in hotspots that compromise overall performance. Additionally, air-cooled systems may lack the precision of more advanced liquid-cooling alternatives, and their design often adds bulk and weight to the battery pack. Noise generation from cooling fans and the potential for dust accumulation further contribute to the list of drawbacks. In high-temperature environments, air cooling may exhibit reduced efficiency, increasing the thermal stress of the cells. The maintenance requirements and energy consumption associated with air-cooled systems underscore the need to carefully consider alternative cooling solutions, especially in applications with demanding thermal conditions.

2.2. Liquid-Cooled BTMS

A liquid-based lithium-ion BTMS involves the use of a liquid, typically as a coolant or heat transfer medium, to regulate the temperature of the battery during operation. In this system, a liquid circulates through the battery pack, absorbing heat generated during the charging and discharging processes. The liquid absorbs the heat and then moves away from the heat source, where it can release the heat to the surroundings through a cooling mechanism. Table A2 (presented in Appendix A) summarizes the existing studies related to liquid-cooled cylindrical BTMS. This table classified the previous studies based on their method, battery type, number of cells, load, pack arrangement, airflow regime, flow inlet, their main focus, and their achievements.

Designing new systems according to the geometry of cylindrical cells is one of the researchers' priorities. Gao et al. [48] introduced an innovative design for a BTMS featuring gradient channels oriented along the flow direction (Figure 6). This design focused on the usefulness of achieving efficient heat transfer within the BTMS by maximizing the contact area between the solid components and the fluid. To address temperature gradient issues along the flow direction, a liquid-cooled BTMS with a gradient channel-based design was proposed. The unique liquid-cooled structure comprised cooling tubes and multiple rows of cells, as illustrated in Figure 6. Cells were symmetrically positioned on both sides of the aluminum tube, facilitating a substantial heat dissipation area. The gradient cooling tube was segmented into two parts along the flow direction, with each large channel in the front corresponding to four small channels in the rear. This design aimed to increase the fluid–solid contact area in the rear segment. Through this gradient design, the heat generated in the cells is effectively transferred to the water, leading to a notable 69.40% enhancement in the temperature uniformity of the battery pack.

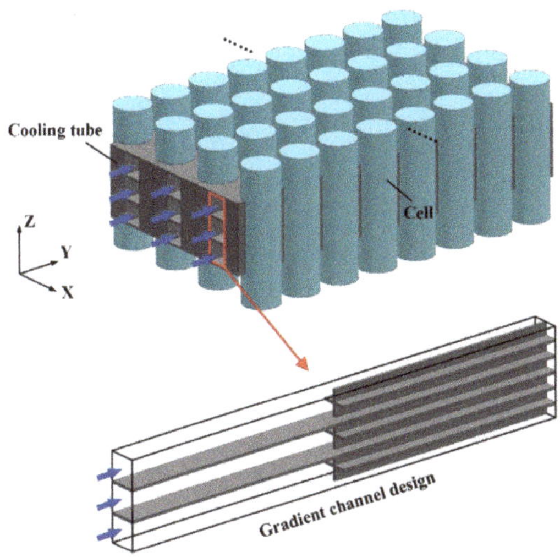

Figure 6. Schematics of a liquid-cooled BTMS with a gradient channel design [48].

Designing suitable channels for better heat dissipation is a key challenge of liquid-cooling systems. Kong et al. [49] explored the efficacy of two liquid-cooling configurations, namely the creation of mini-channel cylinders and a heat sink with channels for lithium-ion battery packs. The study utilized a 18650 lithium-ion cylindrical battery subjected to a 5 C discharge rate. The investigation delved into the impact of variables such as the number of channels, inlet/outlet positions, and mass flow rate on cooling efficiency, considering parameters like maximum temperature and temperature distribution. The findings indicated that augmenting the number of channels led to a reduction in the cell's maximum temperature, albeit with diminishing returns for each incremental increase. Additionally, the most optimal cooling performance was observed when the inlet was positioned closest to the battery end, resulting in enhanced heat dissipation under adiabatic conditions.

The widespread application of mini-channel cooling plates is observed in the liquid cooling of cylindrical cells. Li et al. [50] introduced an innovative liquid-cooling design incorporating mini-channel cold plates distributed vertically along the cells, utilizing water as the cooling medium. The researchers performed CFD simulations to assess the influence of different design parameters on the battery pack. Their specific emphasis was on evaluating the maximum temperature and the maximum temperature difference within the pack, liquid pressure drop, and overall pack weight. The authors highlighted that the incorporation of an aluminum cylindrical enclosure effectively addresses the issue of poor heat conduction through the battery material, significantly improving the cooling rate and temperature uniformity. Their findings indicated that the first design (battery pack with a cylindrical enclosure), featuring only three cooling plates, achieved thermal efficiency comparable to the second design (design without a cylindrical enclosure) with seven cooling plates. Consequently, the cylindrical enclosure proves beneficial in reducing both weight and pumping power. Moreover, a square cross-sectional mini-channel is demonstrated with more effective heat dissipation and reduced flow resistance in comparison to circular and elliptical shapes. The study concluded that converging channels enhance heat removal and temperature uniformity but lead to increased pressure drop, while diverging channels reduce pressure drop at the expense of diminished thermal efficiency.

Researchers are trying to propose new structures for cooling systems. Xia et al. [51] devised an innovative heat sink for the cooling sleeve of a spindle, incorporating a novel network of fractal tree-like channels based on the constructal theory. Previous research indicated that such a heat sink with fractal tree-like channels offers inherent advantages in both heat transfer efficiency and pressure drop. The authors conducted a numerical analysis comparing the heat transfer and pressure drop characteristics of the fractal tree-like channels' heat sink with a conventional helical channel heat sink. The findings conclude that the fractal tree-like channels' heat sink exhibits a lower pressure drop, a more uniformly distributed temperature field, and a higher coefficient of performance compared to the traditional helical channel heat sink. Specifically, the coefficient of performance for the fractal net heat sink is more than twice that of the traditional helical net heat sink.

Parallel channels and serial channels are the conventional configurations of the liquid-cooling system. Wang et al. [52] introduced an innovative modular liquid-cooled system designed for batteries. They conducted numerical simulations and experiments to explore the impact of the mass flow rate of working fluid and cooling modes (parallel and serial) on the thermal characteristics of the battery module (Figure 7). The findings indicated that augmenting the flow rate within a defined range significantly lowers the maximum temperature and improves temperature uniformity in the battery module. However, surpassing a certain flow rate does not show a noticeable enhancement in cooling efficiency. Additionally, parallel cooling, compared to serial cooling, notably enhances the temperature consistency of the battery module.

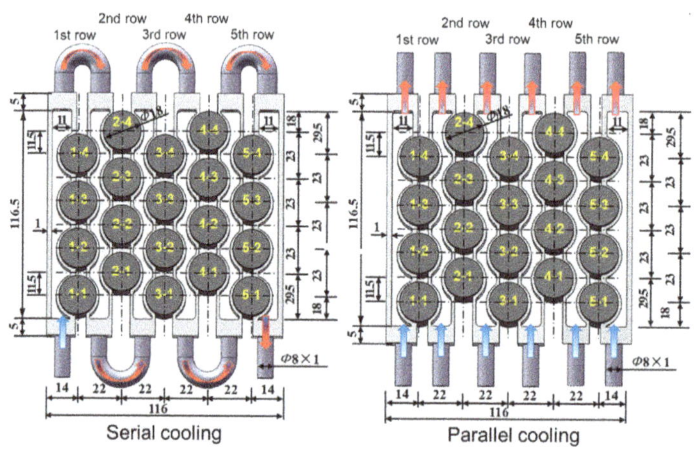

Figure 7. Schematic diagram of the serial cooling and parallel cooling [52].

The placement of the cold plate between the cells or their sides is also one of the issues that is taken into consideration and tested in different situations. Kim et al. [53] conducted experiments involving two battery packs equipped with distinct cold plates to assess and compare the cooling efficiencies of the battery thermal management system (BTMS). The study included an analysis of the internal heat generation within the battery modules, motivated by safety considerations. Figure 8 illustrates the two designed cold plates: Type I and Type II. In the case of Type I, four 18650 batteries were inserted into aluminum fins, with both ends of the fins in direct contact with the plate. Two cooling channels were situated on the two edge sides of the cold plate. In Type II, an individual battery was positioned within a cylindrical aluminum fin, with the fins in contact with an aluminum cold plate. Four cooling channels were incorporated inside the aluminum cold plate, distributed among each battery row, facilitating the direct flow of water into the cooling channels. The fins were meticulously designed to match the outer diameter of the 18650 cells precisely. The findings indicated that Type II exhibited superior cooling efficiency compared to Type I at the same C-rate. Specifically, Type I demonstrated elevated battery temperatures at high C-rates, whereas Type II effectively maintained the battery temperature within the optimal range.

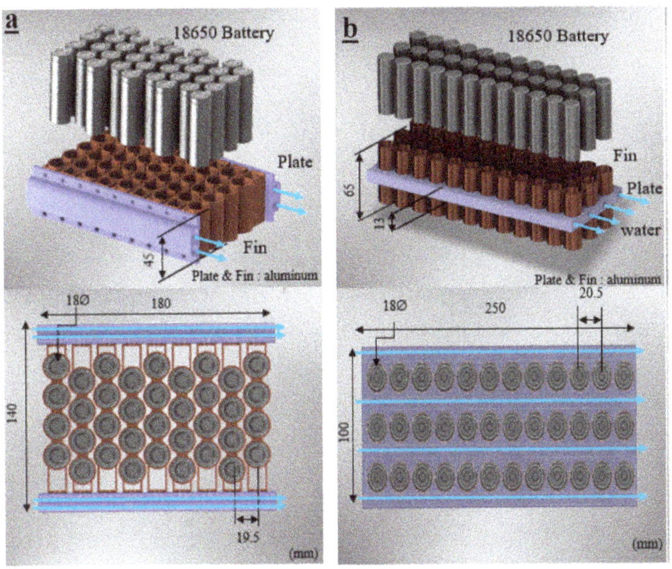

Figure 8. Battery module: (**a**) Type I and (**b**) Type II [53].

Some researchers suggest the implementation of cooling jackets to mitigate the undesired temperature increase and temperature variation in lithium-ion batteries. Sheng et al. [54] designed a liquid-cooling jacket with a cellular structure specifically for cylindrical lithium-ion batteries to mitigate undesired temperature increases and variations (Figure 9). In this approach, numerical investigations were conducted to analyze the impact of fluid flow, channel dimensions, and cooling medium on the thermal behavior of the battery cells. The results revealed that the interlaced flow directions within the developed cellular cooling jacket contributed to achieving a reduced temperature standard deviation and a more uniform thermal profile.

One of the most important design criteria of the cooling system is the ability to provide temperature uniformity in the module. Xu et al. [55] devised an innovative wrench-shaped design for a cylindrical battery module, aiming to specifically address the problem of temperature non-uniformity caused by a continuous rise in coolant temperature along the direction of the coolant flow (Figure 10). The effectiveness of the wrench-shaped cold plate was assessed through a comprehensive evaluation considering temperature rise control, preservation of temperature uniformity, lightweight design, and power consumption. The bifurcation structure incorporated into the wrench-shaped cold plate demonstrated a notably improved heat dissipation effect at the junction between the upstream and downstream segments.

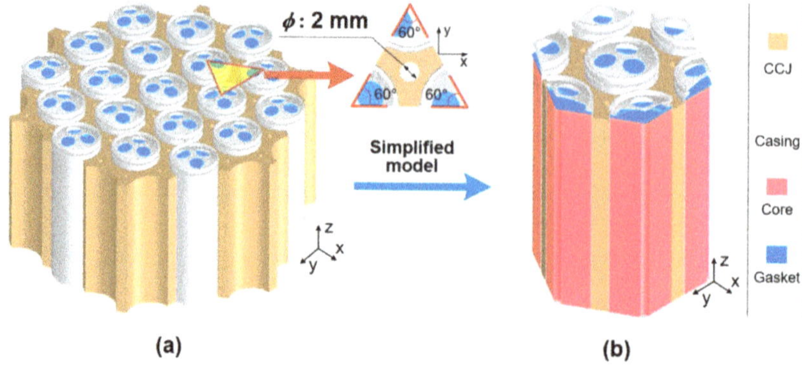

Figure 9. Power battery module. (**a**) Battery module consisting of multiple 21700 cells; (**b**) Simplified CAD model for the module [54].

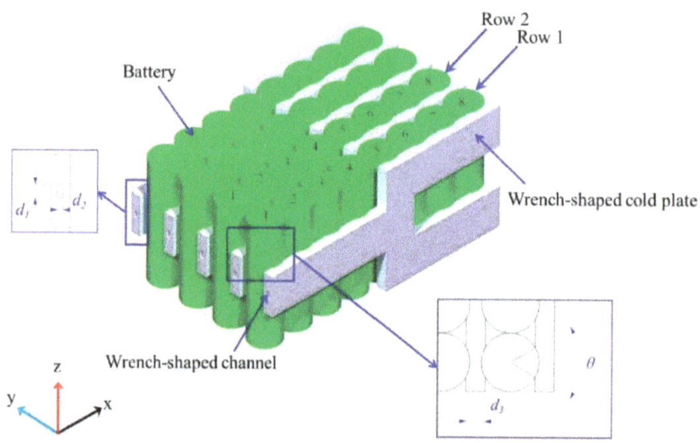

Figure 10. Schematic diagram of the liquid-cooling BTMS [55].

Minimizing the temperature differences poses a greater challenge compared to decreasing the maximum temperature and requires a novel structure for liquid channels. Liu et al. [56] designed an innovative liquid-cooling system that incorporates a vertical tube layout (VLT) along with a combination of gradually increased tube diameter. The research explored the influence of two tube configurations, the diameter of the tube, flow rate, gradient ratio flow velocity, and gradient-incremented tube diameter on the thermal efficiency of the system. The finding indicated that, within a defined range of flow velocities, the battery module featuring a vertical tube layout demonstrates superior performance compared to the horizontal configuration. Moreover, compared to a constant flow velocity and tube diameter, implementing a strategy involving either a gradient ratio flow velocity or a gradually increased tube diameter can reduce temperature differences in the battery, and combining both approaches can further improve temperature uniformity. Tete et al. [57] devised an innovative battery pack design that incorporates cylindrical casings for inserting batteries and utilizes a liquid-cooling medium circulated in its surroundings. Numerical simulations were conducted employing the finite volume method under various operating conditions, including scenarios with no cooling and with water cooling. The observations revealed that the highest heat accumulation occurred at the center of the battery pack. The maximum temperature increase was observed in the cells located at the core of the battery pack, as the coolant that reached this area had already warmed by absorbing heat from cells in the preceding rows.

A key feature that a BTMS desires is to be lightweight. A small, lightweight liquid-cooled BTMS was presented by Lai et al. [58] to manage the maximum temperature and temperature differential of a lithium-ion power battery pack. Their battery pack consists of base plates, liquid-cooled BTMS, an IC control unit, and 18650-type battery cells. Thermal conductive structures (TCSs), a bottom reservoir, and an upper reservoir make up the liquid-cooled BTMS. Engineered polymers are used to make the bottom reservoir and upper reservoir in order to save weight. The cells are arranged in a staggered pattern with a fixed gap between each one. The TCSs are placed inside the cell gaps for a compact size consideration, as seen in Figure 11. Three TCSs surround every cell. Each TCS has a circular liquid channel with a coolant setting in the middle. Each TCS is made with three curved contact surfaces in order to provide efficient thermal contact. By assessing the effects of inner diameter, contact surface height, and contact surface angle on temperature difference, they concluded that choice of inner diameter affects weights the most.

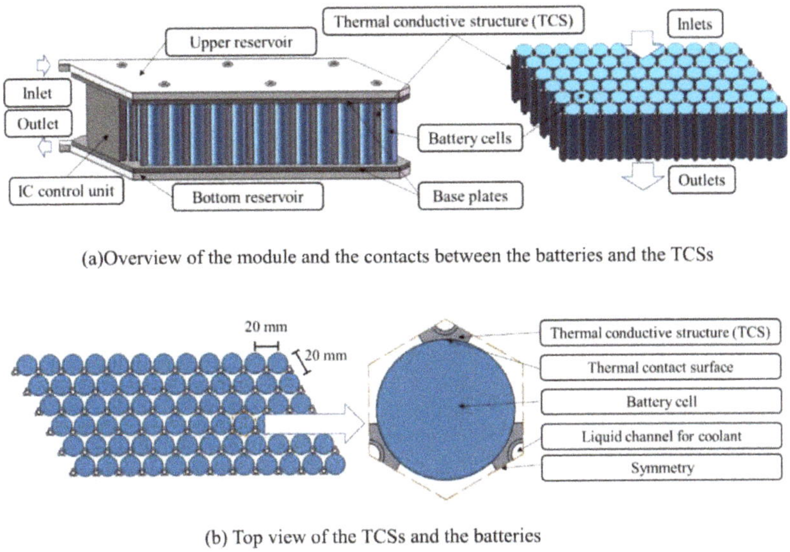

Figure 11. The scheme of the liquid-cooled battery management module [58].

Developing a compact cooling system for cylindrical LIB modules by effectively utilizing the spaces between the cylindrical batteries is a desired approach that will not further decrease the pack volumetric energy density. Shan et al. [59] introduced an innovative liquid-cooling BTMS featuring axially mounted cooling tubes inserted into the gaps between tightly assembled batteries, providing a compact and lightweight solution for cooling cylindrical battery modules. Their work considered the effects of different battery arrangements, leading to the proposed two designs for BTMS with axially mounted cooling tubes (Figure 12). In Design A, a staggered battery arrangement was implemented. The corresponding cooling tubes inserted between the batteries had the same height as the batteries and featured three curved contact surfaces to ensure effective heat contact. To achieve uniform heat dissipation, each battery was surrounded by three cooling tubes, covering half of the battery's side surface. In total, 65 tubes were inserted into the battery module. A circular coolant channel was positioned at the center of the cooling tube, allowing the coolant to flow axially from the top to the bottom of the batteries to dissipate the generated heat. In Design B, an aligned battery arrangement was employed. To adapt to the structural characteristics of the aligned battery module, the cooling tubes inserted between the batteries were designed with four curved contact surfaces, with a curvature radius identical to that of the battery. The height of the cooling tubes matched that of the batteries. They observed that the cooling performance of the BTMS in Design B surpassed that of Design A. However, the BTMS in Design A incurred significantly lower weight and volume costs (0.13 kg and 1.09 L, respectively) compared to the BTMS in Design B (0.72 kg and 1.26 L, respectively), highlighting the superior lightweight and compact characteristics of Design A.

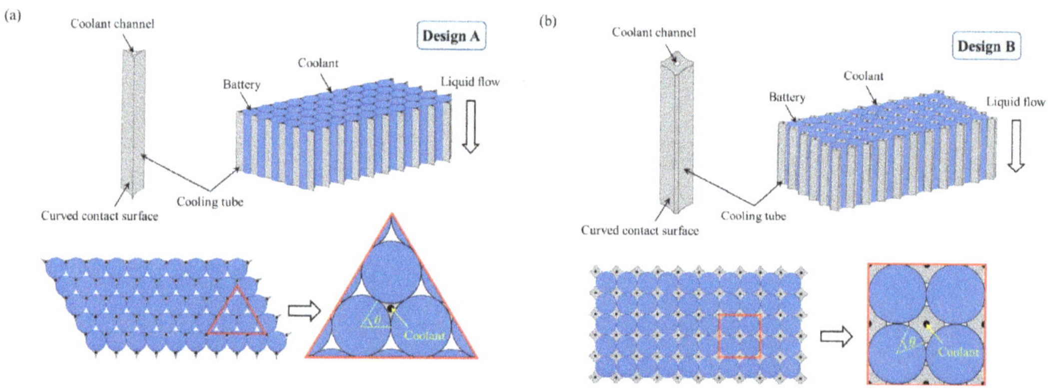

Figure 12. Schematic diagrams of the liquid-cooling battery module with axially mounted cooling tubes: (**a**) Design A with a staggered battery arrangement; (**b**) Design B with an aligned battery arrangement [59].

Pumping power in a BTMS refers to the energy required to circulate a cooling function through a system, and it is the energy consumed by the pump to overcome friction, pressure drops, and other resistances in the fluid circulation system. An efficient BTMS aims to strike a balance between effective heat dissipation and the energy required to circulate the cooling fluid, ensuring optimal performance and energy efficiency. Xiong et al. [60] developed a heat exchanger featuring a structure reminiscent of a spider web, leveraging the benefits associated with a large surface area, high convective heat transfer coefficient, and efficient fluid flow. This design effectively maintains the battery within the appropriate temperature range while minimizing power consumption. When compared to honeycomb-like and traditional helical structures with identical volume fractions and geometric parameters, heat exchangers utilizing the spider-web-like structure exhibit superior thermal and flow characteristics. Liang et al. [61] suggested a unitization-based BTMS solution, incorporating an innovative distributed water-cooling component to address the thermal management challenges of cylindrical LIB packs in EV applications. They aimed to strike a reasonable balance between system cooling performance, power consumption, and lightweight design.

To further facilitate heat transfer, nanofluids have emerged as a promising solution in liquid-cooling systems for BTMS. These nanofluids, consisting of a base fluid and suspended nanoscale particles, exhibit exceptional thermal conductivity, surpassing that of conventional coolants. When integrated into the liquid-cooling systems of LIBs, nanofluids enhance the efficiency of heat transfer, effectively mitigating issues related to overheating during high-power operations. Tousi et al. [62] developed an innovative indirect liquid-cooled BTMS for cooling a battery pack comprising 12 cells. The proposed schematic aimed to control the maximum temperature and thermal non-uniformity, particularly through the use of unidirectional flow. This method intentionally delays the fluid's exit from the channel, maximizing the fluid's cooling efficiency. The novel indirect liquid-cooling channel BTMS was designed to encompass all sides of the batteries, providing a large heat exchanger area to cool the heat generated from all the battery surfaces through heat conduction and dissipation by cooling DI-water and AgO nanofluid in the channel. The study revealed that the addition of AgO nanofluid into the channel significantly contributes to reducing the highest temperature and temperature non-uniformity in the battery pack. However, this method comes with an increased pressure drop in the channel. Transitioning from DI-water to AgO 4% nanofluid results in a 10.1% decrease in the maximum temperature and a 23.95% decrease in the temperature difference values of the battery pack.

Alumina nanofluid is one of the most effective fluids used in liquid-cooling systems. Sarchami et al. [63] devised an innovative BTMS, which incorporates wavy/stair channel liquid cooling and a copper sheath. The wavy-type channel not only aligns with the geometry of LIBs but also offers a substantial contact surface. Additionally, the stair channel is specifically designed to reduce the temperature at the top section of the battery pack. The study revealed that incorporating alumina nanoparticles into the base fluid, i.e., DI-water, enhances the cooling efficiency of the battery pack. When transitioning from DI-water to a 2% volume fraction alumina nanofluid, the peak temperature during charge operation and temperature non-uniformity values decrease by 2.4 K and 0.26 K, respectively. Similarly, for discharge operation, these values were approximately 2.33 K and 0.54 K lower, respectively.

Similar to air cooling, the reciprocating motion in liquid-cooling systems for BTMS offers an effective approach to optimize heat dissipation. In such a system, the cooling fluid undergoes back-and-forth motion within the cooling channels, facilitating efficient heat transfer. This motion can be controlled and adjusted based on the specific thermal requirements of the battery and environmental conditions. The reciprocating flow helps prevent the formation of hotspots by ensuring uniform cooling across the battery cells. Zeng et al. [64] utilized a reciprocating-liquid-flow-based BTMS for cylindrical batteries to address these associated issues (Figure 13). They conducted a comparative analysis among unidirectional flow, cross-direction flow, and reciprocating flow to assess the efficacy of reciprocating flow. Their findings demonstrated that reciprocating flow outperformed both unidirectional and cross-directional flows. The researchers further investigated the impact of velocity and reciprocating period, identifying them as key factors which influenced reciprocating flow. They proposed a comprehensive criterion for BTMS performance, considering maximum temperature, temperature difference, temperature distribution coefficient, and energy consumption, aiming to optimize reciprocating flow. They achieved a significant reduction of 55.3% in temperature difference and a 15.6% decrease in energy consumption.

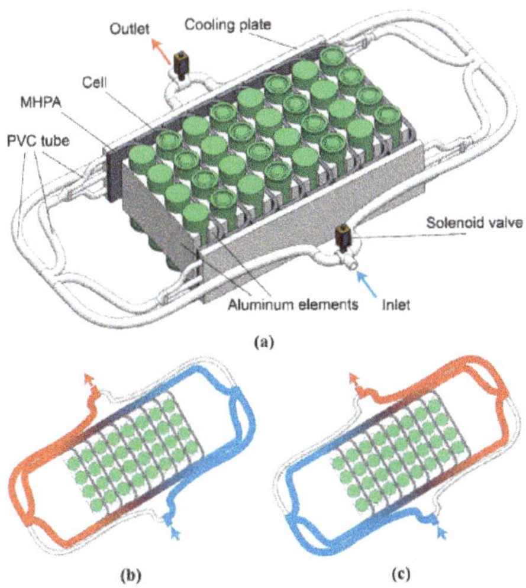

Figure 13. Geometric model of the BTMS with reciprocating flow. (**a**) Schematic diagram of the BTMS. (**b**,**c**) Diagram of reciprocating flow [64].

Optimization in liquid-cooling systems for lithium-ion batteries is critical to enhance battery performance, longevity, and safety. The optimization process involves a careful consideration of factors such as coolant flow rate, heat exchanger design, and pump efficiency. Engineers strive to strike a balance, ensuring that the cooling system effectively dissipates heat from the battery cells while minimizing the associated energy consumption, known as pumping power. Ma et al. [65] suggested an optimization approach for cooling lithium-ion batteries using a triple-step nonlinear method. To overcome the current limitation where the coolant flow rate cannot be precisely aligned with the actual cooling requirements of batteries in thermal management systems, the researchers introduced a triple-step nonlinear approach. They developed a simplified thermal model for lithium-ion batteries employing liquid cooling. The model takes into account factors such as heat generation, heat dissipation mechanisms, and the heat balance equation of the batteries. Wang et al. [66] introduced a combined strategy that integrates single-factor analysis with orthogonal testing to streamline the optimization process for the liquid-cooling structure in extensive battery modules. They illustrated that the contact angle exerts the most substantial influence on cooling performance, while the channel number exhibits a minimal impact in large battery modules with elongated channels. Consequently, for large battery modules, several potential factor combinations are suggested, such as maintaining a fixed contact angle of approximately 70°. The appropriate values for inlet velocity and channel number can vary within the ranges of 0.2–0.5 $m{\cdot}s^{-1}$ and 1–2, respectively, depending on specific applications.

Researchers proposed a method to improve the heat transfer coefficient by introducing baffles into the micro-channels. Xie et al. [67] investigated a micro-channel liquid-cooling system for the EV battery modules. They developed a computational model that couples the heat transfer process from electrochemical–thermal predictions at the cell level to the overall calculation of the battery module. Numerical results demonstrated that the inclusion of baffles in the cooling micro-channels of the battery module could elevate the temperature gradient at the channel wall by enhancing flow mixing. As the height and number of baffles increase, a further improvement in heat transfer effects is achieved for the battery pack, albeit with an accompanying increase in the pressure required to drive the coolant flow.

Although liquid cooling for BTMS has considerable advantages, it has its drawbacks too. These systems can be complex and expensive due to additional components such as pumps and heat exchangers, leading to higher upfront costs and increased maintenance complexity. The risk of leaks poses a significant concern, which results in damage and safety hazards. The weight and size of liquid-cooling systems may be limited in applications where space and weight are critical factors. Compatibility issues during integration, potential energy consumption concerns, the risk of coolant freezing with volume change in cold environments, and environmental impacts further contribute to the considerations that must be weighed against the advantages of efficient heat dissipation.

2.3. PCM-Cooled and PCM-Fin BTMS

PCM (phase-change material) thermal management for LIBs involves the use of PCMs to regulate and control the temperature of the battery cells during various operational conditions. PCMs are substances that can absorb or release thermal energy during the process of changing phases, such as from solid to liquid or vice versa. PCMs with melting points ranging between 20 and 50 °C are commonly employed in research, aligning with the operational temperatures of batteries. The key advantages of PCM usage are linked to their substantial latent heat and their role in temperature regulation. This facilitates the isothermal absorption of heat generated by cells when their temperature surpasses the melting point. Conversely, when the battery temperature is low, the PCM releases heat isothermally. A BTMS based on PCMs thus holds the potential to control battery packs effectively at desired operational temperatures. A PCM-based BTMS exhibits four distinct melting stages during operation. In the initial stage, battery temperature rises significantly with heat conduction dominating, storing heat as sensible heat. The temperature increase slows in the second stage, reaching an equilibrium in the third stage when natural convection and latent heat energy storage offer robust cooling. The fourth stage sees the consumption of latent heat, and heat dissipation occurs through natural convection. Challenges in PCM-based BTMS are associated with the core PCM, characterized by low thermal conductivity, high volume change, low form stability, and flammability (organic PCM). Most PCMs exhibit low thermal conductivities between 0.2 and 0.4 W/(m·K), which are suboptimal for efficient heat transfer [68]. To address this, promising solutions involving creating heat conduction paths and incorporating thermally conductive materials such as porous materials and nano additives into PCMs have been explored. Porous materials absorb PCMs, forming heat conduction paths to enhance thermal conductivity. Nano additives significantly improve thermal conductivity with a minimal mass or volume ratio, although measures must be taken to preserve form stability [69]. Table A3 (presented in Appendix A) summarizes the existing studies related to PCM-cooled cylindrical BTMS. This table classified the previous studies based on their method, battery type, number of cells, load, PCM composite, melting point, thermal conductivity, main focus, and achievements.

PCM-based cooling systems are usually accompanied by fin structures. Fins are thin, extended surfaces that are added to heat transfer surfaces to enhance the efficiency of heat dissipation. The design of fins involves considerations such as length, thickness, spacing, and material properties, all of which impact the overall heat transfer performance. An optimized fin structure increases the surface area available for heat exchange, promotes better airflow, and ultimately improves the cooling efficiency of the system, helping to maintain optimal operating temperatures for the associated equipment. Najafi et al. [70] devised an optimal BTMS by integrating PCM, metal foam, and fins. They assessed the thermal performance under a discharge with a 3C current rate. According to their findings, the BTMS incorporating PCMs, metal foam, and fins exhibited a maximum reduction of 3 K in battery surface temperature compared to a system using pure PCMs. Additionally, this configuration demonstrated a maximum delay of approximately 470 s in the PCM melting process. The results from this optimized BTMS design revealed that the applied fins function as a network of heat sources, facilitating heat dispersion within the system. The use of metal foam contributed to a uniform distribution of heat between the battery and the surrounding environment. Moreover, the analysis of PCM melting indicated that the primary factor contributing to the delay in PCM melting was the integration of metal foam and fins within the BTMS.

Metal fins are favored for their straightforward structures and easy manufacturability, leading to their extensive application in the cooling of LIBs. Sun et al. [71] introduced innovative fin structures comprising longitudinal fins and cylindrical rings, and assessed the impact of various parameters, including the position and number of the cylindrical rings, the number of longitudinal fins, and the heat generation rate, on the thermal management effectiveness of the PCM–fin-based BTMS (Figure 14). Their BTMS configuration included a cylindrical battery, metal fins, PCM, and a nylon housing, with the battery positioned at the center of the cylindrical housing. Metal fins, incorporating a cylindrical ring and four longitudinal fins, enveloped the outer surface of the battery. The ring and fins matched the height of the battery, while the remaining housing sections were filled with paraffin wax. Their findings indicated that the proposed fin structures established a thermally conductive network within the PCM, facilitating the rapid transfer of heat generated in the battery to the PCM and significantly extending the battery's operational duration. Additionally, employing more than one ring increased the heat transfer area between the battery and the PCM but adversely affected the BTMS's thermal management performance due to an excessive reduction of PCMs within the system.

Observing the melting process of PCMs provides an effective means to analyze the thermal characteristics of PCM–fin systems. Chen et al. [72] tried to determine the optimal size of the battery compartment and the number of fins to minimize the maximum battery temperature during the discharge phase (Figure 15). In their investigation, a cylindrical battery was immersed in a chamber filled with a phase-change material (PCM), and several fins of equal length were affixed to the battery. The finding indicated that the battery, fitted with 15 fins, had the most advantageous performance in terms of PCM melting at the onset of the cooling procedure. Around one-third of the cooling duration, the PCM reached its peak melting, measuring 26.16%. Additionally, throughout the entire cooling period, the battery with nine fins exhibited the lowest maximum temperature and the highest volume fraction of liquid PCMs. Beyond this point, an increase in the number of fins led to an elevation in battery temperature.

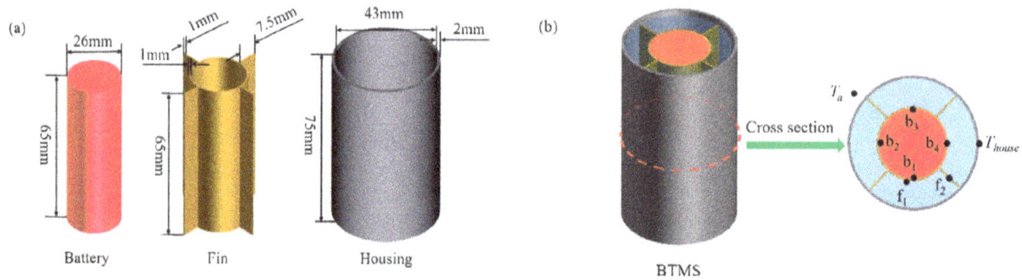

Figure 14. Schematic diagram of the PCM–fin-based BTMS: (**a**) configuration of BTMS and (**b**) locations of thermocouples [71].

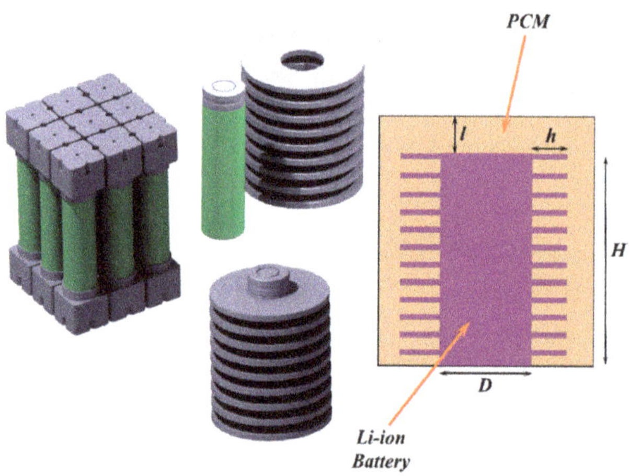

Figure 15. Schematic of the PCM–fin system geometry [72].

Numbers, locations, and shapes of fins can be arranged to reach the optimal thermal performance. Mansir et al. [73] investigated a cylindrical battery pack comprising three cylindrical cells enclosed in a space filled with PCMs. Their objective was to identify the optimal fin shape to maximize heat transfer. In another study, Qi et al. [74] conducted an optimization of the dimensions and spacing of fins installed on a LIB module. Through adjustments in fin height, fin width, and their respective distances, they investigated the resulting values of battery temperature and the quantity of liquid PCMs at various points during charging. The study demonstrated that an increase in fin size led to an augmentation of liquid PCMs during both the average charging period and the full charging time. The same effect was observed when increasing the distance between the fins. Notably, fin height had a more pronounced impact on both the quantity of liquid PCMs and battery temperature compared to alterations in fin width and spacing. Elevating fin height resulted in a reduction of both the maximum and average battery temperature throughout the charging process, including the full charging duration. Moreover, augmenting the distance between fins for fins of average size initially heightened the temperature and subsequently reduced it. Meanwhile, an increase in fin width initially led to a reduction in temperature, followed by an increase.

Mechanical vibration plays a crucial role in the PCM-based BTMS, helping to restrict excessive temperature elevation and enhance the uniformity of temperature in lithium-ion batteries. Zhou et al. [75] examined the impact of mechanical vibration on a PCM-based module for thermal management in a single 26650 cylindrical cell, particularly under high ambient temperature and high discharge rate. Their findings indicated that mechanical vibration improved natural convection heat transfer, mitigating excessive temperature rise and promoting temperature uniformity. Notably, the most significant influence of mechanical vibration on the PCM-based BTMS occurred when the PCM thickness was 12 mm. Additionally, a vibration frequency of 10 Hz yielded the optimal cooling effect, resulting in a 28.05% lower temperature compared to scenarios without vibration. At a vibration frequency of 50 Hz, the temperature uniformity was maximized, achieving a maximum temperature difference of 3.86 K. When the vibration amplitude was 100 mm, the battery exhibited a maximum temperature difference of 1.89 K, corresponding to a 27.74% decrease in maximum temperature.

Enhancing the thermal conductivity of PCMs can be achieved through the addition of expanded graphite or by integrating them into metal foam. Idi et al. [76] developed and optimized a passive thermal management system by utilizing a composite of aluminum foam and paraffin RT27. The primary goal was to maintain the cell temperature below 27 °C during successive charge/discharge cycles. The findings demonstrated that the cell surface temperature was contingent on various factors, including the current rate, duration of charge and discharge cycles, and ambient temperature. The PCM was observed to effectively absorb the heat generated by the cell in a latent form during phase change. However, its limited thermal conductivity imposed constraints on its performance. The incorporation of aluminum foam enhanced the thermal management efficiency of the Li-ion cell. The optimization analysis demonstrated that underestimating the thickness, and consequently the mass of required PCM, resulted in exceedingly high temperatures. Furthermore, it was observed that increasing the volume of the PCM did not markedly influence the surface temperature of the cell.

Incorporating nanoparticles into phase-change materials (PCMs) represents an innovative strategy to enhance their thermal conductivity. Sazvar and Moqtaderi [77] performed a numerical analysis to explore cylindrical battery cooling systems utilizing PCMs. Their emphasis was on enhancing PCM's thermal conductivity through the incorporation of Cu nanoparticles, while also studying the latent heat and melting process of the nano-enhanced phase-change material (NEPCM). Their findings indicated that with increasing volume fractions, the rate of the melting process improved. The most significant enhancement, observed at a volume fraction of 10%, resulted in a 3.9% acceleration of the melting process. In another work, Talele and Zhao [69] investigated the comparative effect between organic paraffin wax PCMs and nano-enhanced PCMs to reduce NMC battery pack peak temperature. Their findings showed that the thermal effectiveness of the battery pack increased with nano-enhanced PCMs despite working at the harshest 2 C-rate and higher ambient temperature till 35 °C.

Some researchers endeavored to develop an analytical model and an exact solution to understand the phase-change process in such systems, aiming to identify the optimal operational conditions. Kermani et al. [78] introduced an innovative analytical model for the cooling process of cylindrical batteries utilizing PCMs. They examined two distinct types of PCMs and explored various scenarios involving different PCM properties and operational conditions. Additionally, they assessed the impact of considering the temperature field of the cells and the natural convection of the liquid PCM. Their findings revealed that PCM1, with a lower melting temperature and higher thermal conductivity compared to PCM2, required a thinner thickness, resulting in a more uniform temperature distribution across the pack. However, PCM2, with a significantly lower density, was preferable when the weight factor of the pack was crucial. In another study, Li et al. [79] introduced a design optimization approach for BTMS incorporating PCMs. The optimization goal was to decrease the PCM mass while meeting two criteria: ensuring that the highest temperature difference in the BTMS remains below a specified limitation, and ensuring the necessary operational time to maintain the battery temperature within a secure range. To overcome the low thermal conductivity of conventional pure PCM, they utilized the EG/PA composite PCM.

BCM-cooled BTMS, while offering certain benefits, comes with its own set of disadvantages. One significant drawback is the limited heat transfer rate of PCMs compared to traditional cooling methods. The relatively low thermal conductivity of these materials can result in slower heat dissipation, potentially leading to temperature imbalances and hotspots within the battery pack. Moreover, the selection of appropriate PCMs involves careful consideration of factors such as melting point and compatibility, adding complexity to system design. The need to integrate PCM solutions into existing battery configurations may require modifications, contributing to increased costs and potential challenges in implementation. Additionally, PCM systems may not be as effective in high-temperature environments, limiting their applicability in certain conditions.

2.4. *Hybrid BTMS*

The hybrid method of thermal management for LIBs combines multiple cooling techniques to optimize performance. Typically, this approach integrates both active and passive cooling methods. Active cooling involves mechanisms such as liquid cooling or air cooling using fans or heat pumps, while passive cooling relies on natural heat dissipation through heat sinks or phase-change materials. By blending these methods, the hybrid approach aims to capitalize on the strengths of each, enhancing the overall efficiency of the thermal management system. This allows for better control over temperature regulation, particularly in high-demand situations or challenging environments. The hybrid method is often designed to provide a balance between the effectiveness of active cooling and the simplicity and reliability of passive solutions, ensuring optimal battery performance, longevity, and safety in diverse operating conditions. Table A4 (presented in Appendix A) summarizes the existing studies related to hybrid cylindrical BTMS. This table classified the previous studies based on the method, battery type, number of cells, load, cooling methods, and achievements.

2.4.1. Combining Air Cooling and Liquid Cooling

Air cooling and liquid cooling are two of the most common cooling methods for the thermal management of lithium-ion batteries. Considering that air cooling alone cannot be effective, it is combined with other systems. In fact, in this type of hybrid system, by adding air cooling to liquid cooling, the heating capacity of the system is increased. Air cooling is usually added to the system along with fins to help reduce the temperature and improve temperature uniformity in areas that are not in contact with liquid cooling. Angani et al. [80] introduced a battery thermal management system incorporating internal Zig-Zag plates. These plates were strategically placed around each cell, maximizing the surface area in contact with the Zig-Zag plates and thereby facilitating improved heat transfer. Two BTMS models were proposed: a bottom plate liquid-cooling system and a hybrid parallel pipe system for battery modules (Figure 16). In the battery module with a bottom plate liquid-cooling system, copper pipes were affixed to the bottom plate for coolant circulation. The heat generated by the cells underwent a three-step process: first, Zig-Zag plates extracted the heat from the cells; second, this heat was conducted to the support plates; and finally, it reached the bottom plate. As the coolant circulated through the pipe, the heat from the bottom plate was carried away. In the battery module with a hybrid parallel pipe cooling system, perforated Zig-Zag plates were employed to allow increased air circulation. Augmenting the airflow velocity reduced the cell temperature by dissipating more heat from the battery. The use of perforated Zig-Zag plates resulted in a reduced battery temperature due to enhanced heat transfer rates compared to nonperforated Zig-Zag plates. A noteworthy increase in the convection heat transfer area ensured more cooling fluid comes into contact with the heat sink. After comparing the results, they demonstrated that the hybrid system achieved more uniform temperature, and a 28% improvement in thermal performance compared to the other system.

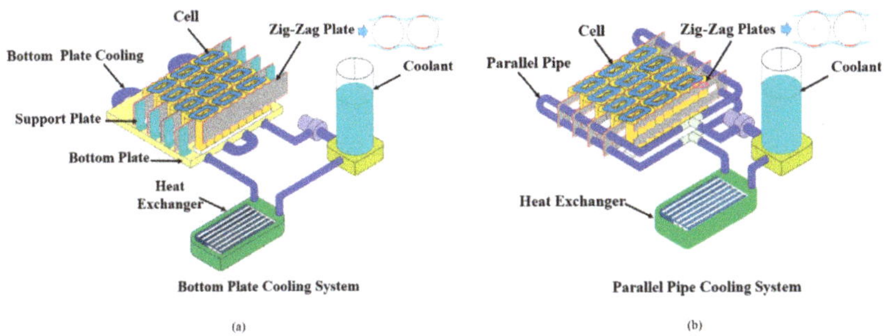

Figure 16. (**a**) Bottom plate liquid-cooling system. (**b**) Hybrid parallel pipe cooling system [80].

In another study, Zhou et al. [81] introduced a thermal management approach that integrated a half-helical coil and air jet impingement for cylindrical LIBs (Figure 17). This method involved transferring heat from the sides of the LIBs through a helical flow and dissipating it from the ends using air jet impingement. Their study revealed that the airflow and coolant flow configurations in the hybrid BTMS significantly impacted the system's heat dissipation. The adoption of a shortened fluid flow path jacket and positioning impinging air jets on both ends of the cylindrical structure resulted in minimizing the maximum temperature and temperature difference to 28.9 °C and 3.6 °C, respectively. Improved cooling performance was evident with an increase in the number of coil turns, attributed to the expanded heat transfer area. Alterations in airflow inlet velocity and inlet temperature had a minor effect on the temperature distribution of lithium-ion batteries. Decreased thermal contact resistance proved advantageous in enhancing the temperature consistency of the lithium-ion battery, with the temperature difference decreasing from 3.6 °C to 2.5 °C as contact resistance decreased from 0.0025 $m^2 \cdot K/W$ to 0.000625 $m^2 \cdot K/W$.

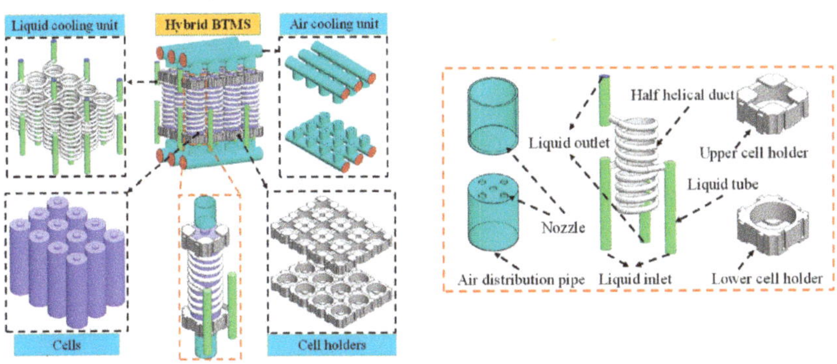

Figure 17. Schematic of a hybrid BTMS based on the half-helical coil coupled with air jet cooling [81].

One of the considerations of cylindrical modules is finding a way to increase the contact surface of the battery with the heat dissipation system. Xin et al. [82] presented a hybrid BTMS that combined air cooling and liquid cooling (Figure 18). Regularly spaced concave surfaces, matching the curvature of the battery's side surfaces, ensured close contact between the heat-conducting blocks (HCB) and the battery. Multiple HCBs were strategically positioned along the axial direction of the battery to promote uniform heat dissipation.

The cooling water circulating through the cooling channel progressively extracts heat from the aluminum block, absorbing it from the battery. In the liquid-cooled BTMS depicted in Figure 18, as the cooling water flows, it continuously absorbs the battery's heat, causing the water temperature to rise gradually and reducing the temperature difference between the battery and the water. Consequently, the heat dissipation of the battery diminishes in the direction of the cooling water flow, leading to a gradual increase in the corresponding battery temperature and the temperature difference among different batteries. Simultaneously, the cooling channel size affects the cooling performance of the battery module. The quantity, design, and arrangement of HCBs also influence the temperature uniformity of the battery and the heat exchange efficiency of the BTMS.

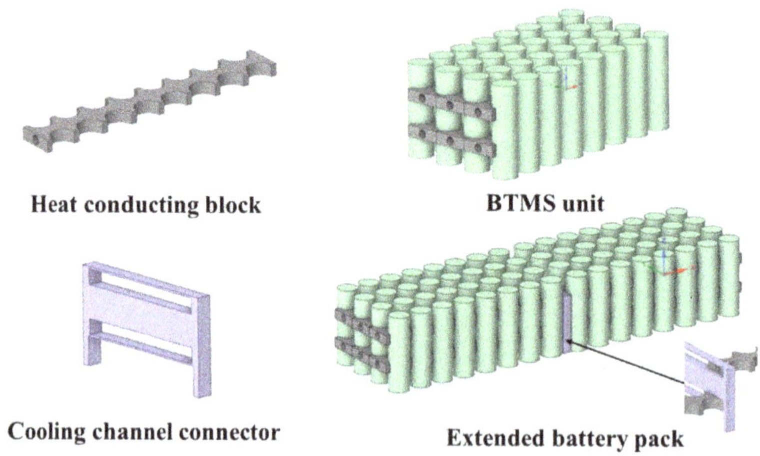

Figure 18. The schematic diagram of the BTMS 3D model [82].

2.4.2. Combining PCM Cooling and Liquid Cooling

The integration of phase-change materials and liquid cooling in a BTMS can combine the benefits of increased cooling capacity with reduced energy consumption. Yang et al. [83] introduced an innovative thermal management system for batteries, resembling a honeycomb structure, which incorporates a cooling plate with a hexagonal design incorporating bionic liquid mini-channels and PCMs. In Figure 19, the integration of the honeycomb-like battery thermal management system with the hexagonal cooling plate and PCMs is illustrated. A pump transports coolant from the water tank to the temperature control module, which subsequently circulates into the battery pack. Figure 19 details the battery module's structure, comprising cooling plates, cells, heat-conducting columns, and tubes. The hexagonal cooling plate incorporates bionic liquid mini-channels inspired by spider webs and honeycomb structures, with evenly distributed through-holes. Heat-conducting columns are positioned within the cell spaces, with their upper and bottom surfaces in contact with the cooling plate and their curved surfaces in contact with the cells. The surplus space in the heat-conducting columns is filled with the PCM. The coolant enters the battery module through the inlet, absorbs the generated heat, and returns to the water tank. In cases where the coolant's heat absorption rate is insufficient compared to the battery's heat generation rate, the PCM acts to absorb the excess heat.

Combining nanofluids with PCMs enhances battery performance during the charging process. Al-Rashed [84] utilized a combination of a hybrid nanofluid (NF) and nano-enhanced PCMs (Figure 20). The fluid within the battery, initially solid, acts as a PCM that undergoes melting due to the heat generated by the battery. Essentially, the battery provides the latent heat of vaporization for the PCM. During the material's solidification, the accumulated heat is transferred to the NF, with NF channels immersed in the PCM. Efficient performance requires careful consideration of the melting and freezing process times. The NF flow within the channels removes heat from the PCM, leading to an increased temperature gradient. This enhancement in battery performance results from heightened velocity attributed to a reduction in the boundary layer thickness.

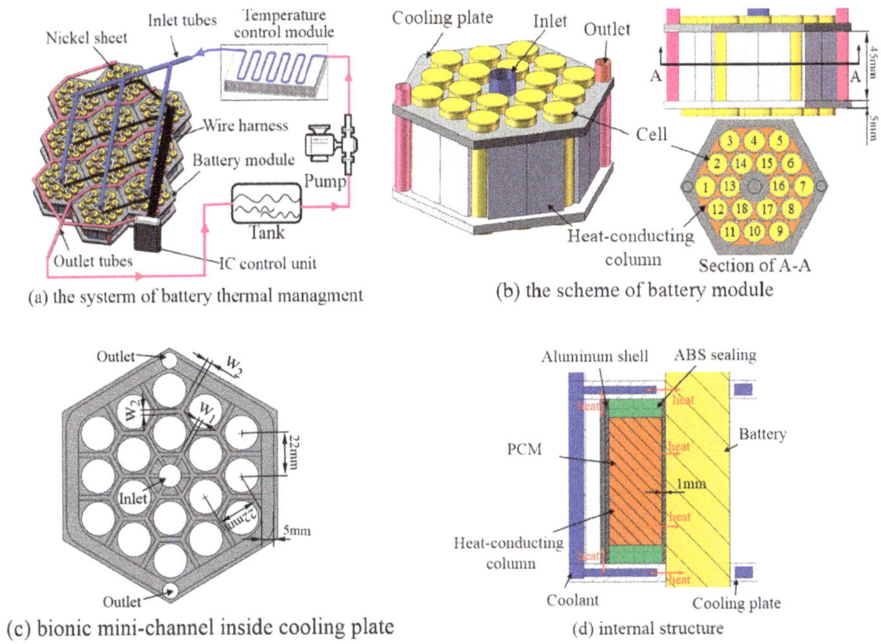

Figure 19. Scheme of the battery thermal management system [83].

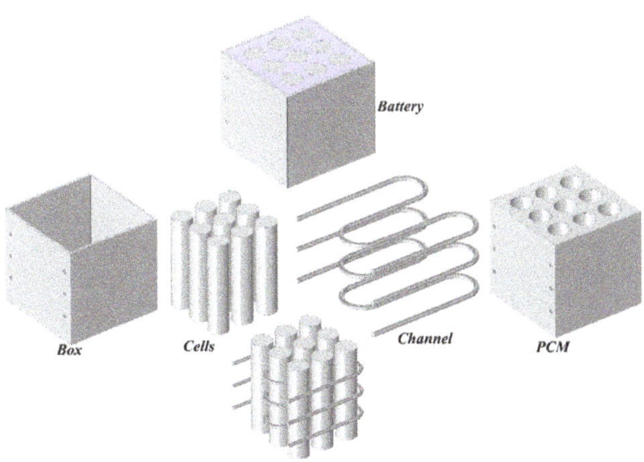

Figure 20. Schematic of a cooling system versus the proposed hybrid cooling system [84].

Combining PCMs and liquid cooling concurrently with a significant flow rate results in a decreased maximum temperature and a more uniform temperature distribution for the battery module, as opposed to employing either PCMs or liquid cooling in isolation. Xin et al. [85] devised a cooling system that incorporates both a composite phase-change material (graphite) and counterflow liquid cooling for a battery module consisting of 25 cylindrical batteries (Figure 21). They examined two distinct structures: the basic case and the composite case, exploring the effects of various parameters, such as thicknesses of the composite phase-change material, fluid directions, mass fractions of expanded graphite, coolant velocities, and coolant temperatures, on the battery module's maximum temperature and temperature uniformity. Their findings revealed a notable reduction in the maximum temperature with a lower coolant inlet temperature, eventually reaching stability. However, this improvement in maximum temperature was accompanied by a decrease in temperature uniformity. In another study, Cao et al. [86] examined the effectiveness of a hybrid thermal management system utilizing water cooling and composite phase PCMs for cylindrical battery cells. The individual contributions of liquid cooling and PCMs were investigated independently. An et al. [87] introduced an innovative thermal management system employing a composite PCM consisting of paraffin (RT44HC) and expanded graphite (EG), coupled with liquid cooling. This system aimed to regulate the rise in temperature and difference of lithium-ion batteries within an optimal range, particularly at a discharge rate of 3 C.

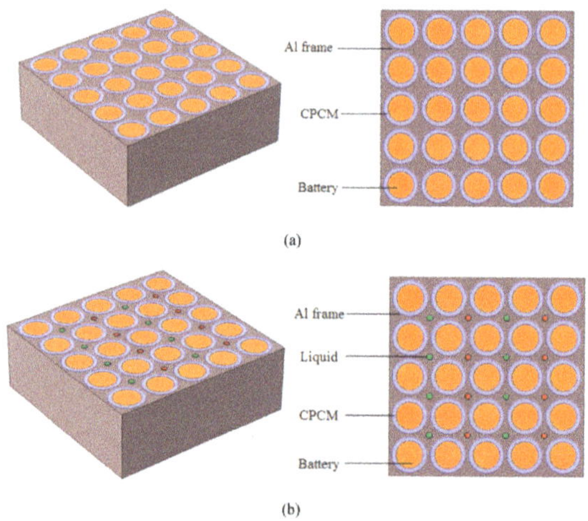

Figure 21. Structures of the battery module with (**a**) only CPCM and (**b**) the hybrid with liquid [85].

2.4.3. Combining Liquid Cooling and Heat Pipe

Liquid cooling results in additional energy consumption and creates temperature gradients within the battery pack. Despite numerous numerical and experimental proposals for cooling structures, working liquids, and strategies to address this issue, the challenge persists. The drawbacks of liquid cooling have spurred research efforts to integrate heat pipes with liquid cooling.

Heat pipes are components designed to dissipate heat efficiently by swiftly transferring thermal energy from one location to another. This process relies on the absorption of thermal energy (latent heat) when a liquid transforms into a gas and its subsequent release when the gas undergoes a phase change back into a liquid. Typically made of highly thermally conductive metals such as copper or aluminum, heat pipes consist of sealed metal pipes enclosing a small quantity of liquid known as a working fluid and a capillary structure (wick). Notably, this system operates without the need for external power or regular maintenance, incurs no operational costs, and demonstrates the capability to function over extended periods.

In order to present an efficient hybrid cooling method, Zheng et al. [88] introduced an innovative hybrid BTMS tailored for cylindrical batteries, incorporating micro heat pipe arrays (MHPA) and liquid cooling (Figure 22). The proposed system comprises Li-ion batteries (Type 18650), aluminum conduction elements, a U-shaped micro heat pipe array (MHPA), and cold plates. Positioned between two rows of batteries, the U-shaped MHPA efficiently draws heat from within the battery pack to the external environment. Liquid channels on either side of the battery pack act as heat exchangers, facilitating the removal of heat as needed. Specially designed aluminum conduction elements are placed between the batteries and the MHPA to maximize surface contact, and thermal grease ensures effective system contact. The heat generated by the batteries is transferred to the MHPA through the aluminum conduction element, enhancing the heat transfer area. Subsequently, the MHPA transfers the heat to the cooling plate through the internal phase-change cycle, and the water in the cooling plate carries the heat away. The U-shaped configuration of the MHPA, resembling a flat heat pipe with independent microscale channels, enables efficient heat transfer. The evaporation section absorbs heat, initiating a phase-change cycle in the internal working fluid. This process transfers heat to the condensation section, ensuring high thermal conductivity and uniform temperature distribution. To optimize the heat exchange area with the cold plates, the MHPA is configured in a U shape.

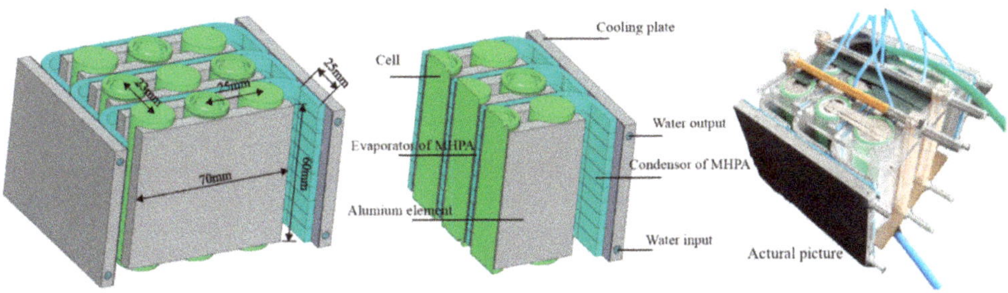

Figure 22. Geometry model of the designed hybrid battery module based on MHPA and liquid cooling [88].

2.4.4. Combining PCM Cooling and Heat Pipes

The thermal performance of a LIB pack can be strongly enhanced by the presence of PCMs and the heat pipe. The PCM exclusively absorbs or releases heat to the cell without participating in heat exchange with the environment, while the heat pipe is involved in exchanging heat with the surroundings. Incorporating a heat pipe alongside a PCM for heat removal can significantly enhance the system's thermal efficiency. When designing and manufacturing the hybrid cooling system, emphasis is placed on achieving lightweight and compact characteristics. The shape and structure of the cooling system are crafted to efficiently utilize the vacant space between tightly assembled cylindrical batteries, ensuring that the assembly process does not introduce additional volume to the battery module. Both the PCM tube and the heat pipe, including the fin structure, are constructed from aluminum, a material known for its lightweight properties and relatively low cost. Peng et al. [89] devised a cylindrical lithium-ion battery module featuring a compact hybrid cooling system integrating PCM and heat pipes. The batteries are closely arranged, and the vacant spaces between them are filled with either heat pipes or PCM tubes, as illustrated in Figure 23. The condensation section of the heat pipe is exposed to the surrounding environment. To ensure electric insulation, the battery module is enveloped in an epoxy board, and additional thermal insulation material is applied to minimize direct heat loss to the surroundings. The study revealed that, during the discharge process, the distribution of the PCM liquid fraction is non-uniform, with the outer layer and top section of the PCM melting first. Variations in heat dissipation rates among batteries, particularly under higher C-rate discharges, lead to temperature disparities among the batteries. Consequently, non-uniform PCM melting occurs, exacerbating the temperature differences between the batteries.

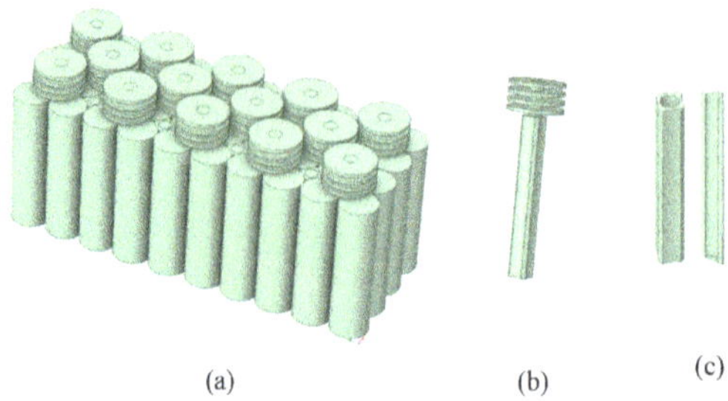

Figure 23. Configurations of (**a**) BTMS and battery module, (**b**) heat pipe, and (**c**) PCM tube [89].

3. Conclusions and Future Prospective

This paper presents a comprehensive review of the thermal management strategies employed in cylindrical lithium-ion battery packs. Lithium-ion batteries are sensitive to temperature fluctuations that can significantly impact their performance, safety, and lifespan. Maintaining an optimal temperature range is crucial to prevent overheating, which can lead to accelerated degradation, reduced capacity, and even safety hazards such as thermal runaway. Additionally, a well-designed thermal management system should also address temperature uniformity, ensuring that all cells within the battery pack experience consistent thermal conditions to avoid imbalances in performance and degradation. In addition, lightweight design is one critical parameter that should be considered.

In addition to all the criteria that should be considered for the thermal management systems of lithium-ion batteries, two important features specific to cylindrical batteries should also be considered. The thermal conductivity of cylindrical cells differs in different directions; therefore, the axial thermal conductivity is much more than the radial thermal conductivity. This feature makes the heat dissipation from the cell to the surrounding environment better through an axis form of heat transfer. The second issue is the geometry of cylindrical packs. Due to the geometry of the cylindrical cells, their placement in the modules comes with problems, such as the space between the cells, which makes it more difficult to remove heat and build a cooling system according to it. The cylindrical form factor imposes limitations on the design and integration of thermal management systems. Designing solutions that fit within the confined space while not compromising safety or performance remains a challenge.

The review covers four major thermal management techniques: air cooling, liquid cooling, PCMs, and hybrid methods. Air-cooling strategies are analyzed for their simplicity and cost-effectiveness, while liquid-cooling systems are explored for their superior heat dissipation capabilities. Phase-change materials, with their latent heat absorption and release properties, are evaluated as potential passive cooling solutions. Additionally, hybrid methods, combining two or more strategies, are discussed for their synergistic effects in achieving optimal thermal management. Figure 24 shows the comparison of different cooling methods of lithium-ion batteries.

Despite many studies, there are still challenges related to the thermal management of cylindrical lithium-ion batteries. Implementing advanced thermal management solutions can increase the overall cost of battery packs. Developing cost-effective strategies without compromising performance is an ongoing challenge. Future thermal management systems aim to enhance the overall performance and efficiency of cylindrical lithium-ion batteries. This includes optimizing heat dissipation to maintain safe operating temperatures, thus improving the battery's lifespan and performance. The utilization of advanced materials, such as novel thermal interface materials and PCMs, to enhance the thermal conductivity and heat dissipation capabilities of the battery systems should be encouraged. Additionally, the development of smart and adaptive thermal management systems that can dynamically respond to changing operating conditions, ensuring optimal thermal performance under varying loads and environmental conditions, merits future investigation. The need to explore innovative cooling techniques to address specific challenges associated with cylindrical battery geometries can still be seen.

	Advantages	Disadvantages
Air cooling	• Low cost • Simple structure • High reliability • Easy maintenance	• Low efficiency • Low cooling capacity • Noise (when using fan)
Liquid cooling	• Better efficiency • Higher cooling capacity • Flexibility in Design • Controllability • Reduced Noise (compared to air cooling)	• Complex structure • Leakage problem • Need a large space • Heavyweight
PCM cooling	• Low cost • Providing temperature uniformity • High efficiency • Simple Structure	• Leakage problem • Low thermal conductivity • Heavyweight • Flammability

Figure 24. Comparison between cooling strategies.

Due to the limited thermal conductivity of air-cooling systems, their widespread adoption is restricted. Therefore, there is a need for a high thermal management system that combines multiple traditional cooling methods. Researchers should prioritize the integration of liquid cooling, PCM cooling, and heat pipes. Hybrid systems, specifically combining liquid cooling and PCM cooling, are recommended due to the higher thermal conductivity of liquids and the greater latent heat of PCMs. While a PCM's standalone thermal conductivity may not be optimal, it can be improved by incorporating a conductive compound. Researchers should explore conductive materials and nanoparticles that enhance PCM conductivity when added. Significantly increasing the thermal conductivity of a PCM could make it the most efficient cooling system when combined with other cooling methods. Although heat pipe cooling is effective, its limited contact area poses a major challenge. To address this, the use of a conductive cold plate is recommended to increase the contact area of both conventional and pulsating heat pipes.

An important future study is the scale-up of thermal management systems. The critical question to raise is: what changes are to be made for battery packs of different sizes including the same type of cells? Take liquid channel cooling for example: to ensure a cell in a pack operates in the same temperature ranges, the amount of heat that has to be carried away by the coolant should be the same. However, as the coolant temperature increases faster in a larger pack, the temperature difference between the cell and local coolant reduces faster; as such, the coolant flow rate has to be increased to ensure a similar operation temperature. However, beyond a certain limit, increasing the coolant flow rate does not actually change the thermal performance; instead, it increases the pumping power. At this point, it might be helpful to increase the number of cooling channels, although geometrical and weight limitations should be considered. Overall, there is no straightforward scale-up strategy for a thermal management system. Parameters such as flow rate, cooling channel number, and others need to be adjusted simultaneously, which further depend on the quantitative prediction and optimization of a BTMS.

Author Contributions: Conceptualization, M.A.-E. and P.Z.; investigation, M.A.-E.; resources, M.A.-E. and P.Z.; data curation, M.A.-E.; writing—original draft preparation, M.A.-E.; writing—review and editing, M.A.-E. and P.Z.; supervision, P.Z.; project administration, P.Z.; funding acquisition, P.Z. All authors have read and agreed to the published version of the manuscript.

Funding: The research was funded by start-up funding at the University of Tennessee Space Institute.

Data Availability Statement: Not applicable.

Acknowledgments: Thanks to the invitation from the special issue editors, with the waiver for open access publication.

Conflicts of Interest: The authors declare no conflicts of interest.

Appendix A

Table A1. Summary of research on air-cooling BTMS (N.M: not mentioned).

Reference	Method	Type	Number	Load	Arrangement	Flow Regime	Airflow	T_{max} (K)	ΔT (K)	Main Focus	Other Objectives
Yang et al. [10]	Exp-Num	18650	37	3	Hexagonal	Turbulent	8–12 m/s	311	1.7	Air distribution plate	Structure parameters
Zhou et al. [36]	Exp-Num	18650	9	3–5	Aligned	Laminar	48 L/min	305.7	3	Air distribution pipe	Inlet pressure, discharge rate
Yang et al. [35]	Exp-Num	18650	1	3	-	Laminar	0.1–1.5 m/s	308	5	Optimization, bionic surface structure	Structure parameter, inlet velocity
Zhou et al. [90]	Exp-Num	18650	1	5	-	Laminar	0.1–1 m/s	310.191	1.978	Optimization, honeycomb cooling structure	Structure parameter
Hasan et al. [91]	Num	18650	52	N.M	Staggered	Turbulent	Re = 1.5–3 × 10^4	324	3.5	Turbulence intensity	Reynolds number
Zhao et al. [92]	Num	18650, 26650, 42110	24	5	Aligned	Turbulent	0.1–10 m/s	307	4		Gap spacing, environment temperatures
Li et al. [93]	Exp-Num	N.M	32	N.M	Aligned	Laminar	0.02–20 m/s	310	3.5		
Kummitha et al. [38]	Num	32700	9	5	Aligned	Turbulent	N.M	301.5	0.26	Cell holders' jacket fins	
Hai et al. [94]	Num	N.M	50	N.M	Aligned		Re = 400	301	0.93		Structure parameter
Mahamud et al. [43]	Num	N.M	8	1–10	Aligned	Turbulent	1 m/s	303	4	Reciprocating flow	Reciprocation period, cell spacing, coolant flow rate
Lu et al. [39]	Num	18650	14	0.5	Stagger-arranged	Laminar	10^{-4} kg/s	303.6	10.5		Channel size, inlet and outlet locations, channel length
Ji et al. [40]	Exp-Num	18650	16	0.6–2	Aligned	Laminar	0.001 m/s	307	2.7	Optimization	Structure parameters
Wang et al. [44]	Num	18650	24	1.5	Aligned	Turbulent	2–5 m/s	304.5	0.97	Optimization	Structure parameters

Table A1. *Cont.*

Reference	Method	Type	Number	Load	Arrangement	Flow Regime	Airflow	T_{max} (K)	ΔT (K)	Main Focus	Other Objectives
Li et al. [95]	Exp-Num	21700	1	0.5	-	N.M		300		Optimization	
Saechan et al. [96]	Num	18650	40	0.5–2	Aligned	Turbulent	1 m/s	315	5.2		Structure parameters
Ruhani et al. [97]	Num	18650	9	1	Aligned	Laminar	Re = 80–140				Reyolds number
Kausthubharam et al. [41]	Num	18650	36	1.4	Aligned	Laminar	0.1–1 m/s	304.2	0.4	Thermal interface material	Structural and operational parameters on
Xun et al. [98]	Num	18650	20	2	Aligned	Laminar	N.M	319	5		Number of cooling channels, channel size
Zhao et al. [99]	Num	18650	71	5	Stagger-arranged	Laminar	0.1 m/s	310	0.7	Multiple short channels	Structural and operational parameters on
Yang et al. [100]	Exp-Num	18650	8	1–4	Aligned	Turbulent	0.87–1.03 m/s	305	2.1		Structure parameters
Fan et al. [101]	Num	18650	32	0.6–4	Aligned, staggered, and cross	Laminar—Turbulent	0.1–10 m/s	305	1.4		Discharge rate, inlet temperature, energy efficiency
Cen et al. [102]	Exp-Num	18650	1	0.5–1.5	-	N.M	N.M	305	1.5		
E et al. [103]	Exp-Num	18650	60	0.5–1	Aligned	Laminar	2 m/s	307.2	2.4		Structure parameters
Chen et al. [104]	Num	18650	50			Laminar		309.2			
Hai et al. [105]	Num	18650	16	N.M	Aligned	Laminar	0.08–0.15 m/s	307.8	N.M		Structure parameters
Wang et al. [106]	Num	18650	25	1–3	Aligned		1 m/s	303	N.M		
Shahid et al. [107]	Exp-Num	18650	32	2	Aligned	Turbulent	N.M	301	1.1		Structure parameters
Zhang et al. [108]	Num	18650	25	3	Aligned	Turbulent	2.89×10^{-3} kg/s	308.8	1.493		Structure parameters
Hasan et al. [109]	Num	N.M	30	N.M	Aligned	Turbulent	Re = 15,000–30,000	303	N.M	Optimization	

Table A1. *Cont.*

Reference	Method	Type	Number	Load	Arrangement	Flow Regime	Airflow	T_{max} (K)	ΔT (K)	Main Focus	Other Objectives
Yang et al. [110]	Num	N.M	6	N.M	Aligned—staggered	Turbulent	0.01326 m^3/s	304.4	N.M	Optimization	Structure parameters
Shahid et al. [111]	Exp-Num	18650	32		Aligned	Turbulent	2–32 m/s	301.2	4.09	Vortex generators	
Peng et al. [112]	Exp-Num	18560	20	2	Aligned	Laminar	1 m/s	301.8	N.M		
Hai et al. [113]	Num	18650	16	N.M	Aligned—non-aligned	N.M	0.03–0.09 m/s	311	N.M		
Marambio et al. [114]	Num-Exp	N.M	25–30	1	Aligned	N.M	2, 2.5 m/s	309	N.M	Predicting the surface temperature	
Qin et al. [115]	Num-Exp	18650	4	1–4	Aligned	Laminar		321.2	4.5	Internal finned structure	
Shahid et al. [111]	Num-Exp	18650	32	2	Aligned		6 m/s	301.2	4.1		
Saw et al. [116]	Num	38120	24	3	Aligned		30 g/s	308.8	4		
Yu et al. [117]	Num-Exp	18650	66	0.5	Staggered		0.8 m/s	309.1	2.8		
Wang et al. [118]	Num-Exp	N.M	30	1	Aligned		8 m/s	55	17.6		
Wang et al. [37]	Num-Exp	N.M	24	1	Aligned		1 m/s	306.7	2.9		

Table A2. Summary of research on liquid-cooling BTMS (N.M: not mentioned).

Reference	Method	Type	Number	Load	Coolant	Inlet Flow	T_{max} (K)	ΔT (K)	Main Focus	Other Objective
Lai et al. [58]	Num	18650	1	5	Water	10^{-4} kg/s	313	4.3137	Weight, thermal conductive structure (TCS)	Mass flow rate, inner diameter, contact surface height, contact surface angle
Gao et al. [48]	Exp-Num	18650	16	2	Water	360 mL/min	310.03	1.96	Gradient channel	Flow direction, number of channel segments, segment lengths, inlet flow rate
Yates et al. [49]	Num	18650	4	5	Water	5×10^{-5} kg/s	313	3.15	Geometric modification	Channel number, hole diameter, mass flow rate, inlet locations
Li et al. [50]	Num	18650	20	1–4	Water	10^{-3} kg/s	303.26	1.984	Geometric modification	Inlet velocity, flow direction, channel cross-section shapes, battery spacing, pressure drop
Xia et al. [51]	Num	N.M	1	Not mentioned	Water	Not mentioned	322.95	2.38	Fractal tree-like channels	Pressure drop
Wang et al. [52]	Exp-Num	18650	20	3	Water	80 mL/min	308.89	4.17	Cooling mode (serial cooling and parallel cooling)	Flow rate, flow direction
Kim et al. [53]	Exp-Num	18650	36	2–10	Water	0.016 kg/s	323.15	N.M	Two different types of cold plates	
Xie et al. [67]	Num	18650	48	2	Water	0.2 m/s	304.01	4.99	Adding baffles	Number, height, and position of baffles
Sun et al. [119]	Num	21700	1	4	Water and ethylene glycol	7.95×10^{-2}	306.44	1.0605	Different channel structures	Channel diameter, inlet velocity
Sheng et al. [54]	Exp-Num	21700	7	1–2	Water	10^{-4} L/s	309.73	1.05	Cellular liquid cooling jacket	Flow rate, channel dimension
Chang et al. [120]	Exp-Num	18650	2	0.5	Water and ethylene glycol	0.65 L/min	308.15	1.67	Reciprocating liquid flow	Flow rate, ambient temperature
Xu et al. [55]	Num	18650	32	1–3	Water	0.1 m/s	30.3.2	3.69	Wrench-shaped	Bifurcation position, pressure drop

Table A2. *Cont.*

Reference	Method	Type	Number	Load	Coolant	Inlet Flow	T_{max} (K)	ΔT (K)	Main Focus	Other Objective
Sarchami et al. [63]	Exp-Num	18650	6	5	Alumina nanofluid	10–120 mL/min	305.13	2.01	Copper sheath	C-rate, alumina nanoparticle concentration, inflow velocity, stair channel geometry
Dong et al. [121]	Num	18650	1	5	Water	1.94×10^{-3} kg/s	305.51	6.98	Double helix structure	Mass flow rate, helix groove pitch, flow diameter
Zhao et al. [99]	Num	18650	71	5	50% water + 50% methanol	0.1 m/s	309.9	2.2	Geometric modification	
Xiong et al. [60]	Num	26650	1	0.5–2	Water	0.4 g/s	302.972	3.858	Bionic flow channel structure	Channel height, channel length, mass flow rate
Shan et al. [59]	Num	18650	10–30	2	Water	0.02 m/s	300.47	2.09	Lightweight design	Diameter of the coolant channel, contact surface angle, flow direction layout
Liang et al. [61]	Num	18650	5	3	Water	6×10^{-7} m^3/s	313.116	4.339	Lightweight design, lower power consumption	Flow rate, inlet water temperature, number of components
Tete et al. [57]	Num	18650	25	0.5–5	Water	0.01 m/s	301.15	0.12	Cell casings	Discharge rate
Tousi et al. [62]	Num	18650 21700	12	1–2	DI-water and AgO nanofluid	0.2–0.28 m/s	306.65	0.67	Using nanofluid	Pressure drop, flow rate, nanoparticle volume fraction, discharge C-rate
Li et al. [122]	Num	18650	40	3	Water	0.03–0.3 m/s	302.4	2.88	Comparing serpentine channel and the U-shaped channel, optimization	Pressure drop
Wang et al. [66]	Num	18650	90	5	Water	0.2–0.5 m/s	302.15	3.3	Optimization	Inlet velocity, channel number, contact angle
Liu et al. [123]	Num	18650	24	3, 5	Water and liquid metal	0.5 m/s	311.85	3.3	Liquid metal	Pumping power, flow velocity, structure

Table A2. *Cont.*

Reference	Method	Type	Number	Load	Coolant	Inlet Flow	T_{max} (K)	ΔT (K)	Main Focus	Other Objective
Yao et al. [124]	Exp-Num	18650	18	3	Water	N.M	311.75	1.9	Bionic spider web channel	Discharge rates, fluid flow arrangements, Reynolds number
Zeng et al. [64]	Exp-Num	18650	32	1–6	Water	0.5 m/s	303.83	0.739	Reciprocating flow	Velocity, reciprocating period
Zhao et al. [125]	Num	18650	71	0.5–5	Water	0.5 m/s	308.15	1		C-rate, flow rate, interfacing area
Esmaeili et al. [126]	Num	18650	16	2	Water	0.00399 kg/s	301.4	0.78	Twisted tapes	Twist pitches, twisted tape edges, intensified ratios
Zhou et al. [127]	Num	18650	9	5	Water	3×10^{-4} kg/s	303.65	4.6	Half-helical duct	Mass flow rate, pitch and number of ducts, flow direction, duct diameter
Rao et al. [128]	Num	18650	24	3	Water	0.05–0.3 m/s	302.6	2.7	Contact surface	Block length, inlet velocity
Zhao et al. [129]	Num	42110	1	N.M	Water	10^{-3} kg/s	300	1.2	Geometric modification	Mini-channel quantity, flow rate, flow direction
Huang et al. [130]	Exp-Num	18650	48	1–5	Water	$1–8 \times 10^{-3}$ kg/s	304		Hysteresis time	Discharge rate, flow rate, heat transfer coefficient
Hasan et al. [131]	Num	18650	52	N.M	SiO_2-water, Al_2O_3 water, ZnO water, CuO water Nanofluid		313		Using nanofluid	Different Reynolds number, pumping power
Bohacek et al. [132]	Exp-Num	18650	96	1	Water	0.1–0.7 L/min	308	4.6	Optimization	Flow rate
Ma et al. [65]	Num		2	1	Water	0.1–1 m/s	305	2	Optimization	
Lv et al. [133]	Exp-Num	21700	25	1–3	Water	0.44 m/s	310.8	4	Spacing	
Liu et al. [56]	Num	26650	25	1.3	Water	0.025–0.4 m/s	308.1	2.25	Geometric modification	Flow velocity, tube diameter
An et al. [134]	Exp	18650	8	1.4	Water	0.05–0.35	308.15	0.7	Increasing contact surface	High-temperature environment, SOC of the battery module

Table A3. Summary of research on PCM-cooling BTMS.

Reference	Method	Type	Number	Load	PCM Composite	Melting Point (K)	Thermal Conductivity (W/(m·K))	T_{max} (K)	ΔT (K)	Main Focus
Idi et al. [76]	Exp-Num	18650	1	1–4.5	Paraffin aluminum foam	300.15–331.15	0.15–0.2	300	1.5	
Lamrani et al. [135]	Num	21700	24	3	RT-31, RT-35 HC, A-32H	300–309	0.2–0.22	305	3	
Kermani et al. [78]	Num	18650	15	2–10	Chloride hexahydrate ($CaCl_2 \cdot 6H_2O$) and paraffin wax	302–309	0.18–1.008	303	2	Optimization, thermophysical properties of the PCM, rate of charge/discharge
Zhou et al. [75]	Num	18650	1	2	*n*-octadecane	301.15	0.152	332.36	1.89	Mechanical vibration frequency, vibration amplitude
Karimi et al. [136]	Exp	18650	1		Paraffin mixture	312–318	0.22	309	3	Nanoparticles
Lv et al. [137]	Exp	18650	24	1, 2.5, 3.5	Paraffin, expandable graphite, low-density polyethylene	317–323	0.16–0.38	321	3.9	
Khaboshan et al. [70]	Num	18650	1	3	*n*-eicosane PCM and metal foam	309.55	0.423	310.6		Fin shapes
Chen et al. [72]	Num	18650	1	0.5–1.7	Paraffin with alkane chain		0.358	315		Number of fins
Bias et al. [138]	Exp-Num	18650	1	3	Paraffin with alkane			314.9	1.5	
Mansir et al. [73]	Num	18650	3	2.4	Paraffin		0.358			Fin arrangement
Liu et al. [139]	Num	18650	1	1–3	Paraffin wax	311.15	0.2	314	1.48	Ambient temperature, fin shape
Qi et al. [74]	Num	18650	1	1–4	Paraffin wax			325		Optimization, fin structure
Li et al. [79]	Num	18650	4		Expanded graphite (eg)/paraffin (pa)	317			2.9	Optimization, mass of PCM
Wang et al. [140]	Num	18650	21			302		311		
Sun et al. [71]	Num	18650	1		Paraffin wax	313–317	0.2			Fin structure

Table A4. Summary of research on hybrid BTMS.

Reference	Method	Type	Number	Load	Cooling Method		Flow Regime	T_{max} (K)	ΔT (K)	Main Focus
Yang et al. [83]	Exp-Num	18650	18	2–4	Liquid cooling	PCM	Laminar	312	3.5	Structure parameters, working parameters, and conditions
Zhou et al. [81]	Exp-Num	18650	1	4	Liquid cooling	Air cooling	Turbulent	303	2	Different schemes, inlet velocity, contact resistance
Xin et al. [82]	Exp-Num	18650	32	3	Liquid cooling	Air cooling	Laminar	303.9	0.58	Heat-conducting blocks, mass flow rate
Angani et al. [80]	Exp-Num	21700	54	1, 1.25	Liquid cooling	Air cooling	Laminar	305	1.4	Structure parameters, airflow
Wang et al. [141]	Num	18650	48		Liquid cooling	Gas cooling	Turbulent	316.5		Airflow, turbulence intensity
Chen et al. [142]	Exp-Num	18650	24	4	Liquid cooling	PCM	Laminar	319.7	2.9	Mass flow rate
Behi et al. [143]	Exp-Num	18650	24	1.5	Heat pipe	Air cooling	Laminar	310.2		
Kong et al. [144]	Exp-Num	21700	6	3	Liquid cooling	PCM	Laminar	307		Cell-to-cell spacing, channel number, coolant velocity
Jiang et al. [145]	Num		9		Liquid cooling	PCM		308		
Khan et al. [146]	Num	18650	1	3	Air cooling	PCM	Laminar	303		Heat transfer coefficient, melting PCM content
Jilte et al. [147]	Num	18650	1	5	Liquid cooling	PCM		307.53		Nanoparticle concentration, structure parameters, melting behavior
Almehmadi et al. [148]	Num	18650	1		Air cooling	PCM	Laminar	304		Flow rate, PCM location
Zeng et al. [88]	Exp-Num	18650	9	1.3	Liquid cooling	Heat pipe		314.18	2.16	Optimization, design parameters
Mashayekhi et al. [149]	Exp-Num	18650	1	1–4	Liquid cooling	PCM		308	0.9	Nanofluid, power consumption, flow rate
Tian et al. [150]	Num	18650	9		Air cooling	PCM		303		Cost of electricity consumption
Gholami Zadeh et al. [151]	Num	18650	12	5	Liquid cooling	PCM		318.1	3.60	Aging analysis
Shahid et al. [152]	Exp-Num	18650	9	1–7	Liquid cooling	PCM	Laminar	302.5	1.8	

Table A4. *Cont.*

Reference	Method	Type	Number	Load	Cooling Method		Flow Regime	T_{max} (K)	ΔT (K)	Main Focus
Wang et al. [153]	Exp	18650	40	2	Heat pipe	PCM		320.8	2.5	
Zhao et al. [154]	Exp	18650	12	5	Heat pipe	PCM		320	5	
Weng et al. [155]	Exp-Num	18650	1	1–4	Air cooling	PCM		303.6	1.6	
Faizan et al. [156]	Exp-Num	18650	1	0.2–2	Liquid cooling	PCM	Laminar	307	1.6	
Cao et al. [86]	Exp-Num	18650	20	3	Liquid cooling	PCM	Laminar	315	1.2	Mass fraction of PCMs
Chavan et al. [157]	Exp-Num	18650	4	1	Liquid cooling	Heat pipe		311.4		
An et al. [87]	Num	18650	25	3	Air cooling	PCM	Laminar	319.4	2	Mass fraction of PCM, channel layout
Peng et al. [89]	Num	18650	4	0.5–2	Heat pipe	PCM		308	1	PCM properties
Zhao et al. [158]	Exp-Num	18650	40	0.96–1.92	Air cooling	PCM	Turbulent	312.53	4.70	Optimization, PCM properties
Li et al. [159]	Exp-Num	18650	40	2–4	Liquid cooling	Air cooling	Turbulent	314	2	Optimization, flow direction
Bamdezh et al. [160]	Num	18650	1	3	Liquid cooling	PCM	Laminar	304	2.7	
Jiang et al. [161]	Num	18650	9	1	Liquid cooling	PCM	Laminar	322		Nanoparticles
Qi et al. [162]	Num	18650	9		Air cooling	PCM	Laminar	305		
Lee et al. [163]	Num	18650	1		Air cooling	PCM		303		
Akkurt et al. [164]	Num	21700	3	3–4	Air cooling	PCM	Laminar	304		
Xin et al. [85]	Num	18650	25	5	Liquid cooling	PCM	Laminar	318.4	3.49	PCM properties
Al-Rashed [84]	Num		9		Liquid cooling	PCM	Laminar	312		Optimization
Cao et al. [165]	Num	18650	40	1–4	Liquid cooling	PCM	Laminar	313	5	Delayed cooling
Junwei et al. [166]	Exp	18650	15	1–6	Liquid cooling	PCM		310.28	3.75	

References

1. Ahmed, A.A.; Nazzal, M.A.; Darras, B.M.; Deiab, I.M. Global Warming Potential, Water Footprint, and Energy Demand of Shared Autonomous Electric Vehicles Incorporating Circular Economy Practices. *Sustain. Prod. Consum.* **2023**, *36*, 449–462. [CrossRef]
2. Halkos, G.E.; Gkampoura, E.C. Reviewing Usage, Potentials, and Limitations of Renewable Energy Sources. *Energies* **2020**, *13*, 2906. [CrossRef]
3. Behabtu, H.A.; Messagie, M.; Coosemans, T.; Berecibar, M.; Fante, K.A.; Kebede, A.A.; Van Mierlo, J. A Review of Energy Storage Technologies' Application Potentials in Renewable Energy Sources Grid Integration. *Sustainability* **2020**, *12*, 10511. [CrossRef]
4. Kebede, A.A.; Kalogiannis, T.; Van Mierlo, J.; Berecibar, M. A Comprehensive Review of Stationary Energy Storage Devices for Large Scale Renewable Energy Sources Grid Integration. *Renew. Sustain. Energy Rev.* **2022**, *159*, 112213. [CrossRef]
5. Chombo, P.V.; Laoonual, Y. A Review of Safety Strategies of a Li-Ion Battery. *J. Power Sources* **2020**, *478*, 228649. [CrossRef]
6. Simon, P.; Gogotsi, Y. Perspectives for Electrochemical Capacitors and Related Devices. *Nat. Mater.* **2020**, *19*, 1151–1163. [CrossRef]
7. Yang, L.; Nik-Ghazali, N.N.; Ali, M.A.H.; Chong, W.T.; Yang, Z.; Liu, H. A Review on Thermal Management in Proton Exchange Membrane Fuel Cells: Temperature Distribution and Control. *Renew. Sustain. Energy Rev.* **2023**, *187*, 113737. [CrossRef]
8. Sun, X.; Zhang, Y.; Zhang, Y.; Wang, L.; Wang, K. Summary of Health-State Estimation of Lithium-Ion Batteries Based on Electrochemical Impedance Spectroscopy. *Energies* **2023**, *16*, 5682. [CrossRef]
9. Zhang, X.; Li, Z.; Luo, L.; Fan, Y.; Du, Z. A Review on Thermal Management of Lithium-Ion Batteries for Electric Vehicles. *Energy* **2022**, *238*, 121652. [CrossRef]
10. Yang, W.; Zhou, F.; Chen, X.; Li, K.; Shen, J. Thermal Performance of Honeycomb-Type Cylindrical Lithium-Ion Battery Pack with Air Distribution Plate and Bionic Heat Sinks. *Appl. Therm. Eng.* **2023**, *218*, 119299. [CrossRef]
11. Mousavi, S.; Zadehkabir, A.; Siavashi, M.; Yang, X. An Improved Hybrid Thermal Management System for Prismatic Li-Ion Batteries Integrated with Mini-Channel and Phase Change Materials. *Appl. Energy* **2023**, *334*, 120643. [CrossRef]
12. Ham, S.H.; Jang, D.S.; Lee, M.; Jang, Y.; Kim, Y. Effective Thermal Management of Pouch-Type Lithium-Ion Batteries Using Tab-Cooling Method Involving Highly Conductive Ceramics. *Appl. Therm. Eng.* **2023**, *220*, 119790. [CrossRef]
13. Liu, C.; Shen, W.; Liu, X.; Chen, Y.; Ding, C.; Huang, Q. Research on Thermal Runaway Process of 18650 Cylindrical Lithium-Ion Batteries with Different Cathodes Using Cone Calorimetry. *J. Energy Storage* **2023**, *64*, 107175. [CrossRef]
14. Kim, E.; Song, J.; Dzakpasu, C.B.; Kim, D.; Lim, J.; Kim, D.; Park, S.; Lee, H.; Kwon, T.S.; Lee, Y.M. Degradation Behavior of 21700 Cylindrical Lithium-Ion Battery Cells during Overdischarge Cycling at Low Temperatures. *J. Energy Storage* **2023**, *72*, 108627. [CrossRef]
15. Sharma, D.K.; Prabhakar, A. A Review on Air Cooled and Air Centric Hybrid Thermal Management Techniques for Li-Ion Battery Packs in Electric Vehicles. *J. Energy Storage* **2021**, *41*, 102885. [CrossRef]
16. Talele, V.; Patil, M.S.; Panchal, S.; Fraser, R.; Fowler, M. Battery Thermal Runaway Propagation Time Delay Strategy Using Phase Change Material Integrated with Pyro Block Lining: Dual Functionality Battery Thermal Design. *J. Energy Storage* **2023**, *65*, 107253. [CrossRef]
17. Liu, X.; Ai, W.; Naylor Marlow, M.; Patel, Y.; Wu, B. The Effect of Cell-to-Cell Variations and Thermal Gradients on the Performance and Degradation of Lithium-Ion Battery Packs. *Appl. Energy* **2019**, *248*, 489–499. [CrossRef]
18. Troxler, Y.; Wu, B.; Marinescu, M.; Yufit, V.; Patel, Y.; Marquis, A.J.; Brandon, N.P.; Offer, G.J. The Effect of Thermal Gradients on the Performance of Lithium-Ion Batteries. *J. Power Sources* **2014**, *247*, 1018–1025. [CrossRef]
19. Song, Z.; Yang, N.; Lin, X.; Pinto Delgado, F.; Hofmann, H.; Sun, J. Progression of Cell-to-Cell Variation within Battery Modules under Different Cooling Structures. *Appl. Energy* **2022**, *312*, 118836. [CrossRef]
20. Raijmakers, L.H.J.; Danilov, D.L.; Eichel, R.A.; Notten, P.H.L. A Review on Various Temperature-Indication Methods for Li-Ion Batteries. *Appl. Energy* **2019**, *240*, 918–945. [CrossRef]
21. Jinasena, A.; Spitthoff, L.; Wahl, M.S.; Lamb, J.J.; Shearing, P.R.; Strømman, A.H.; Burheim, O.S. Online Internal Temperature Sensors in Lithium-Ion Batteries: State-of-the-Art and Future Trends. *Front. Chem. Eng.* **2022**, *4*, 804704. [CrossRef]
22. Zheng, Y.; Che, Y.; Hu, X.; Sui, X.; Stroe, D.I.; Teodorescu, R. Thermal State Monitoring of Lithium-Ion Batteries: Progress, Challenges, and Opportunities. *Prog. Energy Combust. Sci.* **2024**, *100*, 101120. [CrossRef]
23. Zhang, J.; Wu, B.; Li, Z.; Huang, J. Simultaneous Estimation of Thermal Parameters for Large-Format Laminated Lithium-Ion Batteries. *J. Power Sources* **2014**, *259*, 106–116. [CrossRef]
24. Drake, S.J.; Wetz, D.A.; Ostanek, J.K.; Miller, S.P.; Heinzel, J.M.; Jain, A. Measurement of Anisotropic Thermophysical Properties of Cylindrical Li-Ion Cells. *J. Power Sources* **2014**, *252*, 298–304. [CrossRef]
25. Zeng, Y.; Chalise, D.; Lubner, S.D.; Kaur, S.; Prasher, R.S. A Review of Thermal Physics and Management inside Lithium-Ion Batteries for High Energy Density and Fast Charging. *Energy Storage Mater.* **2021**, *41*, 264–288. [CrossRef]
26. Kumar, R.; Goel, V. A Study on Thermal Management System of Lithium-Ion Batteries for Electrical Vehicles: A Critical Review. *J. Energy Storage* **2023**, *71*, 108025. [CrossRef]
27. Gan, Y.; Wang, J.; Liang, J.; Huang, Z.; Hu, M. Development of Thermal Equivalent Circuit Model of Heat Pipe-Based Thermal Management System for a Battery Module with Cylindrical Cells. *Appl. Therm. Eng.* **2020**, *164*, 114523. [CrossRef]
28. Feng, L.; Zhou, S.; Li, Y.; Wang, Y.; Zhao, Q.; Luo, C.; Wang, G.; Yan, K. Experimental Investigation of Thermal and Strain Management for Lithium-Ion Battery Pack in Heat Pipe Cooling. *J. Energy Storage* **2018**, *16*, 84–92. [CrossRef]
29. Liu, W.; Jia, Z.; Luo, Y.; Xie, W.; Deng, T. Experimental Investigation on Thermal Management of Cylindrical Li-Ion Battery Pack Based on Vapor Chamber Combined with Fin Structure. *Appl. Therm. Eng.* **2019**, *162*, 114272. [CrossRef]

30. Alihosseini, A.; Shafaee, M. Experimental Study and Numerical Simulation of a Lithium-Ion Battery Thermal Management System Using a Heat Pipe. *J. Energy Storage* **2021**, *39*, 102616. [CrossRef]
31. Wang, J.; Gan, Y.; Liang, J.; Tan, M.; Li, Y. Sensitivity Analysis of Factors Influencing a Heat Pipe-Based Thermal Management System for a Battery Module with Cylindrical Cells. *Appl. Therm. Eng.* **2019**, *151*, 475–485. [CrossRef]
32. Li, Y.; Bai, M.; Zhou, Z.; Lv, J.; Hu, C.; Gao, L.; Peng, C.; Li, Y.; Li, Y.; Song, Y. Experimental Study of Liquid Immersion Cooling for Different Cylindrical Lithium-Ion Batteries under Rapid Charging Conditions. *Therm. Sci. Eng. Prog.* **2023**, *37*, 101569. [CrossRef]
33. Liu, Y.; Aldan, G.; Huang, X.; Hao, M. Single-Phase Static Immersion Cooling for Cylindrical Lithium-Ion Battery Module. *Appl. Therm. Eng.* **2023**, *233*, 121184. [CrossRef]
34. Song, W.; Bai, F.; Chen, M.; Lin, S.; Feng, Z.; Li, Y. Thermal Management of Standby Battery for Outdoor Base Station Based on the Semiconductor Thermoelectric Device and Phase Change Materials. *Appl. Therm. Eng.* **2018**, *137*, 203–217. [CrossRef]
35. Yang, W.; Zhou, F.; Zhou, H.; Liu, Y. Thermal Performance of Axial Air Cooling System with Bionic Surface Structure for Cylindrical Lithium-Ion Battery Module. *Int. J. Heat Mass Transf.* **2020**, *161*, 120307. [CrossRef]
36. Zhou, H.; Zhou, F.; Xu, L.; Kong, J.; Yang, Q. Thermal Performance of Cylindrical Lithium-Ion Battery Thermal Management System Based on Air Distribution Pipe. *Int. J. Heat Mass Transf.* **2019**, *131*, 984–998. [CrossRef]
37. Wang, T.; Tseng, K.J.; Zhao, J.; Wei, Z. Thermal Investigation of Lithium-Ion Battery Module with Different Cell Arrangement Structures and Forced Air-Cooling Strategies. *Appl. Energy* **2014**, *134*, 229–238. [CrossRef]
38. Kummitha, O.R. Thermal Cooling of Li-Ion Cylindrical Cells Battery Module with Baffles Arrangement for Airflow Cooling Numerical Analysis. *J. Energy Storage* **2023**, *59*, 106474. [CrossRef]
39. Lu, Z.; Yu, X.; Wei, L.; Qiu, Y.; Zhang, L.; Meng, X.; Jin, L. Parametric Study of Forced Air Cooling Strategy for Lithium-Ion Battery Pack with Staggered Arrangement. *Appl. Therm. Eng.* **2018**, *136*, 28–40. [CrossRef]
40. Ji, C.; Wang, B.; Wang, S.; Pan, S.; Wang, D.; Qi, P.; Zhang, K. Optimization on Uniformity of Lithium-Ion Cylindrical Battery Module by Different Arrangement Strategy. *Appl. Therm. Eng.* **2019**, *157*, 113683. [CrossRef]
41. Koorata, P.K.; Chandrasekaran, N. Numerical Investigation of Cooling Performance of a Novel Air-Cooled Thermal Management System for Cylindrical Li-Ion Battery Module. *Appl. Therm. Eng.* **2021**, *193*, 116961. [CrossRef]
42. Lee, N.; Choi, J.; Lee, J.; Shin, D.; Um, S. Prognostic Analysis of Thermal Interface Material Effects on Anisotropic Heat Transfer Characteristics and State of Health of a 21700 Cylindrical Lithium-Ion Battery Module. *J. Energy Storage* **2023**, *72*, 108594. [CrossRef]
43. Mahamud, R.; Park, C. Reciprocating Air Flow for Li-Ion Battery Thermal Management to Improve Temperature Uniformity. *J. Power Sources* **2011**, *196*, 5685–5696. [CrossRef]
44. Wang, Y.; Liu, B.; Han, P.; Hao, C.; Li, S.; You, Z.; Wang, M. Optimization of an Air-Based Thermal Management System for Lithium-Ion Battery Packs. *J. Energy Storage* **2021**, *44*, 103314. [CrossRef]
45. Jaliliantabar, F.; Mamat, R.; Kumarasamy, S. Prediction of Lithium-Ion Battery Temperature in Different Operating Conditions Equipped with Passive Battery Thermal Management System by Artificial Neural Networks. *Mater. Today Proc.* **2022**, *48*, 1796–1804. [CrossRef]
46. Li, A.; Yuen, A.C.Y.; Wang, W.; Chen, T.B.Y.; Lai, C.S.; Yang, W.; Wu, W.; Chan, Q.N.; Kook, S.; Yeoh, G.H. Integration of Computational Fluid Dynamics and Artificial Neural Network for Optimization Design of Battery Thermal Management System. *Batteries* **2022**, *8*, 69. [CrossRef]
47. Wang, Y.; Chen, X.; Li, C.; Yu, Y.; Zhou, G.; Wang, C.Y.; Zhao, W. Temperature Prediction of Lithium-Ion Battery Based on Artificial Neural Network Model. *Appl. Therm. Eng.* **2023**, *228*, 120482. [CrossRef]
48. Gao, R.; Fan, Z.; Liu, S. A Gradient Channel-Based Novel Design of Liquid-Cooled Battery Thermal Management System for Thermal Uniformity Improvement. *J. Energy Storage* **2022**, *48*, 104014. [CrossRef]
49. Yates, M.; Akrami, M.; Javadi, A.A. Analysing the Performance of Liquid Cooling Designs in Cylindrical Lithium-Ion Batteries. *J. Energy Storage* **2021**, *33*, 100913. [CrossRef]
50. Li, W.; Garg, A.; Wang, N.; Gao, L.; Le Phung, M.L.; Tran, V.M. Computational Fluid Dynamics-Based Numerical Analysis for Studying the Effect of Mini-Channel Cooling Plate, Flow Characteristics, and Battery Arrangement for Cylindrical Lithium-Ion Battery Pack. *J. Electrochem. Energy Convers. Storage* **2022**, *19*, 041003. [CrossRef]
51. Xia, C.; Fu, J.; Lai, J.; Yao, X.; Chen, Z. Conjugate Heat Transfer in Fractal Tree-like Channels Network Heat Sink for High-Speed Motorized Spindle Cooling. *Appl. Therm. Eng.* **2015**, *90*, 1032–1042. [CrossRef]
52. Wang, H.; Tao, T.; Xu, J.; Mei, X.; Liu, X.; Gou, P. Cooling Capacity of a Novel Modular Liquid-Cooled Battery Thermal Management System for Cylindrical Lithium Ion Batteries. *Appl. Therm. Eng.* **2020**, *178*, 115591. [CrossRef]
53. Kim, B.R.; Nguyen, T.N.; Park, C.W. Cooling Performance of Thermal Management System for Lithium-Ion Batteries Using Two Types of Cold Plate: Experiment and MATLAB/Simulink-Simscape Simulation. *Int. Commun. Heat Mass Transf.* **2023**, *145*, 106816. [CrossRef]
54. Sheng, L.; Zhang, H.; Su, L.; Zhang, Z.; Zhang, H.; Li, K.; Fang, Y.; Ye, W. Effect Analysis on Thermal Profile Management of a Cylindrical Lithium-Ion Battery Utilizing a Cellular Liquid Cooling Jacket. *Energy* **2021**, *220*, 119725. [CrossRef]
55. Xu, Q.; Xie, Y.; Huang, Y.; Li, X.; Huang, H.; Bei, S.; Wang, H.; Zheng, K.; Wang, X.; Li, L. Enhancement of Thermal Management for Cylindrical Battery Module Based on a Novel Wrench-Shaped Design for the Cold Plate. *Sustain. Energy Technol. Assess.* **2023**, *59*, 103421. [CrossRef]

56. Liu, Z.; Liu, X.; Meng, H.; Guo, L.; Zhang, Z. Numerical Analysis of the Thermal Performance of a Liquid Cooling Battery Module Based on the Gradient Ratio Flow Velocity and Gradient Increment Tube Diameter. *Int. J. Heat Mass Transf.* **2021**, *175*, 121338. [CrossRef]
57. Tete, P.R.; Gupta, M.M.; Joshi, S.S. Numerical Investigation on Thermal Characteristics of a Liquid-Cooled Lithium-Ion Battery Pack with Cylindrical Cell Casings and a Square Duct. *J. Energy Storage* **2022**, *48*, 104041. [CrossRef]
58. Lai, Y.; Wu, W.; Chen, K.; Wang, S.; Xin, C. A Compact and Lightweight Liquid-Cooled Thermal Management Solution for Cylindrical Lithium-Ion Power Battery Pack. *Int. J. Heat Mass Transf.* **2019**, *144*, 118581. [CrossRef]
59. Shan, S.; Li, L.; Xu, Q.; Ling, L.; Xie, Y.; Wang, H.; Zheng, K.; Zhang, L.; Bei, S. Numerical Investigation of a Compact and Lightweight Thermal Management System with Axially Mounted Cooling Tubes for Cylindrical Lithium-Ion Battery Module. *Energy* **2023**, *274*, 127410. [CrossRef]
60. Xiong, X.; Wang, Z.; Fan, Y.; Wang, H. Numerical Analysis of Cylindrical Lithium-Ion Battery Thermal Management System Based on Bionic Flow Channel Structure. *Therm. Sci. Eng. Prog.* **2023**, *42*, 101879. [CrossRef]
61. Liang, G.; Li, J.; He, J.; Tian, J.; Chen, X.; Chen, L. Numerical Investigation on a Unitization-Based Thermal Management for Cylindrical Lithium-Ion Batteries. *Energy Rep.* **2022**, *8*, 4608–4621. [CrossRef]
62. Tousi, M.; Sarchami, A.; Kiani, M.; Najafi, M.; Houshfar, E. Numerical Study of Novel Liquid-Cooled Thermal Management System for Cylindrical Li-Ion Battery Packs under High Discharge Rate Based on AgO Nanofluid and Copper Sheath. *J. Energy Storage* **2021**, *41*, 102910. [CrossRef]
63. Sarchami, A.; Najafi, M.; Imam, A.; Houshfar, E. Experimental Study of Thermal Management System for Cylindrical Li-Ion Battery Pack Based on Nanofluid Cooling and Copper Sheath. *Int. J. Therm. Sci.* **2021**, *171*, 107244. [CrossRef]
64. Zeng, W.; Ma, C.; Hu, S.; Li, S.; Zhang, Y. The Performance Investigation and Optimization of Reciprocating Flow Applied for Liquid-Cooling-Based Battery Thermal Management System. *Energy Convers. Manag.* **2023**, *292*, 117378. [CrossRef]
65. Ma, Y.; Mou, H.; Zhao, H. Cooling Optimization Strategy for Lithium-Ion Batteries Based on Triple-Step Nonlinear Method. *Energy* **2020**, *201*, 117678. [CrossRef]
66. Wang, Y.; Zhang, G.; Yang, X. Optimization of Liquid Cooling Technology for Cylindrical Power Battery Module. *Appl. Therm. Eng.* **2019**, *162*, 114200. [CrossRef]
67. Xie, L.; Huang, Y.; Lai, H. Coupled Prediction Model of Liquid-Cooling Based Thermal Management System for Cylindrical Lithium-Ion Module. *Appl. Therm. Eng.* **2020**, *178*, 115599. [CrossRef]
68. Talele, V.; Patil, M.S.; Panchal, S.; Fraser, R.; Fowler, M.; Gunti, S.R. Novel Metallic Separator Coupled Composite Phase Change Material Passive Thermal Design for Large Format Prismatic Battery Pack. *J. Energy Storage* **2023**, *58*, 106336. [CrossRef]
69. Talele, V.; Zhao, P. Effect of Nano-Enhanced Phase Change Material on the Thermal Management of a 18650 NMC Battery Pack. *J. Energy Storage* **2023**, *64*, 107068. [CrossRef]
70. Najafi Khaboshan, H.; Jaliliantabar, F.; Adam Abdullah, A.; Panchal, S. Improving the Cooling Performance of Cylindrical Lithium-Ion Battery Using Three Passive Methods in a Battery Thermal Management System. *Appl. Therm. Eng.* **2023**, *227*, 120320. [CrossRef]
71. Sun, Z.; Fan, R.; Yan, F.; Zhou, T.; Zheng, N. Thermal Management of the Lithium-Ion Battery by the Composite PCM-Fin Structures. *Int. J. Heat Mass Transf.* **2019**, *145*, 118739. [CrossRef]
72. Chen, H.; Abidi, A.; Hussein, A.K.; Younis, O.; Degani, M.; Heidarshenas, B. Investigation of the Use of Extended Surfaces in Paraffin Wax Phase Change Material in Thermal Management of a Cylindrical Lithium-Ion Battery: Applicable in the Aerospace Industry. *J. Energy Storage* **2022**, *45*, 103685. [CrossRef]
73. Mansir, I.B.; Sinaga, N.; Farouk, N.; Aljaghtham, M.; Diyoke, C.; Nguyen, D.D. Numerical Simulation of Dimensions and Arrangement of Triangular Fins Mounted on Cylindrical Lithium-Ion Batteries in Passive Thermal Management. *J. Energy Storage* **2022**, *50*, 104392. [CrossRef]
74. Qi, X.; Sidi, M.O.; Tlili, I.; Ibrahim, T.K.; Elkotb, M.A.; El-Shorbagy, M.A.; Li, Z. Optimization and Sensitivity Analysis of Extended Surfaces during Melting and Freezing of Phase Changing Materials in Cylindrical Lithium-Ion Battery Cooling. *J. Energy Storage* **2022**, *51*, 104545. [CrossRef]
75. Zhou, Z.; Chen, S.; Luo, M.; Du, W.; Wu, Y.; Yu, Y. Effect of Mechanical Vibration on Phase Change Material Based Thermal Management System for a Cylindrical Lithium-Ion Battery at High Ambient Temperature and High Discharge Rate. *Int. J. Heat Mass Transf.* **2023**, *211*, 124255. [CrossRef]
76. El Idi, M.M.; Karkri, M.; Abdou Tankari, M. A Passive Thermal Management System of Li-Ion Batteries Using PCM Composites: Experimental and Numerical Investigations. *Int. J. Heat Mass Transf.* **2021**, *169*, 120894. [CrossRef]
77. Sazvar, B.; Moqtaderi, H. A Numerical Study on the Capacity Improvement of Cylindrical Battery Cooling Systems Using Nano-Enhanced Phase Change Material and Axisymmetric Stepped Fins. *J. Energy Storage* **2023**, *62*, 106833. [CrossRef]
78. Ranjbar Kermani, J.; Mahlouji Taheri, M.; Shafii, M.B.; Moosavi, A. Analytical Solution, Optimization and Design of a Phase Change Cooling Pack for Cylindrical Lithium-Ion Batteries. *Appl. Therm. Eng.* **2023**, *232*, 120963. [CrossRef]
79. Li, Y.; Du, Y.; Xu, T.; Wu, H.; Zhou, X.; Ling, Z.; Zhang, Z. Optimization of Thermal Management System for Li-Ion Batteries Using Phase Change Material. *Appl. Therm. Eng.* **2018**, *131*, 766–778. [CrossRef]
80. Angani, A.; Kim, H.W.; Hwang, M.H.; Kim, E.; Kim, K.M.; Cha, H.R. A Comparison between Zig-Zag Plated Hybrid Parallel Pipe and Liquid Cooling Battery Thermal Management Systems for Lithium-Ion Battery Module. *Appl. Therm. Eng.* **2023**, *219*, 119599. [CrossRef]

81. Zhou, H.; Niu, J.; Guo, X.; Xu, L.; Song, Z.; Yin, X. Thermal Performance of a Hybrid Thermal Management System Based on the Half Helical Coil Coupled with Air Jet Cooling for Cylindrical Lithium-Ion Battery. *Appl. Therm. Eng.* **2023**, *225*, 120231. [CrossRef]
82. Xin, S.; Wang, C.; Xi, H. Thermal Management Scheme and Optimization of Cylindrical Lithium-Ion Battery Pack Based on Air Cooling and Liquid Cooling. *Appl. Therm. Eng.* **2023**, *224*, 120100. [CrossRef]
83. Yang, W.; Zhou, F.; Liu, Y.; Xu, S.; Chen, X. Thermal Performance of Honeycomb-like Battery Thermal Management System with Bionic Liquid Mini-Channel and Phase Change Materials for Cylindrical Lithium-Ion Battery. *Appl. Therm. Eng.* **2021**, *188*, 116649. [CrossRef]
84. Al-Rashed, A.A.A.A. Thermal Management of Lithium-Ion Batteries with Simultaneous Use of Hybrid Nanofluid and Nano-Enhanced Phase Change Material: A Numerical Study. *J. Energy Storage* **2022**, *46*, 103730. [CrossRef]
85. Xin, Q.; Xiao, J.; Yang, T.; Zhang, H.; Long, X. Thermal Management of Lithium-Ion Batteries under High Ambient Temperature and Rapid Discharging Using Composite PCM and Liquid Cooling. *Appl. Therm. Eng.* **2022**, *210*, 118230. [CrossRef]
86. Cao, J.; Luo, M.; Fang, X.; Ling, Z.; Zhang, Z. Liquid Cooling with Phase Change Materials for Cylindrical Li-Ion Batteries: An Experimental and Numerical Study. *Energy* **2020**, *191*, 116565. [CrossRef]
87. An, Z.; Chen, X.; Zhao, L.; Gao, Z. Numerical Investigation on Integrated Thermal Management for a Lithium-Ion Battery Module with a Composite Phase Change Material and Liquid Cooling. *Appl. Therm. Eng.* **2019**, *163*, 114345. [CrossRef]
88. Zeng, W.; Niu, Y.; Li, S.; Hu, S.; Mao, B.; Zhang, Y. Cooling Performance and Optimization of a New Hybrid Thermal Management System of Cylindrical Battery. *Appl. Therm. Eng.* **2022**, *217*, 119171. [CrossRef]
89. Peng, P.; Wang, Y.; Jiang, F. Numerical Study of PCM Thermal Behavior of a Novel PCM-Heat Pipe Combined System for Li-Ion Battery Thermal Management. *Appl. Therm. Eng.* **2022**, *209*, 118293. [CrossRef]
90. Zhou, J.; Zhang, S.; Yang, H.; Dong, F. Thermal Performance of a Novel Honeycomb Cooling System for Cylindrical Lithium-Ion Batteries. *Heat Transf. Eng.* **2023**, 1–17. [CrossRef]
91. Hasan, H.A.; Togun, H.; Abed, A.M.; Biswas, N.; Mohammed, H.I. Thermal Performance Assessment for an Array of Cylindrical Lithium-Ion Battery Cells Using an Air-Cooling System. *Appl. Energy* **2023**, *346*, 121354. [CrossRef]
92. Zhao, J.; Rao, Z.; Huo, Y.; Liu, X.; Li, Y. Thermal Management of Cylindrical Power Battery Module for Extending the Life of New Energy Electric Vehicles. *Appl. Therm. Eng.* **2015**, *85*, 33–43. [CrossRef]
93. Li, X.; He, F.; Ma, L. Thermal Management of Cylindrical Batteries Investigated Using Wind Tunnel Testing and Computational Fluid Dynamics Simulation. *J. Power Sources* **2013**, *238*, 395–402. [CrossRef]
94. Hai, T.; Abidi, A.; Sajadi, S.M.; Zain, J.M.; Malekshah, E.H.; Aybar, H. Simultaneous Cooling of Plate and Cylindrical Batteries in an Air-Cooled Lithium Battery Thermal Management System, by Changing the Distances of the Batteries from Each Other and the Pack Wall. *J. Taiwan Inst. Chem. Eng.* **2023**, *148*, 104931. [CrossRef]
95. Li, S.; Kirkaldy, N.; Zhang, C.; Gopalakrishnan, K.; Amietszajew, T.; Diaz, L.B.; Barreras, J.V.; Shams, M.; Hua, X.; Patel, Y.; et al. Optimal Cell Tab Design and Cooling Strategy for Cylindrical Lithium-Ion Batteries. *J. Power Sources* **2021**, *492*, 229594. [CrossRef]
96. Saechan, P.; Dhuchakallaya, I. Numerical Study on the Air-Cooled Thermal Management of Lithium-Ion Battery Pack for Electrical Vehicles. *Energy Rep.* **2022**, *8*, 1264–1270. [CrossRef]
97. Ruhani, B.; Abidi, A.; Hussein, A.K.; Younis, O.; Degani, M.; Sharifpur, M. Numerical Simulation of the Effect of Battery Distance and Inlet and Outlet Length on the Cooling of Cylindrical Lithium-Ion Batteries and Overall Performance of Thermal Management System. *J. Energy Storage* **2022**, *45*, 103714. [CrossRef]
98. Xun, J.; Liu, R.; Jiao, K. Numerical and Analytical Modeling of Lithium Ion Battery Thermal Behaviors with Different Cooling Designs. *J. Power Sources* **2013**, *233*, 47–61. [CrossRef]
99. Zhao, C.; Sousa, A.C.M.; Jiang, F. Minimization of Thermal Non-Uniformity in Lithium-Ion Battery Pack Cooled by Channeled Liquid Flow. *Int. J. Heat Mass Transf.* **2019**, *129*, 660–670. [CrossRef]
100. Yang, T.; Yang, N.; Zhang, X.; Li, G. Investigation of the Thermal Performance of Axial-Flow Air Cooling for the Lithium-Ion Battery Pack. *Int. J. Therm. Sci.* **2016**, *108*, 132–144. [CrossRef]
101. Fan, Y.; Bao, Y.; Ling, C.; Chu, Y.; Tan, X.; Yang, S. Experimental Study on the Thermal Management Performance of Air Cooling for High Energy Density Cylindrical Lithium-Ion Batteries. *Appl. Therm. Eng.* **2019**, *155*, 96–109. [CrossRef]
102. Cen, J.; Li, Z.; Jiang, F. Experimental Investigation on Using the Electric Vehicle Air Conditioning System for Lithium-Ion Battery Thermal Management. *Energy Sustain. Dev.* **2018**, *45*, 88–95. [CrossRef]
103. Jiaqiang, E.; Yue, M.; Chen, J.; Zhu, H.; Deng, Y.; Zhu, Y.; Zhang, F.; Wen, M.; Zhang, B.; Kang, S. Effects of the Different Air Cooling Strategies on Cooling Performance of a Lithium-Ion Battery Module with Baffle. *Appl. Therm. Eng.* **2018**, *144*, 231–241. [CrossRef]
104. Chen, H.; Abidi, A.; Mohammad Sajadi, S.; Yuan, Y.; Aybar, H.; Heidarshenas, B. Effect of Splitter Damper on Airflow Conduction for Thermal Management of a Lithium-Ion Battery Cooling System with Plate and Cylindrical Batteries. *J. Taiwan Inst. Chem. Eng.* **2023**, *148*, 104853. [CrossRef]
105. Hai, T.; Said, N.M.; Cheraghian, G.; Zhou, J.; Malekshah, E.H.; Sharifpur, M. Effect of Air Inlet and Outlet Cross Sections on the Cooling System of Cylindrical Lithium Battery with Segmental Arrangement Utilized in Electric Vehicles. *J. Power Sources* **2023**, *553*, 232222. [CrossRef]
106. Wang, T.; Tseng, K.J.; Zhao, J. Development of Efficient Air-Cooling Strategies for Lithium-Ion Battery Module Based on Empirical Heat Source Model. *Appl. Therm. Eng.* **2015**, *90*, 521–529. [CrossRef]

107. Shahid, S.; Agelin-Chaab, M. Development and Analysis of a Technique to Improve Air-Cooling and Temperature Uniformity in a Battery Pack for Cylindrical Batteries. *Therm. Sci. Eng. Prog.* **2018**, *5*, 351–363. [CrossRef]
108. Zhang, F.; Wang, P.; Yi, M. Design Optimization of Forced Air-Cooled Lithium-Ion Battery Module Based on Multi-Vents. *J. Energy Storage* **2021**, *40*, 102781. [CrossRef]
109. Hasan, H.A.; Togun, H.; Mohammed, H.I.; Abed, A.M.; Homod, R.Z. CFD Simulation of Effect Spacing between Lithium-Ion Batteries by Using Flow Air inside the Cooling Pack. *J. Energy Storage* **2023**, *72*, 108631. [CrossRef]
110. Yang, N.; Zhang, X.; Li, G.; Hua, D. Assessment of the Forced Air-Cooling Performance for Cylindrical Lithium-Ion Battery Packs: A Comparative Analysis between Aligned and Staggered Cell Arrangements. *Appl. Therm. Eng.* **2015**, *80*, 55–65. [CrossRef]
111. Shahid, S.; Agelin-Chaab, M. Analysis of Cooling Effectiveness and Temperature Uniformity in a Battery Pack for Cylindrical Batteries. *Energies* **2017**, *10*, 1157. [CrossRef]
112. Peng, X.; Cui, X.; Liao, X.; Garg, A. A Thermal Investigation and Optimization of an Air-Cooled Lithium-Ion Battery Pack. *Energies* **2020**, *13*, 2956. [CrossRef]
113. Hai, T.; Abidi, A.; Abed, A.M.; Zhou, J.; Malekshah, E.H.; Aybar, H. Three-Dimensional Numerical Study of the Effect of an Air-Cooled System on Thermal Management of a Cylindrical Lithium-Ion Battery Pack with Two Different Arrangements of Battery Cells. *J. Power Sources* **2022**, *550*, 232117. [CrossRef]
114. Reyes-Marambio, J.; Moser, F.; Gana, F.; Severino, B.; Calderón-Muñoz, W.R.; Palma-Behnke, R.; Estevez, P.A.; Orchard, M.; Cortés, M. A Fractal Time Thermal Model for Predicting the Surface Temperature of Air-Cooled Cylindrical Li-Ion Cells Based on Experimental Measurements. *J. Power Sources* **2016**, *306*, 636–645. [CrossRef]
115. Qin, P.; Liao, M.; Mei, W.; Sun, J.; Wang, Q. The Experimental and Numerical Investigation on a Hybrid Battery Thermal Management System Based on Forced-Air Convection and Internal Finned Structure. *Appl. Therm. Eng.* **2021**, *195*, 117212. [CrossRef]
116. Saw, L.H.; Ye, Y.; Tay, A.A.O.; Chong, W.T.; Kuan, S.H.; Yew, M.C. Computational Fluid Dynamic and Thermal Analysis of Lithium-Ion Battery Pack with Air Cooling. *Appl. Energy* **2016**, *177*, 783–792. [CrossRef]
117. Yu, X.; Lu, Z.; Zhang, L.; Wei, L.; Cui, X.; Jin, L. Experimental Study on Transient Thermal Characteristics of Stagger-Arranged Lithium-Ion Battery Pack with Air Cooling Strategy. *Int. J. Heat Mass Transf.* **2019**, *143*, 118576. [CrossRef]
118. Wang, Y.W.; Jiang, J.M.; Chung, Y.H.; Chen, W.C.; Shu, C.M. Forced-Air Cooling System for Large-Scale Lithium-Ion Battery Modules during Charge and Discharge Processes. *J. Therm. Anal. Calorim.* **2019**, *135*, 2891–2901. [CrossRef]
119. Sun, Y.; Bai, R.; Ma, J. Development and Analysis of a New Cylindrical Lithium-Ion Battery Thermal Management System. *Chin. J. Mech. Eng.* **2022**, *35*, 1–15. [CrossRef]
120. Chang, G.; Cui, X.; Li, Y.; Ji, Y. Effects of Reciprocating Liquid Flow Battery Thermal Management System on Thermal Characteristics and Uniformity of Large Lithium-Ion Battery Pack. *Int. J. Energy Res.* **2020**, *44*, 6383–6395. [CrossRef]
121. Dong, F.; Cheng, Z.; Zhu, J.; Song, D.; Ni, J. Investigation and Optimization on Cooling Performance of a Novel Double Helix Structure for Cylindrical Lithium-Ion Batteries. *Appl. Therm. Eng.* **2021**, *189*, 116758. [CrossRef]
122. Li, W.; Garg, A.; Xiao, M.; Gao, L. Optimization for Liquid Cooling Cylindrical Battery Thermal Management System Based on Gaussian Process Model. *Therm. Sci. Eng. Appl.* **2021**, *13*, 021015. [CrossRef]
123. Liu, Z.; Wang, H.; Yang, C.; Zhao, J. Simulation Study of Lithium-Ion Battery Thermal Management System Based on a Variable Flow Velocity Method with Liquid Metal. *Appl. Therm. Eng.* **2020**, *179*, 115578. [CrossRef]
124. Yao, F.; Guan, X.; Yang, M.; Wen, C. Study on Liquid Cooling Heat Dissipation of Li-Ion Battery Pack Based on Bionic Cobweb Channel. *J. Energy Storage* **2023**, *68*, 107588. [CrossRef]
125. Zhao, C.; Cao, W.; Dong, T.; Jiang, F. Thermal Behavior Study of Discharging/Charging Cylindrical Lithium-Ion Battery Module Cooled by Channeled Liquid Flow. *Int. J. Heat Mass Transf.* **2018**, *120*, 751–762. [CrossRef]
126. Esmaeili, Z.; Khoshvaght-Aliabadi, M. Thermal Management and Temperature Uniformity Enhancement of Cylindrical Lithium-Ion Battery Pack Based on Liquid Cooling Equipped with Twisted Tapes. *J. Taiwan Inst. Chem. Eng.* **2023**, *148*, 104671. [CrossRef]
127. Zhou, H.; Zhou, F.; Zhang, Q.; Wang, Q.; Song, Z. Thermal Management of Cylindrical Lithium-Ion Battery Based on a Liquid Cooling Method with Half-Helical Duct. *Appl. Therm. Eng.* **2019**, *162*, 114257. [CrossRef]
128. Rao, Z.; Qian, Z.; Kuang, Y.; Li, Y. Thermal Performance of Liquid Cooling Based Thermal Management System for Cylindrical Lithium-Ion Battery Module with Variable Contact Surface. *Appl. Therm. Eng.* **2017**, *123*, 1514–1522. [CrossRef]
129. Zhao, J.; Rao, Z.; Li, Y. Thermal Performance of Mini-Channel Liquid Cooled Cylinder Based Battery Thermal Management for Cylindrical Lithium-Ion Power Battery. *Energy Convers. Manag.* **2015**, *103*, 157–165. [CrossRef]
130. Huang, Y.; Wang, S.; Lu, Y.; Huang, R.; Yu, X. Study on a Liquid Cooled Battery Thermal Management System Pertaining to the Transient Regime. *Appl. Therm. Eng.* **2020**, *180*, 115793. [CrossRef]
131. Hasan, H.A.; Togun, H.; Abed, A.M.; Qasem, N.A.A.; Abderrahmane, A.; Guedri, K.; Eldin, S.M. Numerical Investigation on Cooling Cylindrical Lithium-Ion-Battery by Using Different Types of Nanofluids in an Innovative Cooling System. *Case Stud. Therm. Eng.* **2023**, *49*, 103097. [CrossRef]
132. Bohacek, J.; Raudensky, M.; Karimi-Sibaki, E. Polymeric Hollow Fibers: Uniform Temperature of Li-Ion Cells in Battery Modules. *Appl. Therm. Eng.* **2019**, *159*, 113940. [CrossRef]
133. Lv, Y.; Zhou, D.; Yang, X.; Liu, X.; Li, X.; Zhang, G. Experimental Investigation on a Novel Liquid-Cooling Strategy by Coupling with Graphene-Modified Silica Gel for the Thermal Management of Cylindrical Battery. *Appl. Therm. Eng.* **2019**, *159*, 113885. [CrossRef]

134. An, Z.; Li, D.; Zhang, C.; Luo, Y.; Zhang, J. Behaviours of Thermal Management System with Micro Channels for Cylindrical Lithium-Ion Cells under Fuzzy-PID Control Strategy. *Appl. Therm. Eng.* **2023**, *233*, 121089. [CrossRef]
135. Lamrani, B.; Lebrouhi, B.E.; Khattari, Y.; Kousksou, T. A Simplified Thermal Model for a Lithium-Ion Battery Pack with Phase Change Material Thermal Management System. *J. Energy Storage* **2021**, *44*, 103377. [CrossRef]
136. Karimi, G.; Azizi, M.; Babapoor, A. Experimental Study of a Cylindrical Lithium Ion Battery Thermal Management Using Phase Change Material Composites. *J. Energy Storage* **2016**, *8*, 168–174. [CrossRef]
137. Lv, Y.; Yang, X.; Li, X.; Zhang, G.; Wang, Z.; Yang, C. Experimental Study on a Novel Battery Thermal Management Technology Based on Low Density Polyethylene-Enhanced Composite Phase Change Materials Coupled with Low Fins. *Appl. Energy* **2016**, *178*, 376–382. [CrossRef]
138. Bais, A.R.; Subhedhar, D.G.; Joshi, N.C.; Panchal, S. Numerical Investigation on Thermal Management System for Lithium Ion Battery Using Phase Change Material. *Mater. Today Proc.* **2022**, *66*, 1726–1733. [CrossRef]
139. Liu, J.; Ma, Q.; Li, X. Numerical Simulation of the Combination of Novel Spiral Fin and Phase Change Material for Cylindrical Lithium-Ion Batteries in Passive Thermal Management. *Energies* **2022**, *15*, 8847. [CrossRef]
140. Wang, Z.; Liu, Y.; Mahmoud, M.Z. Simultaneous Application of Active and Passive Methods in Cooling of a Cylindrical Lithium-Ion Battery by Changing the Size of the Elliptical Cavity Filled with Nano Phase Change Materials. *J. Energy Storage* **2022**, *50*, 104693. [CrossRef]
141. Wang, S.; Li, Y.; Li, Y.Z.; Mao, Y.; Zhang, Y.; Guo, W.; Zhong, M. A Forced Gas Cooling Circle Packaging with Liquid Cooling Plate for the Thermal Management of Li-Ion Batteries under Space Environment. *Appl. Therm. Eng.* **2017**, *123*, 929–939. [CrossRef]
142. Chen, X.; Zhou, F.; Yang, W.; Gui, Y.; Zhang, Y. A Hybrid Thermal Management System with Liquid Cooling and Composite Phase Change Materials Containing Various Expanded Graphite Contents for Cylindrical Lithium-Ion Batteries. *Appl. Therm. Eng.* **2022**, *200*, 117702. [CrossRef]
143. Behi, H.; Karimi, D.; Behi, M.; Ghanbarpour, M.; Jaguemont, J.; Sokkeh, M.A.; Gandoman, F.H.; Berecibar, M.; Van Mierlo, J. A New Concept of Thermal Management System in Li-Ion Battery Using Air Cooling and Heat Pipe for Electric Vehicles. *Appl. Therm. Eng.* **2020**, *174*, 115280. [CrossRef]
144. Kong, D.; Peng, R.; Ping, P.; Du, J.; Chen, G.; Wen, J. A Novel Battery Thermal Management System Coupling with PCM and Optimized Controllable Liquid Cooling for Different Ambient Temperatures. *Energy Convers. Manag.* **2020**, *204*, 112280. [CrossRef]
145. Jiang, Y.; Wang, X.; Mahmoud, M.Z.; Elkotb, M.A.; Baloo, L.; Li, Z.; Heidarshenas, B. A Study of Nanoparticle Shape in Water/Alumina/Boehmite Nanofluid Flow in the Thermal Management of a Lithium-Ion Battery under the Presence of Phase-Change Materials. *J. Power Sources* **2022**, *539*, 231522. [CrossRef]
146. Khan, M.N.; Dhahad, H.A.; Alamri, S.; Anqi, A.E.; Sharma, K.; Mehrez, S.; Shamseldin, M.A.; Ibrahim, B.F. Air Cooled Lithium-Ion Battery with Cylindrical Cell in Phase Change Material Filled Cavity of Different Shapes. *J. Energy Storage* **2022**, *50*, 104573. [CrossRef]
147. Jilte, R.; Afzal, A.; Ağbulut, Ü.; Shaik, S.; Khan, S.A.; Linul, E.; Asif, M. Battery Thermal Management of a Novel Helical Channeled Cylindrical Li-Ion Battery with Nanofluid and Hybrid Nanoparticle-Enhanced Phase Change Material. *Int. J. Heat Mass Transf.* **2023**, *216*, 124547. [CrossRef]
148. Almehmadi, F.A.; Alqaed, S.; Mustafa, J.; Jamil, B.; Sharifpur, M.; Cheraghian, G. Combining an Active Method and a Passive Method in Cooling Lithium-Ion Batteries and Using the Generated Heat in Heating a Residential Unit. *J. Energy Storage* **2022**, *49*, 104181. [CrossRef]
149. Mashayekhi, M.; Houshfar, E.; Ashjaee, M. Development of Hybrid Cooling Method with PCM and Al_2O_3 Nanofluid in Aluminium Minichannels Using Heat Source Model of Li-Ion Batteries. *Appl. Therm. Eng.* **2020**, *178*, 115543. [CrossRef]
150. Tian, M.W.; Abed, A.M.; Yan, S.R.; Sajadi, S.M.; Mahmoud, M.Z.; Aybar, H.; Smaisim, G.F. Economic Cost and Numerical Evaluation of Cooling of a Cylindrical Lithium-Ion Battery Pack Using Air and Phase Change Materials. *J. Energy Storage* **2022**, *52*, 104925. [CrossRef]
151. Zadeh, P.G.; Gholamalizadeh, E.; Wang, Y.; Chung, J.D. Electrochemical Modeling of a Thermal Management System for Cylindrical Lithium-Ion Battery Pack Considering Battery Capacity Fade. *Case Stud. Therm. Eng.* **2022**, *32*, 101878. [CrossRef]
152. Shahid, S.; Agelin-Chaab, M. Experimental and Numerical Analysis of a Hybrid Thermal Management Concept at Different Discharge Rates for a Cylindrical Li-Ion Battery Module. *Batteries* **2023**, *9*, 474. [CrossRef]
153. Wang, Y.; Peng, P.; Cao, W.; Dong, T.; Zheng, Y.; Lei, B.; Shi, Y.; Jiang, F. Experimental Study on a Novel Compact Cooling System for Cylindrical Lithium-Ion Battery Module. *Appl. Therm. Eng.* **2020**, *180*, 115772. [CrossRef]
154. Zhao, J.; Lv, P.; Rao, Z. Experimental Study on the Thermal Management Performance of Phase Change Material Coupled with Heat Pipe for Cylindrical Power Battery Pack. *Exp. Therm. Fluid Sci.* **2017**, *82*, 182–188. [CrossRef]
155. Weng, J.; He, Y.; Ouyang, D.; Yang, X.; Chen, M.; Cui, S.; Zhang, G.; Yuen, R.K.K.; Wang, J. Honeycomb-Inspired Design of a Thermal Management Module and Its Mitigation Effect on Thermal Runaway Propagation. *Appl. Therm. Eng.* **2021**, *195*, 117147. [CrossRef]
156. Faizan, M.; Pati, S.; Randive, P. Implications of Novel Cold Plate Design with Hybrid Cooling on Thermal Management of Fast Discharging Lithium-Ion Battery. *J. Energy Storage* **2022**, *53*, 105051. [CrossRef]
157. Chavan, U.; Prajapati, O.; Hujare, P. Lithium Ion Battery Thermal Management by Using Coupled Heat Pipe and Liquid Cold Plate. *Mater. Today Proc.* **2023**, *80*, 382–388. [CrossRef]

158. Zhao, R.; Gu, J.; Liu, J. Optimization of a Phase Change Material Based Internal Cooling System for Cylindrical Li-Ion Battery Pack and a Hybrid Cooling Design. *Energy* **2017**, *135*, 811–822. [CrossRef]
159. Li, M.; Liu, F.; Han, B.; Guo, J.; Xu, Y. Research on Temperature Control Performance of Battery Thermal Management System Composited with Multi-Channel Parallel Liquid Cooling and Air Cooling. *Ionics* **2021**, *27*, 2685–2695. [CrossRef]
160. Bamdezh, M.A.; Molaeimanesh, G.R.; Zanganeh, S. Role of Foam Anisotropy Used in the Phase-Change Composite Material for the Hybrid Thermal Management System of Lithium-Ion Battery. *J. Energy Storage* **2020**, *32*, 101778. [CrossRef]
161. Jiang, Y.; Smaisim, G.F.; Mahmoud, M.Z.; Li, Z.; Aybar, H.; Abed, A.M. Simultaneous Numerical Investigation of the Passive Use of Phase-Change Materials and the Active Use of a Nanofluid inside a Rectangular Duct in the Thermal Management of Lithium-Ion Batteries. *J. Power Sources* **2022**, *541*, 231610. [CrossRef]
162. Qi, X.; Sajadi, S.M.; Mahmoud, M.Z.; Li, Z.; Shamseldin, M.A.; Aybar, H. Study of Circular, Horizontal and Vertical Elliptical Enclosures Filled with Phase Change Material in Thermal Management of Lithium-Ion Batteries in an Air-Cooled System. *J. Energy Storage* **2022**, *53*, 105041. [CrossRef]
163. Lee, J.; Abidi, A.; Mohammad Sajadi, S.; El-Shafay, A.S.; Degani, M.; Sharifpur, M. Study of the Effect of the Aspect Ratio of a Cylindrical Lithium-Ion Battery Enclosure in an Air-Cooled Thermal Management System. *J. Energy Storage* **2022**, *45*, 103684. [CrossRef]
164. Akkurt, N.; Aghakhani, S.; Mahmoud, M.Z.; Tag El Din, E.S.M. The Influence of Battery Distance on a Hybrid Air-Cooled Cylindrical Lithium-Ion Battery Phase Change Material Thermal Management System for Storing Solar Energy. *J. Energy Storage* **2022**, *52*, 104873. [CrossRef]
165. Cao, J.; Ling, Z.; Fang, X.; Zhang, Z. Delayed Liquid Cooling Strategy with Phase Change Material to Achieve High Temperature Uniformity of Li-Ion Battery under High-Rate Discharge. *J. Power Sources* **2020**, *450*, 227673. [CrossRef]
166. Li, J.; Zhang, H. Thermal Characteristics of Power Battery Module with Composite Phase Change Material and External Liquid Cooling. *Int. J. Heat Mass Transf.* **2020**, *156*, 119820. [CrossRef]

batteries

MDPI

Review

Recent Advances in Thermal Management Strategies for Lithium-Ion Batteries: A Comprehensive Review

Yadyra Ortiz, Paul Arévalo *, Diego Peña and Francisco Jurado

Department of Electrical Engineering, University of Jaén, 23700 Jaen, Spain; ymog0001@red.ujaen.es (Y.O.); dppb0002@red.ujaen.es (D.P.); fjurado@ujaen.es (F.J.)
* Correspondence: warevalo@ujaen.es

Abstract: Effective thermal management is essential for ensuring the safety, performance, and longevity of lithium-ion batteries across diverse applications, from electric vehicles to energy storage systems. This paper presents a thorough review of thermal management strategies, emphasizing recent advancements and future prospects. The analysis begins with an evaluation of industry-standard practices and their limitations, followed by a detailed examination of single-phase and multi-phase cooling approaches. Successful implementations and challenges are discussed through relevant examples. The exploration extends to innovative materials and structures that augment thermal efficiency, along with advanced sensors and thermal control systems for real-time monitoring. The paper addresses strategies for mitigating the risks of overheating and propagation. Furthermore, it highlights the significance of advanced models and numerical simulations in comprehending long-term thermal degradation. The integration of machine learning algorithms is explored to enhance precision in detecting and predicting thermal issues. The review concludes with an analysis of challenges and solutions in thermal management under extreme conditions, including ultra-fast charging and low temperatures. In summary, this comprehensive review offers insights into current and future strategies for lithium-ion battery thermal management, with a dedicated focus on improving the safety, performance, and durability of these vital energy sources.

Keywords: thermal management strategies; lithium-ion batteries; extreme conditions thermal management; advanced sensors; machine learning algorithms

1. Introduction

In the current landscape of sustainable mobility, the thermal management of lithium-ion batteries (LIBs) in electric vehicles (EVs) has established itself as an essential field of research, crucial to improving the efficiency and ensuring the safety of these energy systems. Battery thermal management systems (BTMSs) play a key role in this context, as they are decisive in keeping LIBs within an optimal temperature range, thus contributing to optimizing their performance and prolonging their lifetime. Recent research [1] highlights that advances in innovative materials and advanced designs in BTMSs are key for the effective management of the heat generated in the charging and discharging processes, which is especially relevant in high-demand applications such as EVs. However, the field faces significant challenges, mainly related to overheating and temperature variations in LIBs. These problems can compromise both the safety and performance of batteries, accelerating their aging and reducing their energy storage capacity. In extreme cases, these problems can even trigger fire or explosion risks due to thermal runaway (TR) [2]. This situation underlines the imperative need for efficient BTMSs that mitigate these risks, maintaining uniformity in battery temperature and avoiding extreme conditions [3,4]. Studies such as those presented in [5] emphasize the critical importance of integrity in BTMS studies, focusing on battery thermal safety as a key element to prevent overheating-related incidents in advanced battery technologies.

Advances in BTMSs have shown significant benefits in EVs, providing more accurate and uniform temperature regulation, and resulting in improved battery efficiency and reliability. Innovations in phase change materials (PCMs) and other BTMS technologies have improved heat dissipation and TR prevention, increasing the safety and energy density of batteries [1]. The study of heat generation in LIBs has gained importance, particularly in its impact on battery performance and safety. Scholars [6,7] have focused on understanding the interplay between thermo-electrochemical processes within batteries and how variable battery properties affect heat generation internally. Additionally, there has been a notable focus on developing effective thermal management strategies for use in EVs, with innovative methods combining pulsed operations with external liquid circulation, air and liquid cooling, PCMs, and heat pipes, each with its advantages and disadvantages [8–10]. Recently, a hybrid system has been highlighted that combines liquid cooling channels with PCMs, optimizing thermal efficiency and minimizing pressure loss [11]. Despite significant progress in the literature on the thermal management of lithium-ion batteries, critical challenges persist, warranting further in-depth investigation. The optimization of battery efficiency and safety remains a dynamically evolving area of study, given the rapid growth of technologies such as EVs and escalating demands for battery performance and durability. Identifying gaps in the literature underscores the necessity to address specific issues, including managing internal heat generation and implementing practical thermal management strategies in real-world scenarios. Despite advancements, there are still limitations in understanding how the variable properties of batteries impact heat generation and how these issues influence the long-term integrity and safety of batteries in high-demand applications. In this context, this work focuses on filling these gaps in the literature, presenting not only a comprehensive review of current advancements but also proposing innovative contributions that address these specific challenges. The key contributions of this study are summarized as follows:

- The primary contribution of this work lies in its comprehensive approach, addressing not only thermal efficiency to enhance battery performance but also placing significant emphasis on safety. This is achieved through innovative strategies in the design of BTMSs that tackle both overheating and temperature variations, thereby mitigating risks of accelerated aging and potential fire hazards.
- We contribute to the scientific literature by highlighting the essential role of advanced materials and innovative designs in BTMSs. This work provides a thorough review of recent advancements in this regard, emphasizing how these innovations can be crucial for effective thermal management during charging and discharging processes, especially in high-demand applications such as electric vehicles.
- A significant novelty of this review is the emphasis on researching internal heat generation in lithium-ion batteries. Through a detailed analysis of thermo-electrochemical processes and the impact of variable battery properties on heat generation, this work contributes to a better understanding of the fundamentals underlying battery efficiency and safety.
- This paper highlights a comprehensive evaluation of various thermal management strategies used in EVs. From pulsed operations to hybrid systems combining liquid cooling with PCMs, we provide a complete overview of the advantages and disadvantages of each approach, identifying best practices to optimize thermal efficiency and minimize pressure loss.
- We present specific results from a recent hybrid system that combines liquid cooling channels with PCMs. This work not only highlights the theory behind this innovation but also demonstrates its practical application, optimizing thermal efficiency and addressing pressure loss, which is crucial for successful implementation in EVs.

The rest of this study is structured as follows: Section 2 delves into the fundamentals of thermal management of LIBs and the shortcomings of existing systems. Section 3 focuses on advances in cooling methods, while Section 4 discusses sophisticated thermal management models. Section 5, meanwhile, investigates novel technologies in thermal monitoring and regulation. The aim is to highlight the critical role of BTMSs in the sustainable development of EVs and to make valuable contributions to the field of thermal management research, presenting findings relevant to both specialists and researchers in related fields. Finally, Section 6 concludes the paper.

2. Thermal Management in Lithium-Ion Batteries

Thermal management in LIBs is critical to their efficient and safe operation, especially in applications such as EVs and energy storage systems. Maintaining these batteries within an optimal temperature range, typically between 20 °C and 40 °C, is essential to prevent reliability problems [12,13]. There are three types of approaches to thermal management, active, passive, and hybrid systems, each with distinctive characteristics and suitable for different applications and requirements [14]. Active systems employ mechanical or electrical means, such as pumps and fans, to regulate the temperature of the batteries [15,16]. These methods, which include air and liquid cooling, are highly effective in dissipating heat but have the disadvantage of increasing system power consumption and thus reducing the overall efficiency of the battery [17]. Despite their effectiveness, these systems require a more complex design and are usually more expensive [18].

Passive systems, which use technologies such as PCMs and heat pipes, rely on natural processes such as conduction and convection for heat transfer [19]. These methods do not require additional energy, which makes them more efficient in terms of energy consumption and simpler in design. However, they may face challenges in their ability to handle high thermal loads or in situations of extreme temperatures [20]. In addition, certain materials such as PCMs may have issues with low thermal conductivity and risk of leakage after melting [21]. Hybrid systems, which combine aspects of active and passive approaches, seek to balance the advantages of both. For example, the integration of PCMs with air or liquid cooling systems can improve temperature control compared to purely passive methods, without reaching the high energy consumption of fully active systems [22]. These hybrid systems offer a promising solution but require careful design and advanced engineering to achieve an optimal balance between energy efficiency and thermal management effectiveness [23]. The proper choice of thermal management system is essential for LIBs, considering factors such as battery size, lifespan, and charge and discharge rates. Advances in new materials, such as nanometer PCMs, and advanced cooling and heating techniques are improving the efficiency and safety of these systems. These innovations are contributing to the increased adoption of batteries in a variety of applications, reducing costs and encouraging the use of cleaner, more sustainable energies [12]. In addition, the integration and compatibility of these systems with the overall EV or storage system design is a challenge [24]. Numerous research studies have been conducted that have proposed various design improvements to increase the efficiency of BTMSs, as detailed in Table 1.

Table 1. Thermal management systems in batteries: comparative analysis.

Ref.	BTMS Method	Operating Principle	Key Findings	Advantages	Disadvantages
[24]	Active	Uses forced air flow to cool the batteries in a rectangular container.	Modifications to outlet size and shape significantly decrease system temperature, improves cooling uniformity.	Requires no moving parts, improves temperature uniformity.	Limited heat transfer capacity, less effective for high thermal loads.
[25]	Active	Circulates water around the battery pack to dissipate heat.	More effective for thermal management at low cycling rates.	Effective for thermal management at low cycling rates, improves thermal performance.	Not as effective at high cycling rates, may require combination with other systems.

Table 1. *Cont.*

Ref.	BTMS Method	Operating Principle	Key Findings	Advantages	Disadvantages
[26]	Passive	Uses PCMs with applied pressure to enhance heat dissipation.	PCMs show the most promising performance compared to traditional active air/liquid cooling methods.	Maintains stable temperatures without energy consumption, improved performance with pressure.	Increased mechanical complexity and costs due to pressure application.
[27]	Passive	Uses PCMs, such as paraffin, to absorb and release heat during phase change.	Provides more uniform temperature distribution compared to air-cooling and liquid cooling.	Effective thermal management, uniform temperature distribution, paraffin is resistant and safe.	Low thermal conductivity of paraffin, slow thermal response.
[28]	Passive	Proposes a passive BTMS using a tetrahedral lattice porous plate for drone batteries.	Significant reduction in maximum temperature and thermal deviation on the battery surface.	Lightweight, requires no additional equipment, mechanically protects the battery.	Minimal weight increase, challenges in heat management across different operation modes.
[29]	Passive	Based on using PCMs to control temperature through heat absorption and release.	PCM-based BTMSs stand out for their cost-effectiveness and ability to maintain temperature uniformity.	Cost-effective, simple installation, minimal space required, excellent temperature uniformity.	Challenges in PCM application, such as low thermal conductivity and rigidity.
[30]	Hybrid	It combines the high heat absorption of PCMs with the active and localized cooling of thermoelectric coolers (TECs).	Delayed TEC current after PCM reaches 80% melting improves temperature uniformity and energy efficiency.	Improved temperature control, utilizes latent heat of PCMs, active cooling of TEC.	More complex than passive systems, higher cost, TEC requires energy, potential temperature variation.
[31]	Hybrid	Uses active liquid cooling combined with passive cooling materials like PCMs.	Effectively prevents TR propagation; maintains uniform temperature during normal operation.	Effective against TR propagation, maintains thermal uniformity, combines active and passive.	Increased complexity and potential additional costs compared to single systems.
[32]	Hybrid	Integrates liquid cooling systems with passive systems for optimal thermal management.	Considered more viable for future thermal management; effectively cools high-energy/power battery packs.	Combines the advantages of active and passive systems, enhancing overall thermal management.	More complex and expensive than single cooling systems.
[33]	Hybrid	Combination of PCMs with active cooling methods for effective thermal management.	Highlights benefits of integrated solutions, needs further research for higher conductivity PCMs.	Improves thermal uniformity and performance, optimizes temperature.	Challenges in integration, need for high conductivity PCMs, environmental impact of larger PCM volume.
[34]	Hybrid	Uses different techniques like air, liquid, and PCMs to cool batteries.	PCM-RT35 showed the best temperature control ability at ambient temperatures of 20 °C or 30 °C.	PCMs offer a passive approach with high efficiency, good temperature management.	PCMs have limited heat absorption capacity, complexity in managing multiple systems.

Notable challenges are identified in BTMSs, especially under harsh operating conditions. A key limitation lies in the low thermal conductivity of PCMs, which leads to uneven temperature distribution within the battery cells, adversely affecting the performance and efficiency of LIBs [33]. This problem is intensified in extreme situations, such as discharge rates higher than 1 °C or in environments with ambient temperatures higher than 35 °C, where temperature differences of less than 3 °C can be observed between individual cells, significantly impacting the performance and durability of LIBs [35]. Also, substantial limitations are observed in current BTMSs, particularly in fast-loading scenarios and high ambient temperatures, which can result in inefficient thermal management and increase the risk of TR [36]. In addition, air-cooled and passive cooling systems show a limited ability to adapt quickly to variations in thermal load, highlighting the importance of developing more dynamic and adaptive BTMSs to improve battery thermal stability over a wider range of operating conditions [37]. This situation is further complicated by the inadequacy of air-cooled and PCM-based systems in contexts of high ambient temperatures or high charge/discharge rates, where the poor thermal conductivity of these systems compromises the long-term performance of the batteries [38].

3. Innovations in Cooling Approaches for Battery Management Systems

Advances in refrigeration techniques, both single-phase and multiphase, have been significant. Single-phase refrigeration, although simpler in design, faces limitations in its heat transfer capacity compared to multiphase techniques [39,40]. Furthermore, one must not only understand the differences in their effectiveness and where they can be best applied, but also consider how improvements in materials, system design, and implementation strategies can help overcome current obstacles. For example, in single-phase cooling, new nanofluids are being explored as a possible solution to improve thermal conductivity and make heat transfer more efficient. Several studies have examined nanofluids with different combinations of base fluids (such as water, ethylene glycol, and engine oil) and nanoparticles (such as alumina (Al_2O_3), iron oxide (Fe_2O_3), copper oxide (CuO), and titanium dioxide (TiO_2)). Although the potential of engine oil nanofluids with Fe_2O_3 has not yet been fully investigated, a significant improvement in the cooling of the battery model has been observed by increasing the Fe_2O_3 concentration from 2% to 5%, which resulted in a decrease in its temperature. The inclusion of Fe_2O_3 as a nanoparticle not only increased the thermal capacity and heat transfer efficiency but also improved the overall thermal performance of the system compared to using engine oil alone [38].

Another nanofluid containing multi-walled carbon nanotubes (MWCNTs) has also shown promise due to its high thermal conductivity and low density. The thermal uniformity of the battery improves with the use of nanofluids, especially at higher MWCNT concentrations. The maximum thermal deviation drops significantly using the 0.45–0.5% MWCNT nanofluid. Another innovative design is the immersion cooling system, which uses a singular fluid in a static state (Novec-7200) and indicates remarkable efficiency in thermal regulation. This system keeps the maximum cell temperature below 40 °C and ensures that the temperature gradient is maintained within a range of 3 °C [39]. Within the field of multiphase refrigeration, traditional refrigerants such as HFCs (hydrofluorocarbons) and HCFCs (hydrochlorofluorocarbons) are two groups of widely used gases that play a crucial role in refrigeration, proving to be vital for efficient performance in environments with extreme temperatures. However, despite being fundamental components in air-conditioning systems, their effect on the environment has prompted the search for more sustainable options [40].

Therefore, new dielectric fluids with lower boiling points are being explored that promise significant improvements in thermal management under various operating conditions. Among these innovative solutions, the use of fluids such as SF33 stands out for their ability to maintain battery temperatures below 34 °C, representing a notable advance toward more sustainable and efficient cooling practices [41]. In addition, mini-channel cold plate cooling systems containing a two-phase fluid are effective in extracting heat through boiling [42]. The use of specific refrigerants, such as R410a, R134a, and R600a, in direct contact with the battery cells allows a significant improvement in temperature distribution. For example, the use of R600a has achieved impressive heat transfer coefficients, keeping module temperatures within safe and much lower ranges compared to systems without refrigeration [43]. To extend battery lifespan and improve battery safety by effectively optimizing the cooling design, the incorporation of thermal management models becomes indispensable. Further research highlights that choosing the right cooling fluids can markedly increase thermal efficiency, underscoring the vital importance of these models in improving cooling systems [44]. For example, in [45], the authors propose a multi-scale and multiphase model, pioneering in comprehensively simulating the venting process in LIBs during a TR event, from heat and gas generation to particle accumulation. In [46], the authors comment on the NTGK (extreme temperature condition, Tiedemann-Gu-Kim) model, which stands out for its accuracy in simulating the internal electrochemical processes of the batteries, offering realistic results that have been confirmed experimentally, differentiating it from more basic models. Two-phase immersion cooling is the model studied in [47]. This model uses the phase change of the coolant to achieve efficient heat transfer, offering rapid cooling, thermal stability, and energy efficiency, ideal for critical systems such as EV batteries, and maintains the battery temperature below 34 °C.

Another approach analyzed is the pseudo-two-dimensional (P2D) model, which reduces the complexity of the lateral structure in LIBs by focusing on electrode thickness to study ion diffusion and electrochemical dynamics. This method allows an accurate representation of variations in concentrations and electric potentials. Although it is fundamental to understand in detail the behavior of batteries, the intensity of the computational calculations required limits its implementation in instantaneous simulations required by battery management systems [48]. According to other research [39], the electrochemical-thermal model (ECT) is the most complete model for simulating batteries, but its high computational demand limits its practical use. The electrical–thermal (ET) model is preferred in thermal management because of its cost efficiency and accuracy. For thermal degradation studies, the Arrhenius method is used. Equivalent circuit models are valued for their effective approximation of electrochemical behavior with low computational cost. However, thermal management focuses on two strategies for temperature estimation: the first involves the use of artificial intelligence (AI), especially neural networks, and the second relies on models describing the internal electrochemical dynamics. These strategies allow the temperature of batteries to be predicted with high accuracy using power control modules. Artificial neural networks (ANNs), trained with real or simulated data, offer estimates with a minimum error of 1.38% compared to conventional methods, eliminating the need for complex simulations, and positioning itself as an effective tool in thermal optimization [39,40]. Figure 1 shows several types of ANN that can be used to predict battery temperature.

Within these types, Elman-NN networks are suitable to simulate the dynamic thermal behavior of the battery, where the temperature at each instant depends on the previous values. Elman-NNs analyze the variables of mathematical models using hidden layers and a contextual layer, learning the relationship between inputs and outputs with training data [40]. In contrast, LSTMs are key to predicting battery temperature by processing data sequences and addressing thermal complexities. Deep neural networks (DNNs) analyze complex data to predict thermal changes, while convolutional neural networks (CNNs) specialize in detecting and predicting areas of heat from visual data. This set of technologies offers an accurate and advanced solution for efficient thermal management of batteries [41]. The multilayer perceptron (MLP) network is effective in modeling nonlinear interactions, making it ideally suited to address challenges in thermal problems that feature nonlinear physical complexities. It uses meaningful inputs such as heat generation and temperature measurements, providing a rich contextual basis for analysis. MLP excels at predicting temperatures with a margin of error of only 0.8 °C using a single sensor, demonstrating its high efficiency in data-constrained contexts [42].

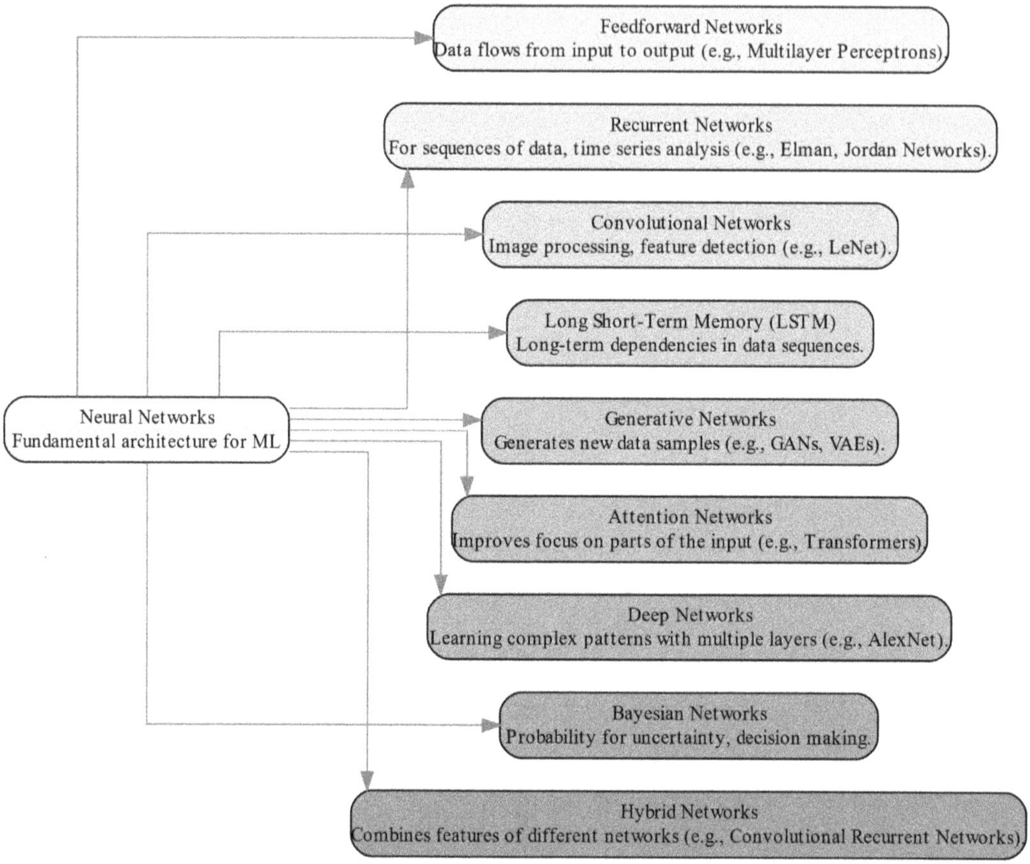

Figure 1. Core structure of machine learning (ML) systems.

On the other hand, model-based approaches delve into the internal reactions and properties to simulate their behavior and battery wear using tools such as the Kalman filter (KF), extended Kalman filter (EKF), unscented Kalman filter (UKF), and particle filtering (PF), among others. These methods seek to pinpoint internal conditions, e.g., load level. Hybrid models, which combine several techniques, are particularly effective. However, challenges, such as accuracy in the representation of deterioration and parameter calibration with limited data, remain complex areas [43]. The KF optimizes estimates of noisy dynamical systems by a recursive process that minimizes the prediction error. It is versatile for linear and nonlinear systems, with its effectiveness depending on the accuracy of the model and noise analysis [44]. There are studies in the literature that address this issue. In [45], the authors perform a comparison between the KF, EKF, and UKF for estimating the battery state of charge. They demonstrated that the UKF was the most accurate with an error of less than 0.3%, outperforming the EKF, which reduced the error to less than 0.5% by accounting for nonlinearities, and the KF, which had an error reduced from 2% to 1.5%. The success of the UKF is due to its effective handling of nonlinearities across sampling points, highlighting its superiority in the accuracy of SOC estimation in lithium-ion batteries. The authors of [46] employed the EKF together with a simple resistance–capacitance (1-RC) model that facilitated the determination of the state of charge (SOC) in batteries. This technique proved to be effective in estimating the overall SOC with an error margin of less than 2%. However, it failed to identify specific variations between individual cells, resulting in notable discrepancies between them. This situation points to the need to dynamically adapt and optimize the model for each cell to obtain an individualized and accurate SOC estimation. The authors of [47] use PF for SOC estimation in batteries by an innovative approach using weighted particles to represent possible states. This method simplifies nonlinear calculations, optimizes performance, and employs the radial simplex sphere principle for efficient sampling, achieving more accurate SOC predictions and corrections.

Therefore, the application of the EKF as an essential component of the algorithm designed for real-time estimation of SOC in LIBs of EVs is validated. Therefore, the prediction and measurement functions for applying the battery model within the EKF are described. The estimation process involves estimating the SOC and battery bias voltage in advance, applying a discrete model that incorporates both prior state and recent inputs, and adding a random error component to reflect uncertainty. Parallel to this, the measurement function bridges the current voltage and current observations to the theoretical battery state, adapting to the nonlinear complexity of the battery model and introducing a random error into the measurements. The KF comes into play by continuously adjusting these initial estimates based on the observed differences between predictions and actual measurements, which effectively refines the accuracy of the battery state estimation [48]. Recent innovations in materials and structures are revolutionizing thermal efficiency, especially in the field of LIBs, a key technology in sectors such as EVs. One of the main innovations is the use of PCMs, which keep the temperature of batteries within a safe and constant range by harnessing the latent heat during their phase transitions [35,49,50]. However, these materials often present the challenge of low thermal conductivity, which is being addressed by incorporating high-conductivity metal matrices and adding metal nanoparticles or porous materials [48,49].

Another area of significant advancement is the development of miniature channel cooling plates, which have been shown to be effective in managing battery pack temperature [51,52]. These plates, often made of aluminum, allow for better heat distribution and more effective temperature control. However, they face challenges related to the complexity and cost of production [51]. Hybrid structures combining passive thermal management and active cooling systems have also emerged as promising solutions, offering greater efficiency, but at the cost of increased weight and complexity. In addition, the use of innovative materials such as graphene has been explored to improve heat dissipation due to their high thermal conductivity [49,53]. Advances in the design of airflow structures and the use of fins on cooling plates have also shown improvements in thermal efficiency [52,54]. However, these solutions may increase airflow resistance and require higher energy consumption. One of the most significant advantages of these innovations is their ability to improve the safety and efficiency of batteries, which is critical in high-demand applications such as EVs [55,56]. However, these technologies still face significant challenges, including the trade-off between improvements in thermal conductivity and production cost, as well as the need to effectively integrate these solutions into large-scale battery designs [50,57]. Although significant progress has been made in improving thermal efficiency through various innovations in materials and structures, challenges remain. These include cost optimization, simplification of manufacturing processes, and effective integration of these solutions into large-scale battery systems. Continued research and development are key to addressing these challenges and taking full advantage of the benefits of these advanced technologies. A summary of innovations in battery thermal management is presented in Table 2.

Table 2. Innovations in Thermal Efficiency.

Ref.	Innovations in Thermal Efficiency	Advantages	Disadvantages	Challenges
[35]	Use of PCMs	They absorb latent heat during phase transition, keeping the battery temperature within a safe range.	Low thermal conductivity, which limits the ability to dissipate heat evenly.	Development of materials with higher thermal conductivity and life cycle.
[51]	Use of miniature channel cooling plates	Increased contact area between the coolant and the cells, which improves heat transfer.	Higher complexity and manufacturing cost.	Optimize channel distribution to reduce pressure loss.
[49]	Phase change composite materials	They combine the advantages of PCMs with those of conductive materials, improving thermal conductivity.	Higher production cost.	Develop composite materials with higher energy density.
[55]	Hybrid system combining heat pipes with evaporative cooling	Improves thermal efficiency in high-current applications.	Increased complexity and manufacturing cost.	Optimize system design to reduce pressure losses.
[58]	Graphene composite structures	Excellent thermal conductivity, which improves heat distribution within the battery.	High production cost.	Develop more efficient production methods.
[56]	Miniature channel design with tilt angles	Reduces pressure losses, improving heat transfer.	Excessively high tilt angles can cause leakage problems.	Optimize the tilt angle for maximum thermal efficiency.
[50]	Passive interfacial thermal regulator based on shape memory alloy	It changes its thermal conductance reversibly, improving battery performance in hot and cold climates.	Challenges related to the development of shape memory alloys with increased thermal cycling and long-term stability.	Optimize device design to facilitate integration into modules and battery packs.
[59]	Hybrid system combining heat pipes with evaporative cooling	Improves thermal efficiency in high-current applications.	Increased complexity and manufacturing cost.	Optimize system design to reduce pressure losses.
[53]	Modular cooling plate design	Greater versatility and adaptability to variable configurations.	Modular designs require joints and connections that can increase the overall thermal resistance.	Achieve large-scale manufacturing of these modular systems in a cost-effective manner.
[57]	System based on liquid cooling of honeycomb structure and phase-change materials	Significantly reduces the maximum temperature and temperature difference in the batteries.	Structural and cooling complexity leads to higher manufacturing costs.	Evaluation under extreme conditions such as actual loading and unloading cycles or thermal packaging situations.
[60]	Modular liquid cooling system	Greater versatility and adaptability to variable configurations.	Modular designs require joints and connections that can increase the overall thermal resistance.	Achieve large-scale manufacturing of these modular systems in a cost-effective manner.
[61]	System based on liquid cooling of honeycomb structure and phase-change materials	Significantly reduces the maximum temperature and temperature difference in the batteries.	Compact structure and uniform heat dissipation.	Evaluate its performance under extreme conditions such as real loading and unloading cycles or thermal packaging situations.

Table 2. *Cont.*

Ref.	Innovations in Thermal Efficiency	Advantages	Disadvantages	Challenges
[54]	Phase change composite materials	Improve thermal uniformity within the battery modules.	Composite materials tend to be more expensive to produce.	Develop composite materials with higher effective thermal conductivity.
[52]	Mini-channel cooling plates with spine-shaped fins	They improve heat transfer performance and reduce thermal gradients.	Horizontal fins cause a significantly higher pressure loss.	Optimize the geometry and arrangement of the fins to achieve the optimum balance between heat transfer and pressure loss.

4. Emerging Technologies in Thermal Monitoring and Control

In recent years, the growing demand for EVs and energy storage systems has driven intense research and development in the field of lithium-ion batteries. A key aspect of this technological evolution is the efficient and safe management of battery temperature, a complex challenge that involves the fusion of advanced sensors, control systems, and risk mitigation, as well as the application of smart technologies and machine learning. The importance of advanced sensors in this area cannot be underestimated. With the incorporation of distributed optical fibers and nanosensors in battery cells, thermal monitoring with high spatial and temporal resolution has been achieved. These sensors provide critical data that are essential to understanding and managing the internal conditions of batteries, enabling the detection of significant temperature variations that could lead to failures [50,62]. In parallel, control systems have significantly evolved thanks to the integration of AI and ML algorithms. These systems not only process the data collected by advanced sensors but also learn from them, continuously improving their ability to predict and mitigate risks. For example, by analyzing thermal behavior patterns, these systems can anticipate and prevent TR incidents, one of the main risks in LIBs [63]. Risk mitigation is enhanced by the development of intelligent technologies and ML. These tools offer an unprecedented ability to analyze and predict battery behavior, enabling more efficient and safer thermal management. The integration of predictive models based on historical data and techniques such as ANNs and reinforcement learning have resulted in more advanced and reliable battery management systems [64,65]. In short, the convergence of these innovative technologies is transforming the way LIBs are monitored, controlled, and safely maintained. This advancement is not only crucial to improve the efficiency and lifetime of these batteries, but also to ensure their safe use in critical applications such as EVs and large-scale energy storage systems [66,67].

4.1. Advanced Sensors

In the field of LIBs, the incorporation of advanced sensors is revolutionizing the way safety and efficiency are monitored and improved. Thin-film RTD (TFRTD) sensors, specifically copper–nickel alloy sensors, have been noted for their ability to be integrated into current collectors, providing fast and accurate internal temperature measurement. These sensors exhibit 82% faster response speeds and 33% higher accuracy compared to external RTDs, which is crucial to avoid overheating and TR [68]. In external short-circuit detection, non-contact magnetoelectric composite sensors, which combine piezoelectric elastomers and magnetostrictive ferrite, have shown high current sensitivity, with an accuracy greater than 99% and a current sensitivity of 0.346–5.975 mV/A. These sensors can distinguish between short circuits and mechanical vibrations, which makes them suitable for applications in EVs [69]. On the other hand, OFDR-based fiber optic sensors offer distributed temperature measurements with a spatial resolution of up to 3 mm. These sensors can identify heat accumulations around positive current tabs during high-rate discharges, revealing non-uniform heat generations even in small cylindrical cells [70]. In smart batteries, the fusion of multiple internal sensors provides a more accurate way to estimate the SOC. For example, expansion force (EF) sensors have been shown to have a more sensitive relationship with SOC compared to voltage and are independent of dynamic current. These sensors can also provide information on battery health status [71].

FBG sensors are noted for their low invasiveness and resistance to electromagnetic interference, with a temperature sensitivity of approximately 10 pm/°C and strain sensitivity of 1–2 pm/με. They are capable of quasi-distributed sensing and thermal mapping within battery packs, making them suitable for estimating the state of charge and state of health, and predicting battery capacity [72]. Finally, fiber optic sensors have been used to monitor strain and temperature variations in individual cells, employing algorithms such as the fast-recursive algorithm to establish nonlinear correlation models between strain signals and key electrical parameters, allowing accurate estimation of the SOC in battery packs [73]. These advances represent major progress in the monitoring and management of lithium-ion batteries, paving the way for safer and more efficient applications, especially in EVs and large-scale energy storage systems. Figure 2 shows the classification of advanced sensors in BTMSs.

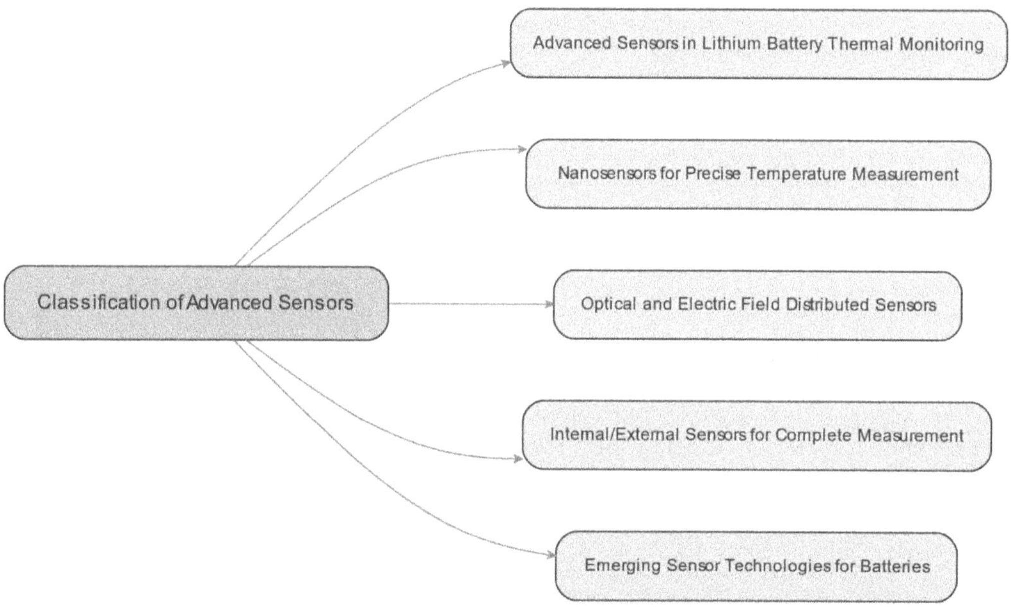

Figure 2. Classification of advanced sensors.

At the heart of this evolution are advanced sensors, whose development continues apace. Emerging technologies such as distributed optical fiber sensors, electric field sensors, and nanosensors offer high spatial and temporal resolution, facilitating detailed thermal mapping of batteries. Particularly, nanosensor-based sensors, composed of materials such as metal oxides and carbon, are capable of measuring temperature in a distributed manner throughout the battery module. Their small size does not affect battery performance and they provide real-time thermal data with high spatial resolution, which is key in smart thermal management [50]. In addition, the use of optical fibers distributed internally in the battery cells allows real-time monitoring of thermal distribution, a significant improvement over conventional sensors. This is vital for detecting temperature variations within the battery and along its length, preventing failures due to temperature differences [62]. The integration of these optical sensors with electric/magnetic field sensors facilitates detailed thermal mapping and allows the detection of variations associated with thermal changes. These dense arrays of sensors collect large volumes of data, essential for accurate and efficient monitoring. Finally, battery monitoring systems include distributed networks of multiple sensors that evaluate thermal and electrical parameters at the cell, module, and pack level, providing a complete picture of the battery status and enhancing the safety and efficiency of battery use [66].

4.2. Application of Intelligent Technologies and Machine Learning

The application of intelligent technologies and ML in battery and power distribution systems has emerged as a critical field in the evolution toward more efficient and safer energy management. In the context of fast charging of lithium-ion batteries, the development of the MSCC-DRL (multi-stage constant current based on deep reinforcement learning) model, which uses deep reinforcement learning, demonstrates a significant advance in reducing charging times while maintaining safety and minimizing battery degradation. In parallel, a systematic review of ML applications in smart distribution systems highlights how these technologies are revolutionizing the planning and operation of power grids, improving the efficiency and effectiveness of these systems [74,75]. In the EV arena, battery technologies and battery management systems have seen remarkable advances. Developments in ANNs for the health management of lithium-ion energy storage batteries, as well as hybrid ML models for thermal modeling and battery diagnostics, are clear examples of how ML is improving the safety, efficiency, and durability of these batteries [76–79]. In addition, the use of deep learning to estimate the state of charge, health, and remaining life of batteries indicates significant progress in intelligent battery management, enabling more accurate and safer operation of battery systems in EVs [79]. The integration of ML models into BMSs has significantly transformed lithium-ion battery management, especially in the context of EVs. The innovative approach presented in [74] illustrates the application of deep reinforcement learning to optimize charging efficiency. This model interacts with the battery environment, allowing the trained agent to autonomously determine the optimal charging profile, thus maximizing battery lifetime. The results show that this approach allows charging batteries in as little as 6–14 minutes, obtaining charging times up to four times shorter than traditional methods. In addition, the agent demonstrates the ability to adapt to variations in parameters such as electrode thickness, optimizing charging autonomously in different conditions.

Study [76] emphasizes how advanced battery and BMS technologies, enriched with ML, have improved safety and efficiency in EVs. Intelligent BMSs, which use ML models, are critical for vital functions such as SOC and cell balancing, thus improving the actual autonomy and safety of EVs. State-of-health (SOH) management of LIBs is another area where ML has had a significant impact. According to [77], ANNs have been successfully applied to predict and monitor SOH, a crucial factor for the safe and efficient operation of energy storage systems. These models can capture complex nonlinear relationships between multiple factors, such as voltage, current, and temperature, and the battery health state. Furthermore, [78] introduces a hybrid ML model for thermal modeling and battery diagnostics. This approach combines mechanistic models with data-driven trade-offs, providing a powerful tool to prevent overheating and ensure safety during battery operation. Deep learning has proven to be a powerful tool for health status estimation thanks to its ability to learn complex relationships between input data and health indicators. Architectures such as feedforward, convolutional, recurrent neural networks (LSTMs and GRUs), and Transformers have been applied with promising results on data-driven models [80–82]. The authors of [79] demonstrate how deep learning is used to accurately estimate the SOC, SOH, and remaining useful life (RUL) of batteries. This approach enables more accurate and effective battery management, which is crucial for the safe and efficient operation of EVs [76,82].

Deep learning (DL) allows these states to be estimated from large sets of historical battery operation data, without the need for complex electrochemical models. Different architectures such as DNNs, recurrent networks (RNNs), LSTMs, and convolutional networks (CNNs) have been successfully applied [78,79]. Other works have applied LSTM networks, GRUs, and autoencoders to estimate the SOH and predict the RUL. For example, the model proposed by [83] based on an autoencoder with a particle filter achieved a mean square error of 12.1457 for the CALCE database, better than the particle filter and Kalman models. Finally, [84] analyzes different AI strategies, including SOH and SOC estimation. These strategies highlight the versatility and effectiveness of ML in various applications within BMSs, thus improving the performance and safety of EVs. Six ML algorithms are analyzed in this study: linear regression, random forest, gradient boosting, light gradient boosting machine, extreme gradient boosting, and support vector machines. The input data include current, temperature, and SOC, while voltage is used as the output [83,85]. The results show that the random forest provides superior performance with an R2 of 0.999 and minimal errors. This shows that ML can accurately estimate the state of lithium-ion batteries.

Therefore, the adoption of ML models in BMSs has opened a path towards more advanced and sustainable energy storage and EV systems. These advances are critical to improving the efficiency, safety, and lifespan of LIBs, marking a milestone in the transition to more efficient and environmentally friendly mobility. Studies [86,87] highlight the importance of these technologies in the future of battery management and electric mobility. Moreover, the integration of AI in LIB management is a growing area of research, with significant applications in EV and energy storage systems. Lithium-ion battery health management, especially in energy storage systems, has gained importance due to the need to manage SOH, SOC, and RUL accurately. ANN models are emerging as effective tools to address these challenges, leveraging their ability to decipher complex and nonlinear relationships between input data and battery health indicators [76,88,89]. On the other hand, accurate monitoring of the internal temperature distribution is crucial for the safety of LIBs. A novel approach to this is a hybrid thermal-neural network (LTNN) model that combines a mechanism-based distributed thermal model with machine learning-based axial thermal gradient compensation. This hybrid LTNN model has been shown to be highly compatible with common state observation methods, providing accurate and spatially resolved internal thermal monitoring and diagnostics for LIBs [78].

In the context of battery management systems (BMSs) in EVs, DL has emerged as a key technique to address battery-related algorithms and operational issues. The use of DL in BMS enables accurate estimation of SOC, SOH, and RUL, which is critical for EV reliability, safety, and performance [82,90]. The role of AI in solving battery management problems also extends to estimating the state of Li-ion batteries. Methods such as random forests, support vector machines (SVMs), and gradient momentum algorithms have demonstrated superior performance in discharge prediction, suggesting that integrating these methods with BMS can significantly improve the performance of EVs [84,91]. In addition, battery safety is a critical issue, especially in EVs and grid-scale storage. Fire incidents have highlighted the importance of battery safety, particularly regarding unpredictable thermal runaway. Machine learning approaches offer new opportunities to predict and prevent battery failures in practical applications, addressing multi-sector and multi-scale challenges [84,90]. AI and ML are playing a crucial role in improving the health and safety management of Li-ion batteries. These technologies offer promising solutions for accurate SOH, SOC, and RUL estimation, advanced thermal management, and failure and safety risk prevention in a variety of applications, including EVs and energy storage systems.

5. Challenges and Solutions in Extreme Conditions

Lithium-ion batteries, crucial in the era of electric mobility, face notable challenges in extreme temperature conditions. These conditions, defined outside the optimal operating range (298.15 K to 323.15 K), significantly impact battery efficiency and safety. At elevated temperatures, the acceleration of electrochemical degradation and the risk of thermal decomposition are primary concerns, while temperatures below the optimal range compromise battery capacity and power [92,93]. The challenge is intensified when considering that heat flow directly affects ionic and electronic conductivity, altering the redox processes at the electrodes and shortening the battery lifetime [94]. To address these difficulties, hybrid thermal management systems have been developed, combining liquid cooling with PCMs. These systems act as thermal buffers, passively absorbing and releasing heat during exothermic and endothermic reactions, thus keeping the temperature within a safe range and extending the battery lifetime [95]. Regulations such as ISO 12405-4 [96] and UN38.3 [97] play a crucial role, establishing maximum temperature limits of 55 °C and 60 °C. In addition, temperatures below 0 °C are considered extreme, although manufacturers often specify stricter ranges to optimize performance and durability. The importance of effective thermal management cannot be underestimated, especially when considering the wide variety of applications for these batteries, from EVs to energy storage in harsh environments [98,99].

In the future, thermal management systems will need to balance high capacity and fast charging with thermal efficiency, especially challenging over a wide temperature range. In addition, fast charging presents additional challenges, where optimization is key to controlling heat and temperature gradients while maintaining safety and performance [100,101]. Energy storage systems in harsh environments require advanced thermal management approaches, such as phase change cooling, to maintain stable performance under extreme conditions [102]. In addition, AI-based controllers, such as ANNs and fuzzy logic, are emerging as solutions to optimize battery safety and lifetime in EVs by dynamically adapting to temperature variations [103]. In conclusion, the effectiveness of thermal management in LIB is vital to overcome the challenges posed by extreme conditions. Continued development of advanced materials, techniques, and regulations is critical to ensure the performance, safety, and long-term viability of these essential technologies [94,98]. Table 3 provides a comprehensive examination of various thermal management technologies employed in BTMSs. The detailed analysis encompasses the technology's description, its relation to thermal leakage, specific benefits, and associated limitations and challenges. The technologies covered include hybrid systems, liquid cooling, PCM active heating systems, phase shift cooling, thermal management with AI, and thermotolerant separators. Each entry sheds light on the unique features, advantages, and considerations of the respective thermal management technology within the realm of BTMSs.

Table 3. Challenges and solutions in BTMSs.

Thermal Management Technology	Detailed Description	Relation to Thermal Leakage	Specific Benefits	Limitations and Challenges
Hybrid Systems Refs. [95,99]	The authors combine the efficiency of liquid cooling with the heat storage capacity of PCMs. They offer a dynamic and adaptive response to temperature variations.	They provide balanced thermal management, absorbing excess heat and releasing it when needed, which is crucial in fast-load or high-demand situations.	They significantly improve battery life and safety by adapting to different operating conditions.	They require a complex design and may have a higher cost.
Liquid Cooling Refs. [94,98,100,101]	It uses a fluid, usually water or a mixture of water and glycol, to efficiently transfer heat from the batteries to a heat exchanger. This technique is especially effective in fast charging or high-power density situations.	Essential to prevent overheating at high temperatures and maintain a stable thermal environment, reducing the risk of TR and battery degradation.	It provides fast and uniform heat dissipation, is scalable, and can be adjusted to different battery sizes and designs.	It can be susceptible to leaks and requires regular maintenance, in addition to an efficient pumping system.

Table 3. *Cont.*

Thermal Management Technology	Detailed Description	Relation to Thermal Leakage	Specific Benefits	Limitations and Challenges
PCM Refs. [49,102–104]	PCMs absorb and release heat during their phase transitions (solid to liquid and vice versa), allowing them to maintain a constant temperature in the battery. They are particularly useful in variable charge and discharge conditions.	They offer passive thermal response, stabilizing the internal temperature of the battery and reducing TR in extreme climates.	They provide high thermal storage capacity with minimal change in temperature, which is ideal for space-constrained applications.	They may have limitations in thermal conductivity and cycle life, as well as challenges in integration with other battery components.
Active Heating Systems Refs. [1,98,105]	These systems use heating elements or strategies such as battery preheating to maintain the optimum temperature in cold environments, thus improving battery response and efficiency.	They are essential to mitigate TR at low temperatures, ensuring that the battery operates efficiently and avoiding problems such as electrolyte crystallization.	Improve performance and safety in cold climates, extending battery life and preventing damage to internal components.	They increase energy consumption and may require additional time before use to reach optimum temperature.
Phase Shift Cooling Ref. [106]	It uses the evaporation and condensation of a refrigerant fluid to effectively absorb and dissipate heat. This method is based on the latent heat of phase change of the refrigerant, offering high heat dissipation capacity.	Efficiently controls temperature under peak load and unload, preventing overheating and excessive thermal runaway.	It offers precise thermal control and is capable of handling high thermal loads, making it suitable for energy-intensive applications.	It requires careful design to ensure the efficiency of the phase change system and can present challenges in refrigerant replenishment.
Thermal Management with AI Refs. [95,105]	It implements AI algorithms to monitor and adjust thermal management in real time, based on usage patterns and environmental conditions.	It enables fast and accurate response to temperature variations, optimizing thermal management to reduce TR and improve efficiency.	Maximizes battery life and performance by continuously adapting to changing conditions, improving safety and efficiency.	It depends on the accuracy of algorithms and data collection and may require constant updates and maintenance.
Thermotolerant Separators Ref. [107]	Advanced separators designed to withstand high temperatures without losing functionality, improving battery stability and safety in extreme heat conditions.	They prevent overheating and reduce TR by maintaining structural and functional integrity at high temperatures, avoiding internal short circuits.	They significantly increase safety in extreme conditions, resisting high temperatures without degrading.	They can increase the cost of battery manufacturing and present challenges in integration with other components.

6. Conclusions

This study on the thermal management of LIBs focuses on fundamental aspects of their sustainable and safe development, particularly in critical applications such as EVs and energy storage systems. This work highlights the cruciality of BTMSs in keeping LIBs within an optimal temperature range, optimizing their performance, and prolonging their lifetime. Current challenges include overheating and temperature variations, which can compromise the safety and performance of batteries, accelerating their aging and reducing their energy storage capacity. Significant innovations in materials and structures have been made that are revolutionizing thermal efficiency in LIBs. The use of PCMs is one such innovation, which helps to maintain battery temperature within a safe and constant range. However, these materials present the challenge of low thermal conductivity, which is being addressed by incorporating high-conductivity metal matrices and the addition of metal nanoparticles or porous materials.

The Integration of smart technologies and ML into battery and power distribution systems has emerged as a critical field. Models such as MSCC-DRL, which uses deep reinforcement learning, are making progress in optimizing charging efficiency and estimating the state of charge, health, and remaining life of batteries, indicating significant progress in intelligent battery management. Looking ahead, thermal management systems will need to balance high capacity and fast charging with thermal efficiency over a wide temperature range. Energy storage systems in harsh environments will require advanced thermal management approaches, and AI-based controllers are emerging as key solutions to optimize

EV battery safety and lifetime by dynamically adapting to temperature variations. Despite significant advances, challenges remain, including cost optimization, simplification of manufacturing processes, and effective integration of these solutions into large-scale battery systems. Continued research and development are critical to address these challenges and maximize the benefits of these advanced technologies.

The significant contribution of this research lies in its innovative approach to the thermal management of LIBs, especially highlighting the adoption of ML models in BMSs. This has paved the way towards more advanced and sustainable energy storage and EV systems, marking a milestone in the transition towards more efficient and environmentally friendly mobility. In summary, this study represents a crucial breakthrough in improving the efficiency, safety, and longevity of lithium-ion batteries, contributing significantly to the future of battery management and electric mobility.

Author Contributions: Conceptualization, Y.O., P.A. and F.J.; data curation, P.A. and F.J.; formal analysis, Y.O., P.A. and F.J.; funding acquisition, P.A.; investigation, Y.O., P.A. and D.P.; methodology, Y.O., P.A., F.J. and D.P.; project administration, F.J.; resources, P.A. and F.J.; software, P.A., Y.O. and F.J.; supervision, F.J.; validation, Y.O., P.A. and F.J.; visualization, D.P.; writing—original draft, Y.O., P.A. and D.P.; writing—review and editing, P.A. and F.J. All authors have read and agreed to the published version of the manuscript.

Funding: This research received no external funding.

Data Availability Statement: Data will be made available on request.

Acknowledgments: The author (Paul Arévalo) thanks the Call for Grants for the Requalification of the Spanish University System for 2021–2023, and the Margarita Salas Grants for the training of young doctors awarded by the Ministry of Universities and financed by the European Union—Next Generation EU.

Conflicts of Interest: The authors declare no conflicts of interest.

References

1. Yue, Q.L.; He, C.X.; Wu, M.C.; Zhao, T.S. Advances in Thermal Management Systems for Next-Generation Power Batteries. *Int. J. Heat. Mass. Transf.* **2021**, *181*, 121853. [CrossRef]
2. Bandhauer, T.M.; Garimella, S.; Fuller, T.F. A Critical Review of Thermal Issues in Lithium-Ion Batteries. *J. Electrochem. Soc.* **2011**, *158*, R1–R25. [CrossRef]
3. Cai, S.; Zhang, X.; Ji, J. Recent Advances in Phase Change Materials-Based Battery Thermal Management Systems for Electric Vehicles. *J. Energy Storage* **2023**, *72*, 108750. [CrossRef]
4. Olabi, A.G.; Maghrabie, H.M.; Hameed, O.; Adhari, K.; Sayed, T.; Youse, B.A.A.; Salameh, T.; Kamil, M.; Abdelkareem, M.A. Battery Thermal Management Systems: Recent Progress and Challenges. *Int. J. Thermofluids* **2022**, *15*, 100171. [CrossRef]
5. Akkurt, N.; Aghakhani, S.; Mahmoud, M.Z.; El Din, E.M.T. The influence of battery distance on a hybrid air-cooled cylindrical lithium-ion battery phase change material thermal management system for storing solar energy. *J. Energy Storage* **2022**, *52*, 1048730. [CrossRef]
6. Richardson, G.; Korotkin, I. Heat Generation and a Conservation Law for Chemical Energy in Li-Ion Batteries. *Electrochim. Acta* **2021**, *392*, 138909. [CrossRef]
7. Wu, W.; Xiao, X.; Huang, X. The Effect of Battery Design Parameters on Heat Generation and Utilization in a Li-Ion Cell. *Electrochim. Acta* **2012**, *83*, 227–240. [CrossRef]
8. Garud, K.S.; Tai, L.D.; Hwang, S.G.; Nguyen, N.H.; Lee, M.Y. A Review of Advanced Cooling Strategies for Battery Thermal Management Systems in Electric Vehicles. *Symmetry* **2023**, *15*, 1322. [CrossRef]
9. Qin, Y.; Xu, Z.; Xiao, S.; Gao, M.; Bai, J.; Liebig, D.; Lu, L.; Han, X.; Li, Y.; Du, J.; et al. Temperature Consistency—Oriented Rapid Heating Strategy Combining Pulsed Operation and External Thermal Management for Lithium-Ion Batteries. *Appl. Energy* **2023**, *335*, 120659. [CrossRef]
10. Chavan, S.; Venkateswarlu, B.; Prabakaran, R.; Salman, M.; Joo, S.W.; Choi, G.S.; Kim, S.C. Thermal Runaway and Mitigation Strategies for Electric Vehicle Lithium-Ion Batteries Using Battery Cooling Approach: A Review of the Current Status and Challenges. *J. Energy Storage* **2023**, *72*, 108569. [CrossRef]
11. Yao, F.; Guan, X.; Chen, Q.; Lin, L. Research on Thermal Management System of Lithium-Ion Battery with a New Type of Spider Web Liquid Cooling Channel and Phase Change Materials. *J. Energy Storage* **2024**, *81*, 110447. [CrossRef]
12. Liu, J.; Yadav, S.; Salman, M.; Chavan, S.; Kim, S.C. Review on Thermal Coupled Battery Models and Parameter Identication or Lithium-Ion Battery Heat Generation in EV Battery Thermal Management System. *Int. J. Heat. Mass. Transf.* **2024**, *218*, 124748. [CrossRef]

13. Khan, N.; Sable, M.J. A Review on Lithium-Ion Battery Thermal Management Systems. *Int. Res. J. Modern. Eng. Technol. Sci.* **2023**, *5*, 2550–2559.
14. Khan, A.; Yaqub, S.; Ali, M.; Ahmad, A.W.; Nazir, H.; Khalid, H.A.; Iqbal, N.; Said, Z.; Sopian, K. A State-of-the-Art Review on Heating and Cooling of Lithium-Ion Batteries for Electric Vehicles. *J. Energy Storage* **2024**, *76*, 109852. [CrossRef]
15. Mahmud, M.; Rahman, K.S.; Rokonuzzaman, M.; Habib, A.K.M.A.; Islam, M.R.; Motakabber, S.M.A.; Channumsin, S.; Chowdhury, S. Lithium-Ion Battery Thermal Management for Electric Vehicles Using Phase Change Material: A Review. *Results Eng.* **2023**, *20*, 101424. [CrossRef]
16. Kumar, R.; Goel, V. A Study on Thermal Management System of Lithium-Ion Batteries for Electrical Vehicles: A Critical Review. *J. Energy Storage* **2023**, *71*, 108025. [CrossRef]
17. Jiang, W.; Feng, G.; Wang, H.; Chang, Z.; Tan, X.; Ji, J.; Zhang, Y.; Zhou, J. Analysis of Hybrid Active-Passive Prismatic Li-Ion Battery Thermal Management System Using Phase Change Materials with Porous-Filled Mini-Channels. *J. Energy Storage* **2024**, *80*, 110144. [CrossRef]
18. Luo, J.; Zou, D.; Wang, Y.; Wang, S.; Huang, L. Battery Thermal Management Systems (BTMs) Based on Phase Change Material (PCM): A Comprehensive Review. *Chem. Eng. J.* **2022**, *430*, 132741. [CrossRef]
19. Moaveni, A.; Siavashi, M.; Mousavi, S. Passive and Hybrid Battery Thermal Management System by Cooling Flow Control, Employing Nano-PCM, Fins, and Metal Foam. *Energy* **2024**, *288*, 129809. [CrossRef]
20. Alaoui, C. Passive/Active BTMS for EV Lithium-Ion Batteries. *IEEE Trans. Veh. Technol.* **2018**, *67*, 3709–3719. [CrossRef]
21. Patel, J.R.; Rathod, M.K. Recent Developments in the Passive and Hybrid Thermal Management Techniques of Lithium-Ion Batteries. *J. Power Sources* **2020**, *480*, 228820. [CrossRef]
22. Ranjbaran, Y.S.; Shojaeefard, M.H.; Molaeimanesh, G.R. Thermal Performance Enhancement of a Passive Battery Thermal Management System Based on Phase Change Material Using Cold Air Passageways for Lithium Batteries. *J. Energy Storage* **2023**, *68*, 107744. [CrossRef]
23. Osmani, K.; Alkhedher, M.; Ramadan, M.; Choi, D.S.; Li, L.K.B.; Doranehgard, M.H.; Olabi, A.G. Recent Progress in the Thermal Management of Lithium-Ion Batteries. *J. Clean. Prod.* **2023**, *389*, 136024. [CrossRef]
24. Chen, J.; Kang, S.; Jiaqiang, E.; Huang, Z.; Wei, K.; Zhang, B.; Zhu, H.; Deng, Y.; Zhang, F.; Liao, G. Effects of Different Phase Change Material Thermal Management Strategies on the Cooling Performance of the Power Lithium Ion Batteries: A Review. *J. Power Sources* **2019**, *442*, 227228. [CrossRef]
25. Li, K.; Yan, J.; Chen, H.; Wang, Q. Water Cooling Based Strategy for Lithium Ion Battery Pack Dynamic Cycling for Thermal Management System. *Appl. Therm. Eng.* **2018**, *132*, 575–585. [CrossRef]
26. Fini, A.S.; Gharehghani, A. Experimental Investigation of Pressure Effect on the PCM Performance in Li-Ion Battery Thermal Management System. *J. Energy Storage* **2024**, *79*, 110273. [CrossRef]
27. Mei, J.; Shi, G.; Liu, H.; Wang, Z.; Chen, M. Investigation on the Optimization Strategy of Phase Change Material Thermal Management System for Lithium-Ion Battery. *J. Energy Storage* **2022**, *55*, 105365. [CrossRef]
28. Son, Y.W.; Kang, D.; Kim, J. Passive Battery Thermal Management System for an Unmanned Aerial Vehicle Using a Tetrahedral Lattice Porous Plate. *Appl. Therm. Eng.* **2023**, *225*, 120186. [CrossRef]
29. Li, Z.; Zhang, Y.; Zhang, S.; Tang, B. Phase Change Materials for Lithium-Ion Battery Thermal Management Systems: A Review. *J. Energy Storage* **2024**, *80*, 110259. [CrossRef]
30. Liu, X.; Yao, L.C.; Su, C.Q.; Xiong, X.; Wang, Y.P. A Hybrid Battery Thermal Management System Coupling with PCM and Optimized Thermoelectric Cooling for High-Rate Discharge Condition. *Case Stud. Therm. Eng.* **2023**, *49*, 103269. [CrossRef]
31. Sun, Z.; Guo, Y.; Zhang, C.; Xu, H.; Zhou, Q.; Wang, C. A Novel Hybrid Battery Thermal Management System for Prevention of Thermal Runaway Propagation. *IEEE Trans. Transp. Electrif.* **2022**, *207*, 3215691. [CrossRef]
32. Kharabati, S.; Saedodin, S. A Systematic Review of Thermal Management Techniques for Electric Vehicle Batteries. *J. Energy Storage* **2024**, *75*, 109586. [CrossRef]
33. Khan, M.M.; Alkhedher, M.; Ramadan, M.; Ghazal, M. Hybrid PCM-Based Thermal Management for Lithium-Ion Batteries: Trends and Challenges. *J. Energy Storage* **2023**, *73*, 108775. [CrossRef]
34. Wang, H.; Guo, Y.; Ren, Y.; Yeboah, S.; Wang, J.; Long, F.; Zhang, Z.; Jiang, R. Investigation of the Thermal Management Potential of Phase Change Material for Lithium-Ion Battery. *Appl. Therm. Eng.* **2024**, *236*, 121590. [CrossRef]
35. Napa, N.; Agrawal, M.K.; Tamma, B. Design of Novel Thermal Management System for Li-Ion Battery Module Using Metal Matrix Based Passive Cooling Method. *J. Energy Storage* **2023**, *73*, 109119. [CrossRef]
36. Afraz, M.V.; Ali Mohammadi, Z.; Karimi, G. A Novel Compact Thermal Management Model for Performance Evaluation of Tesla-like Lithium-Ion Battery Packs. *Energy Convers. Manag.* **2024**, *300*, 117927. [CrossRef]
37. Hwang, F.S.; Confrey, T.; Reidy, C.; Picovici, D.; Callaghan, D.; Culliton, D.; Nolan, C. Review of Battery Thermal Management Systems in Electric Vehicles. *Renew. Sustain. Energy Rev.* **2024**, *192*, 114171. [CrossRef]
38. Rana, J.; Dikshit, M.K. Recent Advancements in Materials for Battery Thermal Management Systems in Li-Ion Battery: A Mini Review. *Mater. Today Proc.* **2023**. [CrossRef]
39. Choi, H.; Lee, H.; Kim, J.; Lee, H. Hybrid Single-Phase Immersion Cooling Structure for Battery Thermal Management under Fast-Charging Conditions. *Energy Convers. Manag.* **2023**, *287*, 117053. [CrossRef]
40. Wang, Y.; Chen, X.; Li, C.; Yu, Y.; Zhou, G.; Wang, C.Y.; Zhao, W. Temperature Prediction of Lithium-Ion Battery Based on Artificial Neural Network Model. *Appl. Therm. Eng.* **2023**, *228*, 120482. [CrossRef]

41. Kurucan, M.; Özbaltan, M.; Yetgin, Z.; Alkaya, A. Applications of Artificial Neural Network Based Battery Management Systems: A Literature Review. *Renew. Sustain. Energy Rev.* **2024**, *192*, 114262. [CrossRef]
42. Xiong, R.; Li, X.; Li, H.; Zhu, B.; Avelin, A. Neural Network and Physical Enable One Sensor to Estimate the Temperature for All Cells in the Battery Pack. *J. Energy Storage* **2024**, *80*, 110387. [CrossRef]
43. Alsuwian, T.; Ansari, S.; Ammirrul Atiqi Mohd Zainuri, M.; Ayob, A.; Hussain, A.; Hossain Lipu, M.S.; Alhawari, A.R.H.; Almawgani, A.H.M.; Almasabi, S.; Taher Hindi, A. A Review of Expert Hybrid and Co-Estimation Techniques for SOH and RUL Estimation in Battery Management System with Electric Vehicle Application. *Expert. Syst. Appl.* **2024**, *246*, 123123. [CrossRef]
44. Xu, L.; Wang, J.; Chen, Q. Kalman Filtering State of Charge Estimation for Battery Management System Based on a Stochastic Fuzzy Neural Network Battery Model. *Energy Convers. Manag.* **2012**, *53*, 33–39. [CrossRef]
45. Rimsha; Murawwat, S.; Gulzar, M.M.; Alzahrani, A.; Hafeez, G.; Khan, F.A.; Abed, A.M. State of Charge Estimation and Error Analysis of Lithium-Ion Batteries for Electric Vehicles Using Kalman Filter and Deep Neural Network. *J. Energy Storage* **2023**, *72*, 108039. [CrossRef]
46. Lee, D.; Lee, P.Y.; Baek, I.; Kwon, S.; Kim, J. Deriving the Optimized Battery Model for Battery Pack and Anomaly Detection Based on the Cloud Battery Management System. *J. Energy Storage* **2023**, *74*, 109338. [CrossRef]
47. Zhou, X.; Zhou, S.; Gao, Z.; Wang, G.; Zong, L.; Liu, J.; Feng, Z.; Ming, H.; Zheng, Y.; Chen, F.; et al. A Statistical Distribution-Based Pack-Integrated Model towards State Estimation for Lithium-Ion Batteries. *eTransportation* **2023**, *19*, 100302. [CrossRef]
48. Kim, M.; So, J. VLSI Design and FPGA Implementation of State-of-Charge and State-of-Health Estimation for Electric Vehicle Battery Management Systems. *J. Energy Storage* **2023**, *73*, 108876. [CrossRef]
49. Jiang, K.; Liao, G.; Jiaqiang, E.; Zhang, F.; Chen, J.; Leng, E. Thermal Management Technology of Power Lithium-Ion Batteries Based on the Phase Transition of Materials: A Review. *J. Energy Storage* **2020**, *32*, 101816. [CrossRef]
50. Fayaz, H.; Afzal, A.; Samee, A.D.M.; Soudagar, M.E.M.; Akram, N.; Mujtaba, M.A.; Jilte, R.D.; Islam, M.T.; Ağbulut, Ü.; Saleel, C.A. Optimization of Thermal and Structural Design in Lithium-Ion Batteries to Obtain Energy Efficient Battery Thermal Management System (BTMS): A Critical Review. *Arch. Comput. Methods Eng.* **2022**, *29*, 129–194. [CrossRef] [PubMed]
51. Deng, T.; Zhang, G.; Ran, Y.; Liu, P. Thermal Performance of Lithium Ion Battery Pack by Using Cold Plate. *Appl. Therm. Eng.* **2019**, *160*, 114088. [CrossRef]
52. Guo, Z.; Xu, Q.; Zhao, S.; Zhai, S.; Zhao, T.; Ni, M. A New Battery Thermal Management System Employing the Mini-Channel Cold Plate with Pin Fins. *Sustain. Energy Technol. Assess.* **2022**, *51*, 101993. [CrossRef]
53. Selvaraj, V.; Vairavasundaram, I. Flyback Converter Employed Non-Dissipative Cell Equalization in Electric Vehicle Lithium-Ion Batteries. *e-Prime Adv. Electr. Eng. Electron. Energy* **2023**, *5*, 100278. [CrossRef]
54. Sharma, D.K.; Prabhakar, A. A Review on Air Cooled and Air Centric Hybrid Thermal Management Techniques for Li-Ion Battery Packs in Electric Vehicles. *J. Energy Storage* **2021**, *41*, 102885. [CrossRef]
55. Murugan, M.; Saravanan, A.; Elumalai, P.V.; Murali, G.; Dhineshbabu, N.R.; Kumar, P.; Afzal, A. Thermal Management System of Lithium-Ion Battery Packs for Electric Vehicles: An Insight Based on Bibliometric Study. *J. Energy Storage* **2022**, *52*, 104723. [CrossRef]
56. Zhang, Y.; Zuo, W.; Jiaqiang, E.; Li, J.; Li, Q.; Sun, K.; Zhou, K.; Zhang, G. Performance Comparison between Straight Channel Cold Plate and Inclined Channel Cold Plate for Thermal Management of a Prismatic LiFePO4 Battery. *Energy* **2022**, *248*, 123637. [CrossRef]
57. Hao, M.; Li, J.; Park, S.; Moura, S.; Dames, C. Efficient Thermal Management of Li-Ion Batteries with a Passive Interfacial Thermal Regulator Based on a Shape Memory Alloy. *Nat. Energy* **2018**, *3*, 899–906. [CrossRef]
58. Zhao, H.; Yang, X.; Wang, C.; Lu, R.; Zhang, T.; Chen, H.; Zheng, X. Progress in Thermal Rectification Due to Heat Conduction in Micro/Nano Solids. *Mater. Today Phys.* **2023**, *30*, 100941. [CrossRef]
59. Behi, H.; Karimi, D.; Jaguemont, J.; Gandoman, F.H.; Kalogiannis, T.; Berecibar, M.; Van Mierlo, J. Novel Thermal Management Methods to Improve the Performance of the Li-Ion Batteries in High Discharge Current Applications. *Energy* **2021**, *224*, 120165. [CrossRef]
60. Wang, H.; Tao, T.; Xu, J.; Mei, X.; Liu, X.; Gou, P. Cooling Capacity of a Novel Modular Liquid-Cooled Battery Thermal Management System for Cylindrical Lithium Ion Batteries. *Appl. Therm. Eng.* **2020**, *178*, 115591. [CrossRef]
61. Yang, T.; Su, S.; Xin, Q.; Zeng, J.; Zhang, H.; Zeng, X.; Xiao, J. Thermal Management of Lithium-Ion Batteries Based on Honeycomb-Structured Liquid Cooling and Phase Change Materials. *Batteries* **2023**, *9*, 287. [CrossRef]
62. Yu, Y.; Vincent, T.; Sansom, J.; Greenwood, D.; Marco, J. Distributed Internal Thermal Monitoring of Lithium Ion Batteries with Fibre Sensors. *J. Energy Storage* **2022**, *50*, 104291. [CrossRef]
63. McKerracher, R.D.; Guzman-Guemez, J.; Wills, R.G.A.; Sharkh, S.M.; Kramer, D. Advances in Prevention of Thermal Runaway in Lithium-Ion Batteries. *Adv. Energy Sustain. Res.* **2021**, *2*, 20000059. [CrossRef]
64. Ghalkhani, M.; Habibi, S. Review of the Li-Ion Battery, Thermal Management, and AI-Based Battery Management System for EV Application. *Energies* **2023**, *16*, 185. [CrossRef]
65. Boujelbene, M.; Goodarzi, M.; Ali, M.A.; Shigidi, I.M.T.A.; Pashameah, R.A.; Homod, R.Z.; Alzahrani, E.; Safaei, M.R. Machine-Learning Optimization of an Innovative Design of a Li-Ion Battery Arrangement Cooling System. *J. Energy Storage* **2023**, *58*, 106331. [CrossRef]
66. Ghaeminezhad, N.; Wang, Z.; Ouyang, Q. A Review on Lithium-Ion Battery Thermal Management System Techniques: A Control-Oriented Analysis. *Appl. Therm. Eng.* **2023**, *219*, 119497. [CrossRef]

67. Zheng, Y.; Che, Y.; Hu, X.; Sui, X.; Stroe, D.I.; Teodorescu, R. Thermal State Monitoring of Lithium-Ion Batteries: Progress, Challenges, and Opportunities. *Prog. Energy Combust. Sci.* **2024**, *100*, 101120. [CrossRef]
68. Ling, X.; Zhang, Q.; Xiang, Y.; Chen, J.S.; Peng, X.; Hu, X. A Cu/Ni Alloy Thin-Film Sensor Integrated with Current Collector for in-Situ Monitoring of Lithium-Ion Battery Internal Temperature by High-Throughput Selecting Method. *Int. J. Heat. Mass. Transf.* **2023**, *214*, 124383. [CrossRef]
69. Liu, Q.; Huang, Q.; Guo, H.; Zhang, X.; Wang, K.; Zhang, Q.; Xu, H.; Xu, J.; Tashiro, Y.; Li, Z.; et al. Contactless Sensor for Real-Time Monitoring of Lithium Battery External Short Circuit Based on Magnetoelectric Elastomer Composites. *J. Power Sources* **2024**, *589*, 233776. [CrossRef]
70. Guo, Z.; Briggs, C.; Vincent, T.A.; Gulsoy, B.; Sansom, J.E.H.; Marco, J. Global Thermal Image of Cylindrical 21700 Li-Ion Batteries with Distributed Optical Fibre Sensor. *J. Power Sources* **2024**, *592*, 233980. [CrossRef]
71. Mao, S.; Han, X.; Lu, Y.; Wang, D.; Su, A.; Lu, L.; Feng, X.; Ouyang, M. Multi Sensor Fusion Methods for State of Charge Estimation of Smart Lithium-Ion Batteries. *J. Energy Storage* **2023**, *72*, 108736. [CrossRef]
72. Chen, D.; Zhao, Q.; Zheng, Y.; Xu, Y.; Chen, Y.; Ni, J.; Zhao, Y. Recent Progress in Lithium-Ion Battery Safety Monitoring Based on Fiber Bragg Grating Sensors. *Sensors* **2023**, *23*, 5609. [CrossRef]
73. Liu, S.; Li, K.; Chong, B.; Chen, Y. State-of-Charge Estimation of Li-Ion Battery Packs Based on Optic Fibre Sensor Measurements. *Transp. Res. Procedia* **2023**, *70*, 388–397. [CrossRef]
74. El Ouazzani, H.; El Hassani, I.; Barka, N.; Masrour, T. MSCC-DRL: Multi-Stage Constant Current Based on Deep Reinforcement Learning for Fast Charging of Lithium Ion Battery. *J. Energy Storage* **2024**, *75*, 109695. [CrossRef]
75. Matijašević, T.; Antić, T.; Capuder, T. A Systematic Review of Machine Learning Applications in the Operation of Smart Distribution Systems. *Energy Rep.* **2022**, *8*, 12379–12407. [CrossRef]
76. Waseem, M.; Ahmad, M.; Parveen, A.; Suhaib, M. Battery Technologies and Functionality of Battery Management System for EVs: Current Status, Key Challenges, and Future Prospectives. *J. Power Sources* **2023**, *580*, 233349. [CrossRef]
77. Zou, Y.; Lin, Z.; Li, D.; Liu, Z.C. Advancements in Artificial Neural Networks for Health Management of Energy Storage Lithium-Ion Batteries: A Comprehensive Review. *J. Energy Storage* **2023**, *73*, 109069. [CrossRef]
78. Wei, Z.; Li, P.; Cao, W.; Chen, H.; Wang, W.; Yu, Y.; He, H. Machine Learning-Based Hybrid Thermal Modeling and Diagnostic for Lithium-Ion Battery Enabled by Embedded Sensing. *Appl. Therm. Eng.* **2022**, *216*, 119059. [CrossRef]
79. Hossain Lipu, M.S.; Ansari, S.; Miah, M.S.; Meraj, S.T.; Hasan, K.; Shihavuddin, A.S.M.; Hannan, M.A.; Muttaqi, K.M.; Hussain, A. Deep Learning Enabled State of Charge, State of Health and Remaining Useful Life Estimation for Smart Battery Management System: Methods, Implementations, Issues and Prospects. *J. Energy Storage* **2022**, *55*, 105752. [CrossRef]
80. Bai, G.; Su, Y.; Rahman, M.M.; Wang, Z. Prognostics of Lithium-Ion Batteries Using Knowledge-Constrained Machine Learning and Kalman Filtering. *Reliab. Eng. Syst. Saf.* **2023**, *231*, 108944. [CrossRef]
81. Samanta, A.; Williamson, S.S. A Comprehensive Review of Lithium-Ion Cell Temperature Estimation Techniques Applicable to Health-Conscious Fast Charging and Smart Battery Management Systems. *Energies* **2021**, *14*, 5960. [CrossRef]
82. Zhao, J.; Feng, X.; Pang, Q.; Wang, J.; Lian, Y.; Ouyang, M.; Burke, A.F. Battery Prognostics and Health Management from a Machine Learning Perspective. *J. Power Sources* **2023**, *581*, 233474. [CrossRef]
83. Jiao, M.; Wang, D.; Qiu, J. A GRU-RNN Based Momentum Optimized Algorithm for SOC Estimation. *J. Power Sources* **2020**, *459*, 228051. [CrossRef]
84. Khawaja, Y.; Shankar, N.; Qiqieh, I.; Alzubi, J.; Alzubi, O.; Nallakaruppan, M.K.; Padmanaban, S. Battery Management Solutions for Li-Ion Batteries Based on Artificial Intelligence. *Ain Shams Eng. J.* **2023**, *14*, 102213. [CrossRef]
85. Mamo, T.; Wang, F.K. Long Short-Term Memory with Attention Mechanism for State of Charge Estimation of Lithium-Ion Batteries. *IEEE Access* **2020**, *8*, 94140–94151. [CrossRef]
86. Surya, S.; Rao, V.; Williamson, S.S. Comprehensive Review on Smart Techniques for Estimation of State of Health for Battery Management System Application. *Energies* **2021**, *14*, 4617. [CrossRef]
87. Zhao, J.; Feng, X.; Pang, Q.; Fowler, M.; Lian, Y.; Ouyang, M.; Burke, A.F. Battery Safety: Machine Learning-Based Prognostics. *Prog. Energy Combust. Sci.* **2024**, *102*, 101142. [CrossRef]
88. Yang, Q.; Xu, J.; Li, X.; Xu, D.; Cao, B. State-of-Health Estimation of Lithium-Ion Battery Based on Fractional Impedance Model and Interval Capacity. *Int. J. Electr. Power Energy Syst.* **2020**, *119*, 105883. [CrossRef]
89. Nawaz, M.; Ahmed, J.; Abbas, G. Energy-Efficient Battery Management System for Healthcare Devices. *J. Energy Storage* **2022**, *51*, 104358. [CrossRef]
90. Park, K.; Choi, Y.; Choi, W.J.; Ryu, H.Y.; Kim, H. LSTM-Based Battery Remaining Useful Life Prediction with Multi-Channel Charging Profiles. *IEEE Access* **2020**, *8*, 20786–20798. [CrossRef]
91. Hannan, M.A.; Lipu, M.S.H.; Hussain, A.; Saad, M.H.; Ayob, A. Neural Network Approach for Estimating State of Charge of Lithium-Ion Battery Using Backtracking Search Algorithm. *IEEE Access* **2018**, *6*, 10069–10079. [CrossRef]
92. Zhang, F.; Zhai, L.; Zhang, L.; Yi, M.; Du, B.; Li, S. A Novel Hybrid Battery Thermal Management System with Fins Added on and between Liquid Cooling Channels in Composite Phase Change Materials. *Appl. Therm. Eng.* **2022**, *207*, 118198. [CrossRef]
93. Chen, X.; Zhou, F.; Yang, W.; Gui, Y.; Zhang, Y. A Hybrid Thermal Management System with Liquid Cooling and Composite Phase Change Materials Containing Various Expanded Graphite Contents for Cylindrical Lithium-Ion Batteries. *Appl. Therm. Eng.* **2022**, *200*, 117702. [CrossRef]

94. Huo, Y.; Rao, Z.; Liu, X.; Zhao, J. Investigation of Power Battery Thermal Management by Using Mini-Channel Cold Plate. *Energy Convers. Manag.* **2015**, *89*, 387–395. [CrossRef]
95. Hasan, M.K.; Mahmud, M.; Ahasan Habib, A.K.M.; Motakabber, S.M.A.; Islam, S. Review of Electric Vehicle Energy Storage and Management System: Standards, Issues, and Challenges. *J. Energy Storage* **2021**, *41*, 102940. [CrossRef]
96. *ISO 12405-4:2018*; Electrically Propelled Road Vehicles—Test Specification for Lithium-Ion Traction Battery Packs and Systems—Part 4: Performance Testing. International Organization for Standardization: Geneva, Switzerland, 2018. Available online: https://bit.ly/3wxWXrL (accessed on 19 January 2024).
97. UN38.3 Certification for Lithium Batteries. Electrically Propelled Road Vehicles-Test Specification for Lithium-Ion Traction Battery Packs and Systems. Available online: https://bit.ly/3TgkNRD (accessed on 19 January 2024).
98. Wang, Y.; Zhang, X.; Li, K.; Zhao, G.; Chen, Z. Perspectives and Challenges for Future Lithium-Ion Battery Control and Management. *eTransportation* **2023**, *18*, 100260. [CrossRef]
99. Akbarzadeh, M.; Jaguemont, J.; Kalogiannis, T.; Karimi, D.; He, J.; Jin, L.; Xie, P.; Van Mierlo, J.; Berecibar, M. A Novel Liquid Cooling Plate Concept for Thermal Management of Lithium-Ion Batteries in Electric Vehicles. *Energy Convers. Manag.* **2021**, *231*, 113862. [CrossRef]
100. Yue, Q.L.; He, C.X.; Sun, J.; Xu, J.B.; Zhao, T.S. A Passive Thermal Management System with Thermally Enhanced Water Adsorbents for Lithium-Ion Batteries Powering Electric Vehicles. *Appl. Therm. Eng.* **2022**, *207*, 118156. [CrossRef]
101. Ren, Y.; Yu, Z.; Song, G. Thermal Management of a Li-Ion Battery Pack Employing Water Evaporation. *J. Power Sources* **2017**, *360*, 166–171. [CrossRef]
102. Shahjalal, M.; Shams, T.; Islam, M.E.; Alam, W.; Modak, M.; Hossain, S.B.; Ramadesigan, V.; Ahmed, M.R.; Ahmed, H.; Iqbal, A. A Review of Thermal Management for Li-Ion Batteries: Prospects, Challenges, and Issues. *J. Energy Storage* **2021**, *39*, 102518. [CrossRef]
103. Shen, Z.G.; Chen, S.; Liu, X.; Chen, B. A Review on Thermal Management Performance Enhancement of Phase Change Materials for Vehicle Lithium-Ion Batteries. *Renew. Sustain. Energy Rev.* **2021**, *148*, 111301. [CrossRef]
104. Samimi, F.; Babapoor, A.; Azizi, M.; Karimi, G. Thermal Management Analysis of a Li-Ion Battery Cell Using Phase Change Material Loaded with Carbon Fibers. *Energy* **2016**, *96*, 355–371. [CrossRef]
105. Hossain Lipu, M.S.; Hannan, M.A.; Karim, T.F.; Hussain, A.; Saad, M.H.M.; Ayob, A.; Miah, M.S.; Indra Mahlia, T.M. Intelligent Algorithms and Control Strategies for Battery Management System in Electric Vehicles: Progress, Challenges and Future Outlook. *J. Clean. Prod.* **2021**, *292*, 126044. [CrossRef]
106. Chen, M.; Zhang, Y.; Xing, G.; Chou, S.L.; Tang, Y. Electrochemical Energy Storage Devices Working in Extreme Conditions. *Energy Environ. Sci.* **2021**, *14*, 3323–3351. [CrossRef]
107. Li, Y.; Yu, L.; Hu, W.; Hu, X. Thermotolerant Separators for Safe Lithium-Ion Batteries under Extreme Conditions. *J. Mater. Chem. A Mater.* **2020**, *8*, 20294–20317. [CrossRef]

batteries

MDPI

Article

Li-Ion Battery Thermal Characterization for Thermal Management Design

Aron Saxon, Chuanbo Yang *, Shriram Santhanagopalan, Matthew Keyser and Andrew Colclasure

National Renewable Energy Laboratory, 15013 Denver West Parkway, Golden, CO 80401, USA; aron.saxon@nrel.gov (A.S.); shriram.santhanagopalan@nrel.gov (S.S.); matthew.keyser@nrel.gov (M.K.); andrew.colclasure@nrel.gov (A.C.)

* Correspondence: chuanbo.yang@nrel.gov; Tel.: +1-720-618-4954

Abstract: Battery design efforts often prioritize enhancing the energy density of the active materials and their utilization. However, optimizing thermal management systems at both the cell and pack levels is also key to achieving mission-relevant battery design. Battery thermal management systems, responsible for managing the thermal profile of battery cells, are crucial for balancing the trade-offs between battery performance and lifetime. Designing such systems requires accounting for the multitude of heat sources within battery cells and packs. This paper provides a summary of heat generation characterizations observed in several commercial Li-ion battery cells using isothermal battery calorimetry. The primary focus is on assessing the impact of temperatures, C-rates, and formation cycles. Moreover, a module-level characterization demonstrated the significant additional heat generated by module interconnects. Characterizing heat signatures at each level helps inform manufacturing at the design, production, and characterization phases that might otherwise go unaccounted for at the full pack level. Further testing of a 5 kWh battery pack revealed that a considerable temperature non-uniformity may arise due to inefficient cooling arrangements. To mitigate this type of challenge, a combined thermal characterization and multi-domain modeling approach is proposed, offering a solution without the need for constructing a costly module prototype.

Keywords: lithium-ion battery; battery thermal efficiency; battery thermal management; battery thermal model; battery thermal management design

1. Introduction

Battery electric vehicles (BEVs) improve renewable energy integration and enhance grid energy efficiency, leading to reduced greenhouse gas emissions. Accelerating BEV adoption needs to address both the range and charge anxieties of BEV users. These forces point toward development in extreme-fast-charge (XFC) enabled battery design and increased specific energy and volumetric energy density. Fast charge rates shall be comparable to refueling a petroleum-fueled vehicle and the range should exceed vehicles on the road today [1]. The increase in energy densities and charging capability results in an increase in heat generation, which requires upsizing the battery thermal management system (BTMS) [2].

The need for BTMSs arises from the sensitivity of lithium-ion batteries to their operating temperatures [3]. If the cells become too hot, the life cycle of the battery is significantly decreased. Inversely, if the cell temperature is too low, the power performance and the available capacity drop significantly. In designing a full vehicle pack, it is imperative to account for all the heat sources within the battery pack and to couple the form factor with the appropriate thermal management system.

Today, batteries have excellent thermal efficiency—generally around 98% under normal cycling conditions found within BEVs. Any inefficiencies, although small at the cell level, stack upon themselves once integrated into a battery pack. For instance, a pack with a rated power output of 150 kW is expected to generate 3 kW of heat load on the system when considering that same 98% heat efficiency. The inclusion of a cooling system designed to reject 3 kW of heat load results in an addition of approximately 5 kg to the battery system [4]. This calculation does not include the cooling requirement under extreme fast charging, which could be as high as 15 kW [5]. Assuming the thermal management mass scales with system power and a battery pack with a specific energy of 300 Wh/kg at the cell level, a 15 kW management system accounts for 5% of the mass of the batteries within the pack. This example does not account for the impact on the system volume.

The advent of newer lithium-ion chemistries, such as high-nickel NMCs [6,7] and silicon-based anodes [8], alongside efforts to achieve higher energy densities, drives the need for continually more robust thermal management systems. The higher the energy density, the larger the volumetric heating from inefficiencies. Even with improvements in efficiency, the cooling load will increase because of the decrease in the available heat transfer area, resulting in an increase in the heat flux to be rejected by battery systems. An accurate estimate of heat generation rate versus operating condition (e.g., state of charge or C-rate) is needed to effectively design a thermal management unit that can keep up with this demand.

Characterizing heat generation sources is a crucial step in designing a BTMS [9]. The reliability of heat source characterization relies on a fundamental understanding at various levels within a battery pack. In this study, we provide a concise review of the different heat generation terms within a battery pack. Additionally, we detail the operating parameters, such as the C-rate and temperature, as well as the interconnect design, which play crucial roles in influencing the amount of heat generated. Furthermore, we offer examples demonstrating how heat measurements in isothermal battery calorimeters during electrochemical cycling aid in overcoming practical design challenges. Finally, we illustrate how to match the design of thermal management units with target battery performance.

2. Methods

All heat generation measurements discussed in this article were collected at the National Renewable Energy Laboratory's Battery Thermal and Life Analysis Laboratory. In these experiments, a battery sample is placed within an isothermal calorimeter and cycled within the heat chamber via an external power supply, as illustrated in Figure 1. The heat chamber is submerged within a liquid isothermal bath. The heat dissipated from a battery sample is measured using heat flux sensors attached to the surface of the heat chamber. The total thermal energy is calculated by temporally integrating the measured heat flux profile. Due to the thermal lag between the sample surface and the sensors in the calorimeter, it is important to note that the measured heat flux may not accurately represent the time-resolved heat flux at the cell surface, especially under high power loads. A more detailed explanation of the calorimeter technique is provided in Isothermal Battery Calorimeters [10]. "Isothermal" is used to describe the temperature condition of the full system that is used to measure heat generated by the device under test. The full system is controlled to ±0.001 °C, which allows ±5 mW of baseline heat stability. This setup, while isothermal at the full system scale, still allows the device under test to experience a temperature increase.

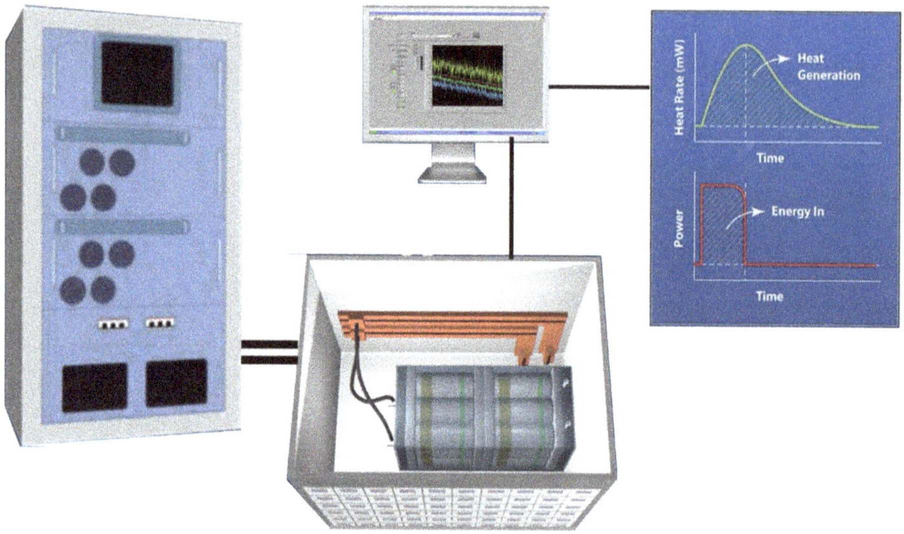

Figure 1. Illustration of battery thermal characterization using isothermal battery calorimeters at National Renewable Energy Laboratory. The battery in the test chamber of the isothermal battery calorimeter (bottom). In the graph (right), the red curve shows electrical power input to the battery, and the green curve shows the measured heat signature.

By integrating battery thermal characterization at cell level with system-level simulations using battery computer-aided engineering (CAE) tools, such as multi-scale multi-domain battery modeling [11,12], the thermal performance of a battery system can be assessed. These combined experimental and modeling approaches can accelerate battery prototype development and reduce development cost. A design trade-off analysis of BTMSs can be conducted without building a costly system. More importantly, these approaches are effective for quantifying spatial thermal and electrochemical inhomogeneities that are difficult and expensive to measure in experiments.

3. Results and Discussion

3.1. Cell Heat Generation

The main source of heat within a battery pack is the cells themselves. Heat generation in a battery cell arises from two components: the electrode sandwich and current collectors, which can be expressed as

$$q_{cell} = \int q'''_{electrode\ sandwish} dV + \int q'''_{current\ collectors}\, dV \tag{1}$$

where q is the heat in the unit of W and q''' is volumetric heat in the unit of W/m^3.

The electrical heat produced by both positive and negative current collectors follows Ohm's law and can be calculated using Equation (2) [1,13].

$$q'''_{current\ collectors} = \left(\sigma|\nabla\varnothing|^2\right)_{pos} + \left(\sigma|\nabla\varnothing|^2\right)_{neg} \tag{2}$$

where σ represents electrical conductivity and $\varnothing$ represents the local electric potential.

For the electrochemical heat generated by the electrode sandwich, Bernardi et al. [14] proposed a comprehensive equation that considers electrochemical reactions, joule heating, phase changes, and mixing effects. Mixing heat and phase change heat are typically small and can be neglected [15,16]. Therefore, the volumetric heat generated by the sandwich can be simplified to Equation (3).

$$q'''_{electrode\ sandwish} = i'''(U - V) - i'''\left(T\frac{\partial U}{\partial T}\right) \tag{3}$$

where i''' is the volumetric reaction current in the unit of A/m^3; U is the open circuit voltage; V is the sandwich volage between current collectors; and T is the temperature in K. The first term corresponds to irreversible overpotential heat due to ohmic losses in the sandwich, while the second term represents the reversible entropic heat.

Joule heating from the electrolyte and resistance within the cell matrix are the largest contributors to the heat signature [17]. Ohmic heating sourced from the ion flux within the electrolyte is the largest source of cell heat originating from the materials within the cell during normal operation. The heat sourced from the electrolyte can be attributed to low electrolyte conductivity/diffusivity at higher C-rates and formation/growth of the solid electrolyte interface (SEI). Matrix ohmic heating is the heat generated within the solid portion of the lithium-ion cell. Matrix ohmic heating has historically been neglected due to the relatively larger electrical conductivity compared to the electrolyte ionic conductivity. In the drive to bolster specific energy, cell designs have increased the electrode thickness from around 40 μm to over 100 μm [18]. This maximization of active material in turn impacts the relative amount of inactive current conductors and increases the amount of heating from the current conductors. These heat sources are the resistances at interfaces between cell components and ohmic losses of the current-conducting substrate. Electrolyte and matrix ohmic heating are dependent on factors such as the C-rate, cell size, and cell age. The heating rates scale with the increase in each of these parameters. Advanced diagnostic techniques, such as the micro calorimeter developed by Santhanagopalan et al. [19], measure the time-resolved thermal performance of small battery samples while conducting an electrochemical impedance spectroscopy (EIS) characterization. This enables the identification of the heat contribution of each individual component while also relating their bulk contribution.

The chemical reactions within the cell generate both reversible and irreversible heat. The reversible heating source is a result of the entropic changes in the lattice structure when lithium is inserted into or pulled from the structure. While there are some irreversible entropic losses, a reversible reaction dominates the heat signature at low cycle rates—usually those under C/10 charge or discharge [17]. The irreversible heating is attributed to the charge transfer overpotential during the oxidation and redox at the interfaces between the electrolyte and active cell materials. Under normal operating conditions, the reversible heat sources contribute to 10% of the total cell heating, whereas the irreversible losses account for around 40% [17].

Individual cell energy inefficiencies are correlated to transport and kinetic properties within the cell. At the cell level, this correlation can be illustrated through the comparison between energy and power cells. An energy cell designed to prioritize meeting an energy requirement will typically have higher active material loading, which will have adverse heat effects due to the longer path the ions have to travel during intercalation. This longer pathway results in irreversible heat losses. Similarly, the total effective thermal conductivity of these higher electrode loading cells is also lower, resulting in increased levels of heat not dispersing throughout the cell. In contrast, a power cell designed for high current outputs to meet a power target will have a lower material loading and thinner electrodes, leading to lower irreversible losses due to ion transport. Although these cells have a better thermal signature, they suffer from a total energy standpoint due to having less active material and a larger fraction of the total mass/volume being dedicated to substrates that do not participate in the cell reaction. Designing a system that minimizes heat losses within the battery requires an understanding and quantification of each heat source.

3.2. Cell Heat Generation with C-Rate and Temperature

The heat generated by the cell increases as the cell's cycled C-rate increases. This is mostly because of ohmic joule heating, which increases proportionally to the square of C-rates. The joule heating can be simply modeled using Ohm's law:

$$q = I^2R \tag{4}$$

where q is the heating rate; I denotes the cell current; and R is the total cell resistance. In this case, the resistance captures both ionic diffusion resistances within the active material and electrolyte, as well as the electrical resistances of the electron's path.

The local temperature during operation has an inverse effect on cell heat generation. As the temperature decreases, the ionic conductivity of the electrolyte decreases due to the increased viscosity of the electrolyte, creating more resistance to the ionic flow. Conversely, the battery operates more efficiently at higher temperatures, but the trade-off for increased performance is usually a decrease in battery life. Thermal management systems work to balance this trade-off between performance and life degradation.

Figure 2a illustrates the heat generation of a 12 Ah LMO/LTO pouch cell across multiple temperatures at various C-rates, with the cutoff voltages set at 2.8 V and 1.6 V. At a rate of 1C, the heat generation rate was approximately 0.25 W at 30 °C, 0.50 W at 0 °C, and 1.5 W at −15 °C. This effect was also evident at the higher rate of 4C. The heat generation tripled from 2 W to 6 W between 30 °C and 0 °C, and doubled from 6 W to around 13 W between 0 °C and −15 °C. This behavior of the cell indicates that there was a resistance spike between 30 °C and 0 °C, likely due to a rapid decrease in the ionic conductivity between those temperatures. Over this same temperature change, the cell discharge capacity was inversely proportional to the cell resistance.

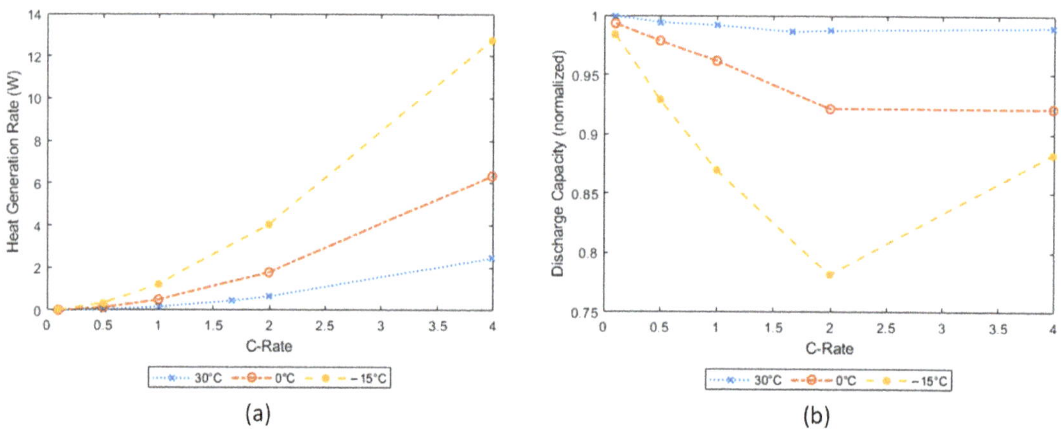

Figure 2. (**a**) Graph illustrating how heat generation changes with both temperature and discharge rate. (**b**) Graph demonstrating the decrease in available electrical energy as the cycle temperature decreases from 30 °C to 0 °C to −15 °C.

Figure 2b demonstrates this decrease in available energy by showing how the available capacity vs. the C-rate of the cell changes. Comparing the change in discharge capacities against the amount of heat generation at the different temperatures and C-rates also serves to evaluate transport limitations within the cell. Comparing Figure 2a,b, we can see that the amount of heat generation was not always proportional to the improvement in cell capacity. In addition, we must note that although these results are specific to the cell studied under this example, the analysis procedure outlined here broadly applies to multiple chemistries and/or cell designs. In fact, these heat signals can be used to benchmark performance across different cell formats with the same chemistry, or vice versa.

Next, we aimed to understand how heat generation is impacted by both temperature and discharge rate. For an ideal ohmic resistor for a given ambient temperature, the heat generation rate should scale with the square of the C-rate. At 30 °C, the heat generation rate was 0.18 W at 1C, 0.67 W at 2C, and 2.5 W at 4C. When normalized for the currents (i.e., when the heat generation rate is divided by the square of the C-rate), we see that the resistance at 4C was approximately 87% of that at 1C. This is attributed to the enhanced transport properties heating within the cell at higher currents. Subsequently, the ambient temperature was lowered from 30 °C to 0 °C and the experiment was repeated. At 0 °C, the average heat rate increased from 1C to 4C and was nearly three times what it was at 30 °C. A similar trend was observed moving from 0 °C to −15 °C. The 4C heat rate increased from 2.5 W at 30 °C to 6 W at 0 °C and to around 13 W at −15 °C. From these results, it is evident that the capacity at 0 °C decreased as the C-rate increased. At −15 °C this correlation was amplified. The cell's discharge capacity fell to a point at −15 °C where the 2C capacity was only approximately 77% of the available capacity at 30 °C, indicating a need for an external heat source to minimize the performance loss at low temperatures.

3.3. *Evolution of Heat Generation in Battery Formation Cycling*

Heat generation rates change with cell aging, leading to a decrease in capacity and an increase in resistance [20]. However, there is currently a lack of publicly available experimental quantification regarding the effect of aging on battery heat generation. Several factors can influence changes in cell resistance and heat generation, including the formation of the solid–electrolyte interface, changes to wetting properties of the electrodes and/or the separator membrane, degradation of the electrolyte additives, gas generation, mechanical degradation, and lithium plating.

Cells often undergo formation cycles, where the potential within the different cell components equilibrate and the surface layers on the electrodes mature before the cell capacities level off. Cell formation is a widely empirical, tightly controlled process. The formation parameters, such as ambient temperature, durations, and voltages (or currents), of the charge and discharge are often determined from an extensive experimental design process.

Figure 3 shows an increase in the heating rates with increasing current (i.e., C-rates) passing through 4.4 Ah cylindrical LFP/graphite cells with cutoff voltages of 3.6 V and 2.0 V. The cell samples underwent constant-current constant-voltage (CCCV) charging with a cutoff current of C/20. During discharge, the cell samples were discharged at specified C-rates with a cutoff voltage of 2.0 V. In this study, these measurements were repeated for 30 cycles after the first formation cycle. As noted, there was an increase in the slope of the heat-generation rate versus the charging rate (C-rate) past the C/2 currents for the initial cycle. This increase leveled off with cell cycling, indicating that the limitations were likely from surface limitations because of inactive electrode materials and limited electrolyte-wettability, rather than inherently poor conductivity in the bulk electrolyte at the higher currents. With formation aging, the surface impedances at the electrodes leveled off, resulting in lower heat generation rates. It is also worth noting that the difference between the heating rates during charge and discharge was slightly higher at the C/3 and C/2 rates for the initial cycle than at 1C. This indicates that the heat generation rate during the initial charge step helps to quantify limitations at the electrode surface.

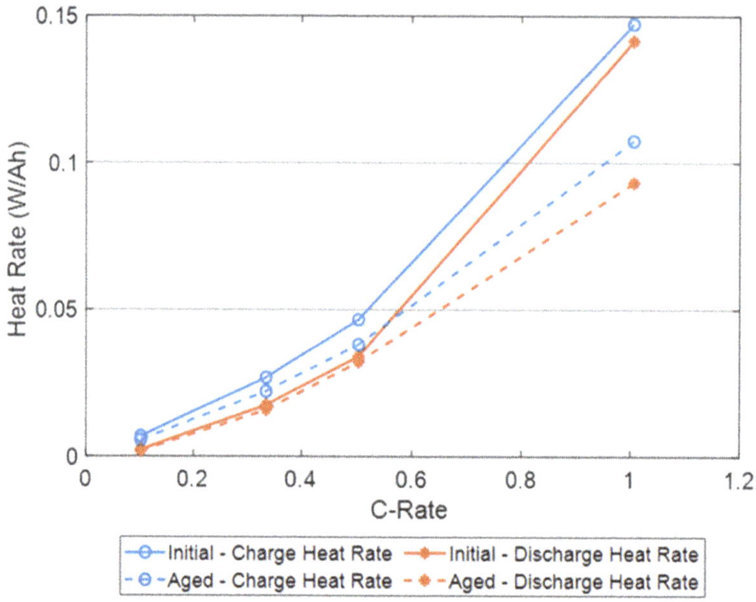

Figure 3. Illustration of heat generation decreasing after formation cycling of a cylindrical LFP/graphite cell. In some cases, the cell efficiency can increase (i.e., lower heat generation rate) following a break-in period. In this cell, the 1C heat rate decreases from approximately 0.15 to 0.1 W/Ah following the break-in period.

3.4. Module and Full Pack Interconnect Design

A previous experiment conducted in-house revealed the possibility of additional heating occurring when a cylindrical 4.5 Ah LFP/graphite cell was incorporated into a module compared to what was measured at the cell level. The module was designed for a hybrid electrical vehicle application, and the cell was intended for high-power applications. The power-to-energy ratio, defined as the maximum rated power over a known duration divided by the available energy in the battery, is greater than 10 [4]. Figure 4 compares the heat generation rates from the single LFP cell in a module, with heat generation rates normalized on a per-cell basis, to that of an identical but stand-alone cell at different discharge currents.

The increase in heat between the stand-alone cell and cells in the module is attributed to the module interconnects. The equivalent discharge current at the relevant hybrid electric vehicle application for this cell is approximately 8C. At this current rate, the module generates approximately 20% more heat per cell than the stand-alone cell. We can determine that this additional heating is a result of interconnects by applying Ohm's law. In this case, the average interconnect resistance is approximately 0.25 Ω.

Accounting for the total heat source is critical when designing a BTMS. In this case, scaling the expected total heat based solely on each cell's heat signature would result in a thermal management system unable to keep up with the heat generated, resulting in higher operating temperatures than designed and a decrease in expected life and possibly creating an unsafe thermal runaway condition if subjected to prolonged use.

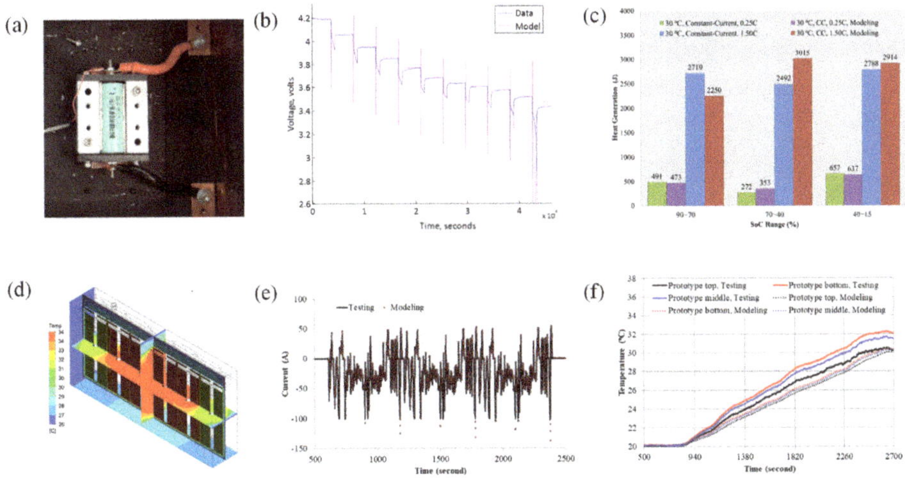

Figure 6. A bottom–up approach combining cell characterization and 3D modeling for BTMS design: (**a**) a cell under calorimeter testing; (**b**) validation of model electrical performance; (**c**) validation of model heat generation prediction; (**d**) temperature distribution within the protype calculated using the 3D model; (**e**) comparison of current response between testing and modeling; and (**f**) discrepancies in prototype surface temperatures between testing and modeling.

Once validated, the cell-level model is upscaled for 3D simulations of a battery prototype composed of six individual cells and a battery housing. Under the US06 driving cycle power loading, the prototype's current response predicted using the model is nearly identical with the experimental measurement, as shown in Figure 6e. Additionally, the simulated spatial temperature distribution on the prototype surface closely matches that measured using the thermocouple during US06 cycling. Figure 6f shows that the maximum discrepancy is less than 2.5 °C, indicating that heat transfer pathways from the cells to prototype ambient through the housing components are accurately presented.

As shown in Figure 6d, the simulation quantified the temperature difference between the center cells and corner cells, which is an essential input for temperature gradient control. The results suggest the design load needs to be limited or the heat dissipation performance in the axial direction of jelly rolls should be enhanced using the BTMS.

4. Conclusions

Precise heat signature measurement within batteries has vast implications for BTMS designs, which profoundly shapes battery pack performance, longevity, and safety. The thermal performance of battery cells is influenced by an intricate interplay of variables, including cell chemistry, cell design, and operational parameters such as the C-rate and temperature range. This paper introduces a calorimetry characterization methodology aimed at dissecting the various components contributing to heat generation. It leverages isothermal battery calorimeters to capture heat signatures across various operating conditions.

The calorimetry characterizations have revealed several key factors affecting heat generation within a battery pack:

- C-rate: the heat generated by a Li-ion cell increases as the cell's C-rate increases. Especially at high C-rates, the heat is proportional to the square of the C-rates.
- Temperature: Temperature also plays a significant role, affecting energy efficiency due to changes in cell resistance. Notably, the discharged capacity could be increased at subzero temperatures at higher discharge C-rates. This is because of substantial self-heating at higher C-rates.

- Cycling: The calorimetry measurement captured the reduction of heat generation rates during formation and subsequent aging cycles. This reduction is attributed to the formation of solid electrolyte interphase (SEI) layers, which can enhance cell performance, particularly at higher C-rates.
- Module interconnects: This study underscores the importance of considering the heat contribution from interconnects when designing BTMSs. Interconnects can considerably increase heat generation, as demonstrated in the case of a LiFePO4 (LFP) module for a hybrid electric vehicle, where interconnects led to approximately 20% more heat per cell compared to standalone cells at application current rates of around 8C. Measuring heat signatures from single cells and comparing them against the heat signatures in modules or packs provides information on the engineering limitations associated with how the cells are packaged.

BTMSs serve multiple roles, not only managing heat generated to control cell temperatures but also minimizing temperature variations within battery packs. The presence of significant temperature heterogeneities can exert detrimental effects on battery performance and longevity. With single-sided cooling, up to a 12 °C temperature difference could be generated within a real 5 kWh battery pack designed for a plug-in hybrid electric vehicle.

The optimization of cooling strategies can be effectively addressed using a bottom–up approach, which integrates battery thermal characterization and cutting-edge battery computer-aided engineering tools. As exemplified in the presented case study, the approach provided accurate predictions of thermal performance in a battery module prototype.

Author Contributions: Conceptualization, A.S. and S.S.; methodology, M.K. and A.S.; software, C.Y.; validation, A.S. and C.Y.; formal analysis, A.S. and C.Y.; investigation, A.S., C.Y. and M.K.; resources, A.S.; data curation, A.S. and M.K.; writing—original draft preparation, A.S. and C.Y.; writing—review and editing, C.Y., A.C. and S.S.; visualization, A.S., M.K., and C.Y.; supervision, M.K.; project administration, M.K.; funding acquisition, M.K. All authors have read and agreed to the published version of the manuscript.

Funding: This work was supported by U.S. DOE EERE Vehicle Technologies Office under Contract No. DE-AC36-08GO28308.

Data Availability Statement: Data are contained within the article.

Acknowledgments: This work was authored by the National Renewable Energy Laboratory(Golden, CO, USA), operated by Alliance for Sustainable Energy, LLC, for the U.S. Department of Energy (DOE) under Contract No. DE-AC36-08GO28308. Funding provided by U.S. Department of Energy Office of Energy Efficiency and Renewable Energy Vehicle Technologies Office. The authors acknowledge Brian Cunningham, Samuel Gillard, and Haiyan Croft from the U.S. DOE for their support in this research. The views expressed herein do not necessarily represent the views of the DOE or the U.S. Government. The U.S. Government retains and the publisher, by accepting the article for publication, acknowledges that the U.S. Government retains a nonexclusive, paid-up, irrevocable, worldwide license to publish or reproduce the published form of this work, or allow others to do so, for U.S. Government purposes.

Conflicts of Interest: The authors declare no conflict of interest.

References

1. Gnann, T.; Funke, S.; Jakobsson, N.; Plötz, P.; Sprei, F.; Bennehag, A. Fast charging infrastructure for electric vehicles: Today's situation and future needs. *Transp. Res. Part D Transp. Environ.* **2018**, *62*, 314–329. [CrossRef]
2. Liu, Y.; Zhu, Y.; Cui, Y. Challenges and opportunities towards fast-charging battery materials. *Nat. Energy* **2019**, *4*, 540–550. [CrossRef]
3. Zeng, Y.; Chalise, D.; Lubner, S.D.; Kaur, S.; Prasher, R.S. A review of thermal physics and management inside lithium-ion batteries for high energy density and fast charging. *Energy Storage Mater.* **2021**, *41*, 264–288. [CrossRef]
4. United States Advanced Battery Consortium. USABC Li-Ion Battery Thermal Management System Requirements. Available online: http://uscar.org/download/246/energy-storage-system-goals/12843/thermal-management-system-goals.pdf (accessed on 6 October 2021).

5. Keyser, M.; Pesaran, A.; Li, Q.; Santhanagopalan, S.; Smith, K.; Wood, E.; Ahmed, S.; Bloom, I.; Dufek, E.; Shirk, M.; et al. Enabling fast charging–Battery thermal considerations. *J. Power Sources* **2017**, *367*, 228–236. [CrossRef]
6. Tian, X.; Guo, R.; Bai, Y.; Li, N.; Wang, X.; Wang, J.; Wu, C. High-Performance High-Nickel Multi-Element Cathode Materials for Lithium-Ion Batteries. *Batteries* **2023**, *9*, 319. [CrossRef]
7. Deng, D. Li-ion batteries: Basics, progress, and challenges. *Energy Sci. Eng.* **2015**, *3*, 385–418. [CrossRef]
8. Eshetu, G.G.; Zhang, H.; Judez, X.; Adenusi, H.; Armand, M.; Passerini, S.; Figgemeier, E. Production of high-energy Li-ion batteries comprising silicon-containing anodes and insertion-type cathodes. *Nat. Commun.* **2021**, *12*, 5459. [CrossRef]
9. Pesaran, A.; Keyser, M.; Burch, S. An Approach for Designing Thermal Management Systems for Electric and Hybrid Vehicle Battery Packs. In Proceedings of the 4th Vehicle Thermal Management Systems, London, UK, 24–27 May 1999.
10. Keyser, M.; Smith, K. Battery Thermal Modeling and Testing. NREL/PR-5400-50916, Arlington, Virginia. 2011. Available online: https://www.nrel.gov/docs/fy11osti/50916.pdf (accessed on 1 March 2024).
11. Kim, G.-H.; Smith, K.; Lawrence-Simon, J.; Yang, C. Efficient and Extensible Quasi-Explicit Modular Nonlinear Multiscale Battery Model: GH-MSMD. *J. Electrochem. Soc.* **2017**, *164*, A1076–A1088. [CrossRef]
12. Panchal, S.; Dincer, I.; Agelin-Chaab, A.; Fraser, R.; Fowler, M. Transient electrochemical heat transfer modeling and experimental validation of a large sized $LiFePO_4$/graphite battery. *Int. J. Heat Mass Transf.* **2017**, *109*, 1239–1251. [CrossRef]
13. Kim, U.S.; Shin, C.B.; Kim, C.-S. Effect of electrode configuration on the thermal behavior of a lithium-polymer battery. *J. Power Sources* **2008**, *180*, 909–916. [CrossRef]
14. Bernardi, D.; Pawlikowski, E.; Newman, J. A General Energy Balance for Battery Systems. *J. Electrochem. Soc.* **1985**, *132*, 5–12. [CrossRef]
15. Bandhauer, T.M.; Garimella, S.; Fuller, T.F. A Critical Review of Thermal Issues in Lithium-Ion Batteries. *J. Electrochem. Soc.* **2011**, *158*, R1. [CrossRef]
16. Gu, W.B.; Wang, C.Y. Thermal-Electrochemical Modeling of Battery Systems. *J. Electrochem. Soc.* **2000**, *147*, 2910–2922. [CrossRef]
17. Srinivasan, V.; Yang, C.Y. Analysis of Electrochemical and Thermal Behavior of Li-Ion Cells. *J. Electrochem. Soc.* **2003**, *150*, A98. [CrossRef]
18. Quinn, J.B.; Waldmann, T.; Richter, K.; Kasper, M.; Wohlfahrt-Mehrens, M. Energy Density of Cyclindrical Li-Ion Cells: A Comparison of Commercial 18650 to the 21700 Cells. *J. Electrochem. Soc.* **2018**, *165*, A3284–A3291. [CrossRef]
19. Santhanagopalan, S.; Keyser, M.; Saxon, A.; Ireland, J. Methods and Devices for Electrochemical System Analysis. U.S. Patent 10,107,696, 23 October 2018.
20. Shen, W.; Wang, N.; Zhang, J.; Wang, F.; Zhang, G. Heat Generation and Degradation Mechanism of Lithium-Ion Batteries during High-Temperature Aging. *ACS Omega* **2022**, *7*, 44733–44742. [CrossRef]
21. Smith, K.; Saxon, A.; Keyser, M.; Lundstrom, B.; Cao, Z.; Roc, A. Life prediction model for grid-connected Li-ion battery energy storage system. In Proceedings of the 2017 American Controls Conference (ACC), Seattle, WA, USA, 24–26 May 2017; pp. 4062–4068.
22. Ansys, Inc. *Ansys Fluent Theory Guide*; Release 2021R2; Ansys, Inc.: Canonsburg, PA, USA, July 2021.
23. Doyle, M.; Fuller, T.F.; Newman, J. Modeling of Galvanostatic Charge and Discharge of the Lithium/Polymer/Insertion Cell. *J. Electrochem. Soc.* **1993**, *140*, 1526–1533. [CrossRef]
24. Kim, U.S.; Yi, J.; Shin, C.B.; Han, T.; Park, S. Modeling the Dependence of the Discharge Behavior of a Lithium-Ion Battery on the Environmental Temperature. *J. Electrochem. Soc.* **2011**, *158*, A611. [CrossRef]
25. Wang, Q.; Jiang, B.; Li, B.; Yan, Y. A critical review of thermal management models and solutions of lithium-ion batteries for the development of pure electric vehicles. *Renew. Sustain. Energy Rev.* **2016**, *64*, 106–128. [CrossRef]

batteries

MDPI

Article

Effects of Non-Uniform Temperature Distribution on the Degradation of Liquid-Cooled Parallel-Connected Lithium-Ion Cells

Takuto Iriyama, Muriel Carter, Gabriel M. Cavalheiro, Pragati Poudel, George J. Nelson and Guangsheng Zhang *

Department of Mechanical & Aerospace Engineering, The University of Alabama in Huntsville, Huntsville, AL 35899, USA
* Correspondence: gz0002@uah.edu

Abstract: Our previous work on an air-cooled stack of five pouch-format lithium-ion (Li-ion) cells showed that non-uniform temperature can cause accelerated degradation, especially of the middle cell. In this work, a stack of five similar cells was cycled at a higher C-rate and water-cooled to create a larger temperature gradient for comparison with the air-cooled stack. It was hypothesized that the larger temperature gradient in the water-cooled stack would exacerbate the degradation of the middle cell. However, the results showed that the middle cell degraded slightly slower than the side cells in the water-cooled stack. This trend is opposite to that in the air-cooled stack. This difference could be attributed to the combined effects of a smaller temperature rise and larger temperature gradient in the water-cooled stack than in the air-cooled stack. Post-mortem analysis of cycled cells and a fresh cell showed that the degradation mainly came from the anode. Increased lithium plating and decreased porosity in the side cells are possible mechanisms for the faster degradation compared with the middle cell. It was also found that all the cells in the water-cooled stack experienced a phenomenon of capacity drop and recovery after a low C-rate reference performance test and extended rest. This phenomenon can be attributed to lithium diffusion between the anode active area and the anode overhang area.

Keywords: Li-ion battery; degradation; temperature distribution; liquid cooling; anode overhang

1. Introduction

Li-ion batteries have been widely used in various applications. However, their durability still needs to be further improved, especially for electric vehicle (EV) and grid-scale energy storage applications. There are many mechanisms of degradation, such as through growth of the solid–electrolyte interphase (SEI) layer, lithium plating, electrode particle fracture, and transition metal dissolution [1–3]. Most of these degradation mechanisms are temperature dependent [2]. High temperatures can accelerate the SEI layer growth and transition metal dissolution. Low temperatures can increase the risk of lithium plating and electrode particle fracture.

Due to the heat generation during operation [4,5] and the low thermal conductivity of Li-ion cell components [6–8], temperature distributions in Li-ion cells are generally non-uniform across the cell thickness [9–14], especially in large-format cells during higher-rate operation. Due to the dependence of Li-ion battery degradation on temperature, the non-uniform temperature distribution could cause non-uniform and accelerated degradation. Hunt et al. compared the effects of surface cooling and tab cooling on the degradation of single Li-ion cells [15]. The research showed that strong surface cooling accelerated degradation, which was attributed to large temperature gradients across the cell thickness. Carter et al. [16] studied the influence of a non-uniform temperature distribution by imposing a temperature gradient externally and also observed accelerated degradation. Some other experimental results showed that non-uniform temperature distribution can lead to non-uniform resistance and current distribution [17–19], which would also accelerate degradation because a higher current can exacerbate particle cracking and lithium plating.

In most of the above-mentioned experimental studies on the degradation effects of non-uniform temperature distribution, the temperature differences were applied externally. To confirm that non-uniform temperature distribution due to heat generation inside large Li-ion cells can cause non-uniform and accelerated degradation, Cavalheiro et al. [20] tested a stack of five parallel-connected pouch cells for 2215 cycles and compared their degradation with a baseline cell. The temperature distributions, individual cell degradation, and current distributions of the stack were measured simultaneously. It was found that all five cells in the stack experienced a higher temperature rise and degraded faster than the baseline cell. In particular, the capacity retention of the middle cell in the stack, which experienced the largest temperature rise, decreased to 51% after 2215 cycles. In comparison, the capacity retention of the side cells was still above 82%, and that of the baseline cell was nearly 88%. These results clearly supported the hypothesis that a non-uniform temperature distribution causes non-uniform and accelerated degradation. To confirm the degradation mechanisms of the cells in the work by Cavalheiro et al. [20], Gonzalez Malabet et al. [21] performed electrochemical impedance spectroscopy (EIS) and post-mortem analysis of the cycled cells for comparison with a fresh cell. While the EIS implied that the main degradation mechanism for the middle cell was SEI layer growth, post-mortem analysis using X-ray diffraction (XRD), optical microscopy, and scanning electron microscopy (SEM) with energy dispersive X-ray spectroscopy (EDS) revealed the presence of lithium plating, especially in the middle cell. The analysis also implied that fast SEI layer growth in the middle cell induced lithium plating and caused abrupt failure of the middle cell near the end of the cycling. The analysis was consistent with a previous modeling study [22].

While the studies by Cavalheiro et al. [20] and Gonzalez Malabet et al. [21] confirmed that a non-uniform temperature distribution during operation can cause non-uniform and accelerated degradation, the Li-ion cells in the studies were air-cooled using natural convection. Compared with air cooling, liquid cooling has a much stronger cooling capability and is much more widely used in EV applications [23]. But stronger cooling would lead to a larger temperature gradient and exacerbate the effects of a non-uniform temperature distribution. Therefore, it is hypothesized that the degradation of liquid-cooled Li-ion cells will be more non-uniform than that of air-cooled cells. To test this hypothesis and to enhance our understanding of the effects of non-uniform temperature distribution that is more relevant for practical applications, a stack of five parallel-connected Li-ion cells, similar to those in our previous study [20], are cycled under water-cooled conditions in this study. The discharging rate during cycling is set to twice that in the previous study to further increase the temperature gradient. Surprisingly, however, the water-cooled stack of Li-ion cells shows very different behaviors to the air-cooled stack. More details are described in the following sections. Section 2 describes the experimental setup and test protocol. Section 3 describes and discusses the experimental results. Finally, Section 4 briefly summarizes the new findings.

2. Experimental

2.1. Experimental Setup

As schematically shown in Figure 1a, five pouch-format 3-Ah Li-ion cells (30107-0, Tenergy, Fremont, CA, USA) were stacked together and placed between a pair of cooling plates (ATS-CP-1001, Advanced Thermal Solutions, Norwood, MA, USA). The dimension of each cell is 102 mm (L) by 50 mm (W) by 6 mm (H). The pair of cooling plates was connected to a water circulator (Refrigerated/Heated Bath Circulator, Thermo Fisher Scientific, Waltham, MA, USA) to keep the water temperature at a set value (set to 24 °C in this study to be comparable to our previous study with the air-cooled stack [20]). Six T-type thermocouples (5TC-TT-T-30-72, Omega Engineering Inc., Norwalk, CT, USA) were used to measure temperature distributions across the thickness of the stack, labelled as T1 through T6 from the top to the bottom. As shown in Figure 1b, the sides of the cells were thermally insulated by polystyrene foam to reduce heat loss. The five pouch cells were connected in parallel and then connected to a battery tester (LBT 21084, Arbin Instruments, College Station, TX, USA). Shunt resistors (PLV7AL, 2 ± 0.005% mΩ, Precision Resistor Company Inc., Clearwater, FL, USA) were used for the measurement of individual cell currents (I1, I2, I3, I4, I5), while the total stack current (I_{total}) was controlled and measured by the battery tester.

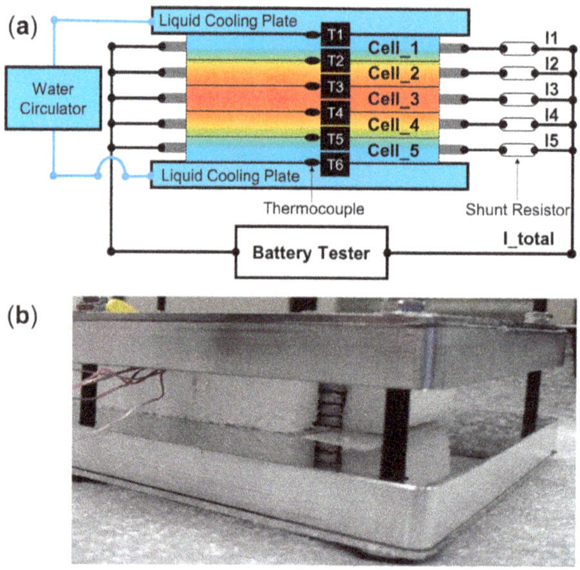

Figure 1. (**a**) Schematic of a water-cooled stack of five Li-ion cells connected in parallel (reused with permission from The Electrochemical Society [24]); (**b**) A photo of an experimental stack.

2.2. Selection of Experimental Li-Ion Cells

To ensure comparable performance of cells in the stack, five cells were selected from the same batch. The selected cells had a similar high-frequency resistance (7.9 ± 0.35 mΩ) as measured by a resistance tester (BATTERY HiTESTER 3561, HIOKI USA, Dallas, TX, USA), similar DC internal resistance (19.2 ± 0.7 mΩ, as determined by the open-circuit voltage and the cell voltage after 10 s of 2C discharge [25]), and similar discharge capacity (3.34 ± 0.03 Ah, according to 1C discharge testing).

2.3. Test Protocol

As shown in Figure 2a, the stack was first rested for 40 min, then charged at a constant current of 1C (corresponding to 15 A) up to a voltage of 4.2 V, followed by a constant voltage step of 4.2 V until the current decreased to C/20 (corresponding to 0.75 A). After a rest of 20 min, the stack was discharged at a constant current of 2C (corresponding to 30 A) until a voltage of 2.8 V was reached. After another rest of 40 min, the stack was charged again in a new cycle. Reference performance testing (RPT) was performed after every 50 or 100 cycles. The charging protocol during RPT was the same (1C, 4.2 V max, C/20 cut-off), but the discharging rate was C/3 (corresponding to 5 A) so that the cells could be fully discharged. After each RPT, the stack experienced an extended rest (at least 4 h) before the resumption of normal cycling.

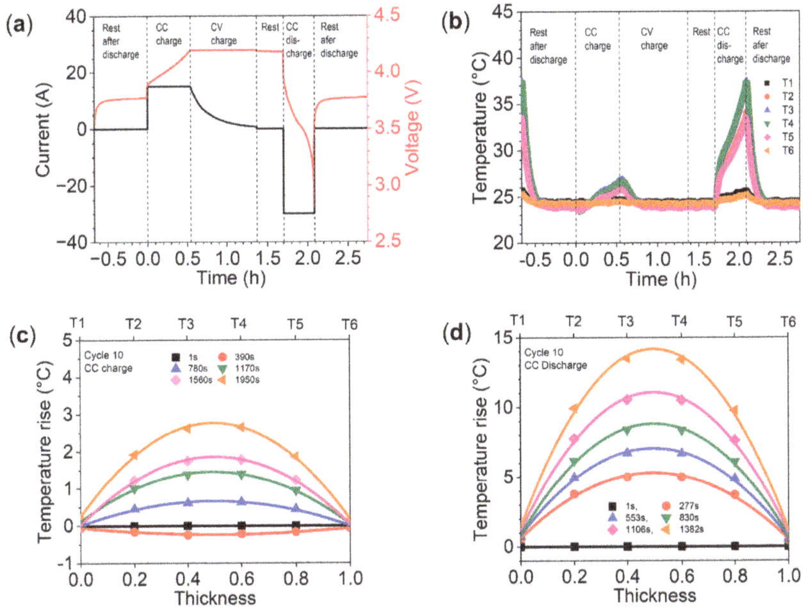

Figure 2. Representative test protocol and temperature profiles during cycle 10. (**a**) Stack voltage and current; (**b**) Variation in local temperatures; (**c**) Temperature distribution during CC charging; (**d**) Temperature distribution during discharging (reused with permission from The Electrochemical Society [24]).

2.4. Post-Mortem Analysis

Cell 3, cell 4, and cell 5 in the stack were disassembled for post-mortem analysis. A fresh cell was also disassembled as a reference. The cells were first fully discharged and then disassembled in an argon-filled glovebox (LABstart pro, MBRAUN USA, Stratham, NH, USA). Electrode layers, including 15 anode layers and 14 cathode layers in each cell, were separated from the tabs. Individual electrode layers were washed twice using 99% dimethyl carbonate (DMC, Thermo Scientific Chemicals, Waltham, MA, USA) and dried before being moved out of the glovebox for further examination. Pictures of dried electrodes were taken using a phone camera. Some regions of the electrodes were further examined using a JEOL 7000 field-emission scanning electron microscope (JEOL Ltd., Tokyo, Japan) at the University of Alabama Analytical Research Facility. The FE-SEM system has an Oxford energy dispersive spectroscopy (EDS) system attached, which was used for the analysis of elemental content in the electrodes.

3. Results and Discussion

3.1. Temperature Distribution Behaviours

As a representation, Figure 2b shows the variation in local temperatures in the stack during cycle 10. It can be seen that the local temperatures T2 to T5 decreased slightly at the beginning of charging. This slight decrease can be attributed to the negative entropic heat generation, which is discussed in Section 3.2. Then, the local temperatures T2 to T5 started to increase because the heat generation became positive at higher SOC levels. After a few minutes into the constant voltage (CV) charging stage, local temperatures began to decrease due to the decline in charging current and the heat generation rate. The temperatures went back to 24 °C shortly after the end of charging. Local temperatures T1 and T6 showed little change during this period because they were located between the outside cells and the cooling plates that were kept at a constant temperature. All the local temperatures increased to higher levels and more rapidly during discharging than during charging, which can be attributed to the higher current during discharging (30 A) than that during charging (15 A). Even T1 and T6 showed obvious increases during discharging.

The temperature distribution across the thickness of the stack is better seen in Figure 2c,d, which show the rise in local temperatures at different times of charging and discharging. T1 and T6 show very small changes due to their contact with the cooling plates, while local temperatures closer to the center of the stack show larger changes due to heat generation inside the cells. Note that the largest temperature difference occurs between T1 and T2 as well as T5 and T6. The differences between T2 and T3 as well as T4 and T5 are smaller but still noticeable. There is not an obvious temperature difference between T3 and T4 due to the symmetry of their locations although they increased the most. These results of the temperature rise and temperature difference indicate that side cells (cell 1 and cell 5) experienced the smallest average temperature rise but the largest temperature gradient. In comparison, the middle cell (cell 3) experienced the largest temperature rise but the smallest temperature gradient.

3.2. Heat Generation Analysis

In order to understand the temperature behavior of the stack, heat generation was analyzed. Irreversible heat generation due to ohmic heating and reversible heat generation due to entropy change were considered. Other sources of heat generation, such as mixed heat generation, were assumed to be negligible [26,27]. Then the heat generation rate q_{gen} during the charging of a Li-ion cell could be estimated from Equation (1):

$$q_{gen} = q_{gen,irr} + q_{gen,rev} = |I(V - U)| + IT\left(\frac{\partial U}{\partial T}\right) \tag{1}$$

where I is the current and its value is positive during charging, V is the cell voltage, U is the open-circuit voltage (OCV), T is the absolute temperature in the Kelvin scale, and $\left(\frac{\partial U}{\partial T}\right)$ is the entropic coefficient. The term $|I(V - U)|$ represents the irreversible heat generation rate $q_{gen,irr}$, which is always positive. The term $IT\left(\frac{\partial U}{\partial T}\right)$ represents the reversible heat generation rate $q_{gen,rev}$, which can be positive or negative depending on the signs of the current and the entropic coefficient.

The open-circuit voltage at different SOC was estimated as the average of the charging voltage and discharging voltage at C/30 (0.1 A), which nearly overlapped due to the small current. The results are shown in Figure 3a. The entropic coefficient was obtained with a baseline cell using a procedure similar to that in an earlier study [13]. The cell was first fully charged at 25 °C and then discharged at C/30 to different SOC levels (~5% interval). At each SOC level, the cell was rested for at least four hours for its OCV to become stable, then the cell was heated from 25 °C to 40 °C or cooled from 40 °C to 25 °C and rested for another four hours to reach a new stable OCV. Then, the entropic coefficient was determined by assuming a linear change in the OCV between 25 °C and 40 °C. The variation in the entropic coefficient with SOC levels is shown in Figure 3b. It can be seen that the entropic coefficient is negative for a wide range of SOCs, especially below a 40% SOC. Then, according to Equation (1), the entropic heat generation at the beginning of charging is negative, which can offset the positive irreversible heat generation and thus cause the cell temperature to decrease slightly, as shown in Figure 2b. Then, during discharging when the current is negative, the entropic heat generation is mostly positive. Combined with the always positive irreversible heat generation and the higher discharging current rate, the heat generation during discharging would be much higher than that during charging.

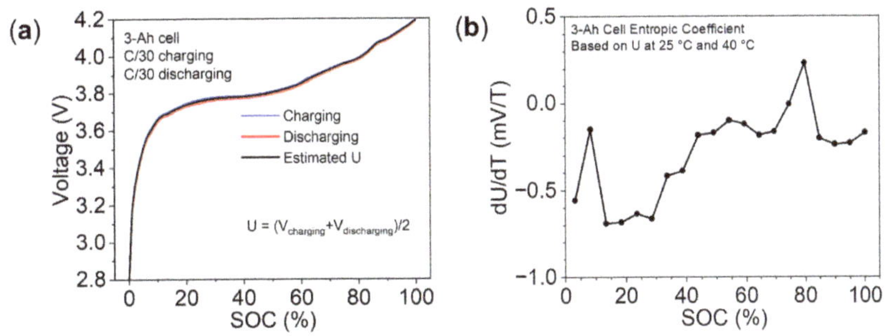

Figure 3. (**a**) Variation in cell voltage during C/30 charging and C/30 discharging as well as the estimated OCV; (**b**) Estimated entropic coefficient at different SOC levels.

3.3. Degradation of Cells in the Stack

Figure 4a shows the capacity retention of cells in the stack over 1600 regular cycles (2C discharging), which indicates the usable capacity during 2C discharging. Figure 4b shows the capacity retention determined by RPT (C/3 discharging), which indicates the retention of true capacity during low-current discharging. The gap between cycles 1100 and 1200 was due to a failure in data recording by the battery tester. The abnormally higher capacity between cycles 1300 and 1400 was due to insufficient water and insufficient cooling of the water circulator, which happened during the winter break. Note that these malfunctions did not change the general trend of degradation or the conclusion of this work.

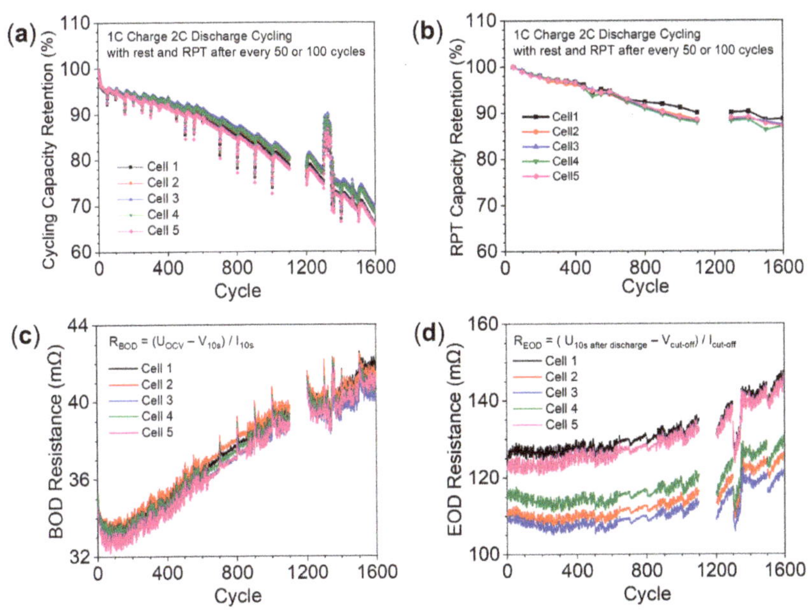

Figure 4. (**a**) Capacity retention of the stack during regular cycling (data for cycles 1101 to 1200 were missing due to a data acquisition malfunction; the abnormal higher capacity between cycles 1301 and 1400 was due to insufficient cooling; the data between cycle 1 and 1000 were reused with permission from The Electrochemical Society [24]); (**b**) Capacity retention of the stack during RPT; (**c**) Variation in BOD resistance with cycle number; (**d**) Variation in EOD resistance with cycle number.

Three interesting phenomena can be observed from Figure 4a,b. First, the capacity retention in regular cycles was significantly lower than that in RPT cycles. This trend is expected. It indicates that the cells in the stack suffered power fade as well as capacity fade. Power fade is associated with an increase in cell internal resistance, which reduces the usable capacity during high-current discharging. To confirm this, the variations in DC resistance at the beginning of discharge (BOD) and at the end of discharge (EOD) with cycle number are shown in Figures 4c and 4d respectively. Note that the BOD resistance was determined by using the difference between the open-circuit voltage (U_{OCV}, ~4.18 V) and the cell voltage at 10 s of discharging (V_{10s}) divided by the individual cell current (I_{10s}). The EOD resistance was determined by using the difference between the open-circuit voltage at 10 s after discharge ($U_{10s\ after\ discharge}$) and the cell cut-off voltage ($V_{cut\text{-}off}$; 2.8 V in this study) divided by the individual cell current at cut-off ($I_{cut\text{-}off}$). It can be seen that the resistance of the cells indeed increased significantly with cycling. Note that the EOD resistance of all the cells was much higher than the BOD resistance, which can be attributed to the electrolyte concentration gradient and the high diffusion resistance near the end of discharge [28]. The BOD resistance was similar among the five cells because the temperature distribution was uniform at the beginning. The EOD resistance was very different, with cell 3 resistance being the lowest while those of cells 1 and 5 were the highest. The difference can be attributed to cell 3 having the highest temperature, while cells 1 and 5 had the lowest temperatures at the end of discharge, as shown in Figure 2d. Second, the middle cell (cell 3) showed slightly slower degradation than the outside cells (cells 1 and 5). This trend can be more clearly seen in Figure 5a, which shows the detailed capacity retention during cycles 1590 to 1600. This trend is unexpected because it is the opposite of what was observed in our earlier study with air-cooled stacks [20], in which the middle cell degraded the quickest. More discussion on this phenomenon is presented in Section 3.4. Third, the capacity retention of every cell experienced a significant drop after each RPT cycle, but this recovered in a few cycles and then decreased again. This phenomenon is more clearly shown in Figure 5b, with details before and after the RPT at around cycle 1500. More discussion on this phenomenon is presented in Section 3.5.

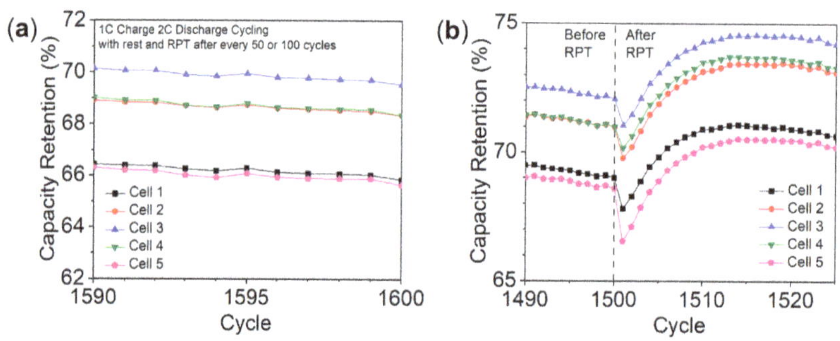

Figure 5. (**a**) Details of the capacity retention from cycles 1590 to 1600. (**b**) Details of the capacity retention before and after RPT.

3.4. Non-Uniform Degradation Due to Temperature Differences

As mentioned above in Figure 4a, the middle cell in the stack degraded slightly more slowly than the outside cells. This trend is opposite to the trend observed in our earlier study of an air-cooled stack. A few factors could influence the non-uniform degradation of cells in the stack. The first factor is the non-uniform current distribution across the stack due to a non-uniform temperature distribution, but this was ruled out as discussed below. As shown in Figure 6, cell 3 always had the highest current due to having the highest temperature. Li-ion cells degrade faster at a higher C-rate. Therefore, non-uniform current distribution is unlikely to be the cause of slower degradation for cell 3. The second factor is non-uniform lithium plating. Considering that cell 3 had the highest temperature during cycling, it would have a lower risk of lithium plating than the side cells. Post-mortem analysis in Section 3.6 indicated a greater presence of lithium plating in cell 5 than in cell 3, which supported this explanation. The third factor is the temperature gradient across the cell. An earlier work by Carter et al. [16] showed that the temperature gradient across the thickness of Li-ion cells can accelerate degradation. As shown in Figure 7, the temperature gradient is smallest in the middle cell and largest in the side cells. Note that the temperature gradient of the side cells in this water-cooled stack (~10 °C) is much larger than that of the side cells in the previous air-cooled stack (~4 °C) [20]. Therefore, the larger temperature gradient across the side cells in this water-cooled stack could cause faster degradation than for the middle cell. The fourth factor is the maximum temperature of the middle cell. As shown in Figure 7, the maximum temperature of the middle cell stayed below 44 °C in this water-cooled stack. In comparison, the value ranged from 44 °C to 50 °C in our earlier air-cooled stack. The reduction in the maximum temperature of the middle cell would slow down degradation. Further investigation is needed in the future to examine these factors more closely.

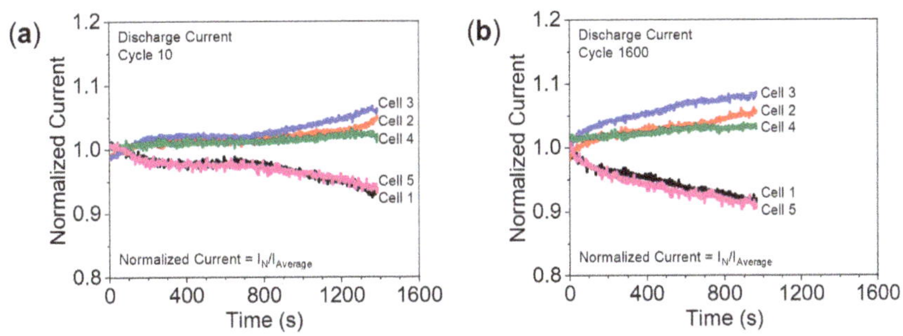

Figure 6. Current of each cell during discharge in (**a**) cycle 10 and (**b**) cycle 1600. The current was normalized by dividing individual cell current by the average current of the cells in the stack (6 A).

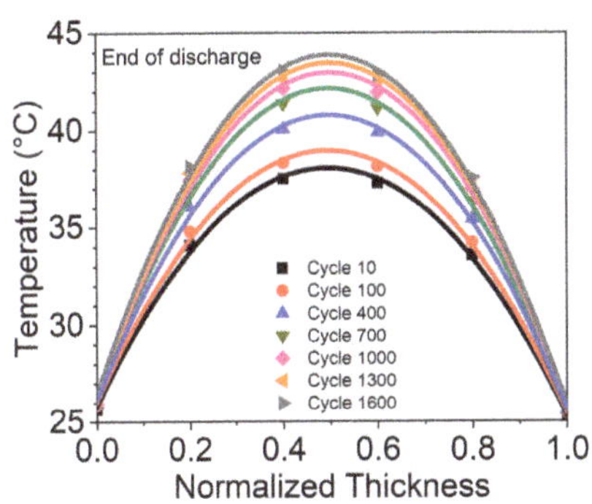

Figure 7. Temperature distribution at the end of discharge in different cycles.

3.5. The Phenomenon of Capacity Drop and Recovery after RPT

As mentioned earlier and as clearly shown in Figure 5b, all the cells in the stack experienced a capacity drop and recovery after an RPT cycle. The discharge rate was C/3 during the RPT cycle, which is much lower than that during regular cycling. Therefore, it is possible that the phenomenon of capacity drop and recovery was related to the RPT cycles. To understand this phenomenon, individual 3-Ah cells from the same batch as cells in the stack were tested. These individual cells were cycled similarly to the cells in the stack using a constant current, constant voltage charge, and 2C discharge. However, the charging rate during cycling, the rest time after RPT, and the starting SOC for the reference performance test were adjusted.

Figure 8a shows the results of a cell with different charging rates during cycling. The cell was charged at 1C, C/2, C/3, and C/4 for 20 cycles before one RPT cycle. The rest time after RPT was kept at 2 h. It can be seen that the phenomenon of capacity drop and recovery exists in all cases. This result indicates that the charging rate or lithium plating was not the cause of the capacity drop and recovery.

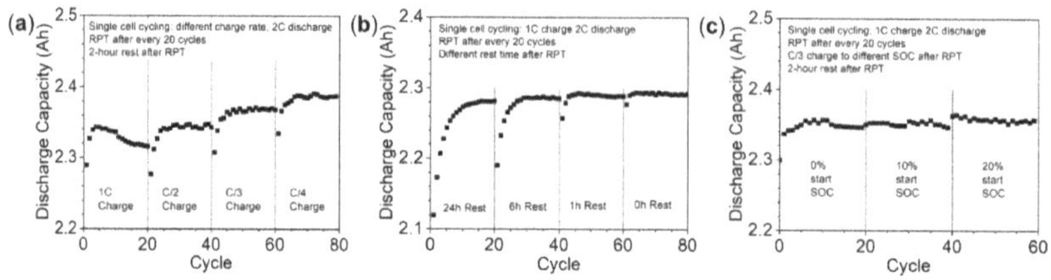

Figure 8. Results of individual cell testing (**a**) with different charging C-rates, (**b**) with different rest times after RPT, and (**c**) with different SOCs after RPT.

Figure 8b shows the results of a cell with different rest times after RPT. The charging rate is 1C. It can be seen that rest time after RPT influences the capacity drop and recovery. Generally, longer rest time is associated with more capacity drop.

Figure 8c shows the results of a cell with different SOCs after RPT. In the 0% SOC case, the cell was simply discharged at the C/3 rate to 2.8 V during the RPT cycle and rested for two hours before regular cycling. In the 10% and 20% starting-SOC cases, the cell was first discharged to 0% SOC in the RPT cycle, then charged to 10% SOC or 20% SOC at the C/3 rate, and then rested for two hours before regular cycling. It can be seen that the phenomenon of capacity drop and recovery was less obvious in the 10% SOC case and completely disappeared in the 20% SOC case. The discharge capacity actually increased after resting in the 20% SOC case, which was opposite to the trend observed in the 0% SOC case. This result clearly showed that the 0% SOC after the C/3 RPT cycle was related to the phenomenon of capacity drop and recovery of cells in the water-cooled stack. During regular cycling, the cells were discharged at 2C to 2.8 V, and the average SOC would be higher than 20% SOC.

Based on the results in Figure 8b,c, it is clear that the low SOC and rest time after each RPT cycle are related to the phenomena of capacity drop and recovery. A literature search indicates that anode overhang effects [29] are the likely causes of the phenomenon. Anode overhang refers to the area of the anode outside the coverage of the cathode. The anode electrode sheet is typically manufactured to be slightly larger than the cathode sheet in Li-ion cells to avoid over-lithiation and lithium plating at the anode edge. Depending on the SOC of a Li-ion cell, the lithium concentration in the anode active area (covered by the cathode) and anode overhang area (outside the cathode) can be different. The concentration difference can cause diffusion between the active area and the overhang area, which can influence the usable discharge capacity. During the regular 1C charge/2C discharge cycling, the SOC of the cell varied from 100% SOC to ~30% SOC, so the SOC in the anode overhang would be in that range. When the cells experienced a C/3 discharge during the RPT cycles, the SOC in the anode active area would reach ~0%, which is significantly lower than that in the anode overhang area. This difference would cause diffusion of lithium from the overhang area to the active area. After extended rest, the diffusion would cause the lithium concentration in the overhang area to be much lower than that during regular cycling. Then, after the first charge after RPT, the lithium concentration in the anode active area would be much higher than that in the anode overhang, causing lithium diffusion to the anode overhang. The diffusion would continue during rest and discharge, reducing the amount of lithium that can be transported to the cathode, thereby causing the reduction in usable capacity as compared with that before RPT. Diffusion from the anode active area to the anode overhang would still occur during the following cycles until a new equilibrium is reached. However, the amount of diffusion would be less, and therefore the "reduced capacity" would be less, and the dischargeable capacity during cycling would recover until capacity degradation started to dominate again.

Based on the above explanation, the phenomenon of capacity drop and recovery would disappear if there were no RPT. To test this hypothesis, the stack was further cycled with rests of more than 24 h after every 100 cycles but without RPT. The capacity retention from cycles 1601 to 1900 is shown in Figure 9. The phenomenon of capacity drop and recovery indeed disappeared, suggesting that the above explanation based on anode overhang effects was reasonable. Note that the capacity increase after rest is commonly observed in Li-ion cell cycling with extended rest, which can be attributed to relaxation effects [30]. Lithium diffusion from the anode overhang area to the active area during rest with a higher starting SOC may have also played a role in the capacity increase [29].

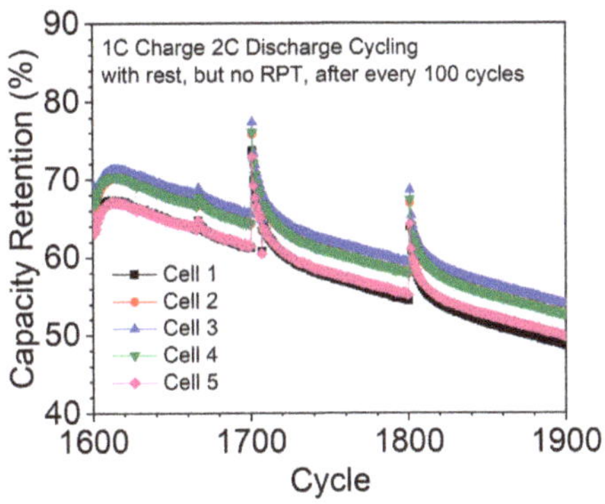

Figure 9. Capacity retention of cells with extended rest but without RPT after 1700 and 1800 cycles.

3.6. Post-Morterm Analysis

Cells 3, 4, and 5 in the stack, as well as a fresh cell, were fully discharged and then disassembled in an argon-filled glovebox for further analysis. Each cell has 15 double-coated anode layers and 14 double-coated cathode layers. Figure 10 shows pictures of representative cathode layers, while Figure 11 shows representative anode layers. A complete set of anode pictures is shown in Supplementary Figures S1–S8. All the cathode layers from the cycled cells and the fresh cell look very similar. This similarity is further supported by the SEM images in Figure 12. In comparison, the anode layers from the cycled cells look very different from each other and from the fresh cell. This implies that the degradation of the cycled cells mainly came from the anode, which is consistent with our earlier work of the air-cooled stack [20,21]. Therefore, the following discussion focuses on the differences of the anode layers.

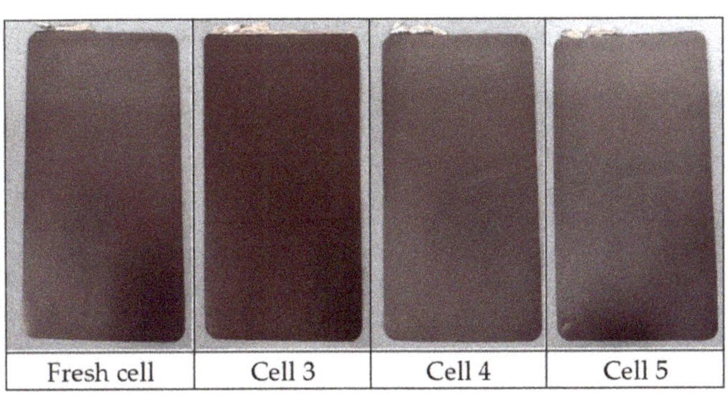

Figure 10. Comparison of the cathode layer of a fresh cell, cell 3, cell 4, and cell 5. Each cell has 14 cathode layers. There was no obvious difference among the different cathode layers in a cell.

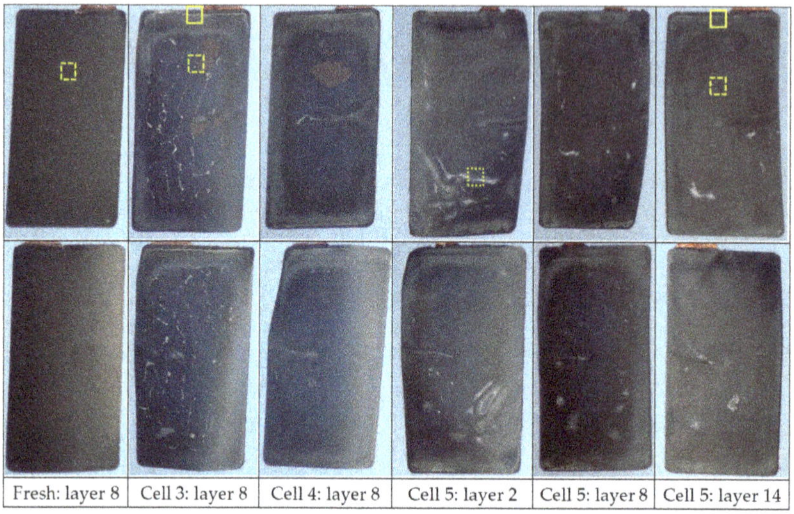

Figure 11. Front side (top row) and back side (bottom row) of anode layers from a fresh cell, cell 3, cell 4, and cell 5. Each cell has 15 anode layers. For cells 3, 4, and 5, the front side faces the top of the stack, and the back side faces the bottom of the stack in Figure 1a. The yellow squares indicate regions for SEM or EDS analysis.

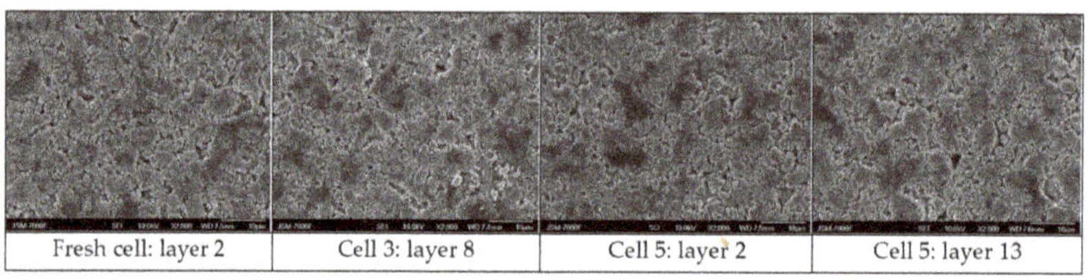

Figure 12. SEM images of the front side of the cathode from a fresh cell, cell 3, and cell 5 at 2000× magnification.

Figure 11 and Supplementary Figures S1–S8 show a few interesting differences in the anode layers. First, some anode coating was tightly attached to the separators and delaminated from the Cu foil in cell 3 and cell 4 during disassembly. This phenomenon was not observed in cell 5. This difference indicates that the anodes were easier to delaminate from the Cu foil in cell 3 and cell 4 than in cell 5. This is consistent with earlier studies that attributed anode delamination to high temperatures [31,32]. Second, the color of the anode layers is lighter near the edge compared with that in the center. The color difference is most obvious for cell 3 and least obvious for cell 5. According to an earlier study [33], the light-colored edge is associated with anode overhang effects, which can lead to lithium plating at the cell edge. Third, there is significantly more light-colored deposition on the anode surface in cell 5 than in cell 3 and cell 4, which indicates that there is more lithium plating in cell 5. Fourth, there is some white deposition on most of the anode surface in cell 3. This white deposition looks unusual and very different from the deposition generally associated with lithium plating in terms of color and distribution. Most of this white deposition seemed to appear after evaporation of the liquid electrolyte during cell disassembly. This phenomenon is also noticeable in cells 4 and 5 but in much smaller amounts. Fifth, there is a tendency for deposition in the grooves caused by thermocouples between the cells, which indicates the effects of non-uniform compression on lithium plating [34]. Smaller thermocouples should be used in future research to reduce this effect.

It is worth comparing the anode layers in Figure 11 with those in the air-cooled stack [21]. The side cell showed increased lithium plating compared with that in the air-cooled stack, which can be attributed to the stronger cooling capability of the liquid, which maintained a lower temperature. The lithium plating of the side cells in the water-cooled stack may have contributed to the slightly faster degradation than for the middle cell.

Figure 13 shows SEM images of the front side of the anodes in the upper middle areas without deposition from a fresh cell, cell 3, and cell 5 at different magnifications. The anode surface looks smoother and less porous in the cycled cells compared with the fresh cell, indicating increased buildup on the anode surface of the cycled cells. Cell 5 seems to have a little more buildup on the graphite particles and lower porosity compared with cell 3, which may have contributed to the slightly faster degradation. However, further investigation is needed to determine the difference in chemical compositions of the buildup and the mechanisms.

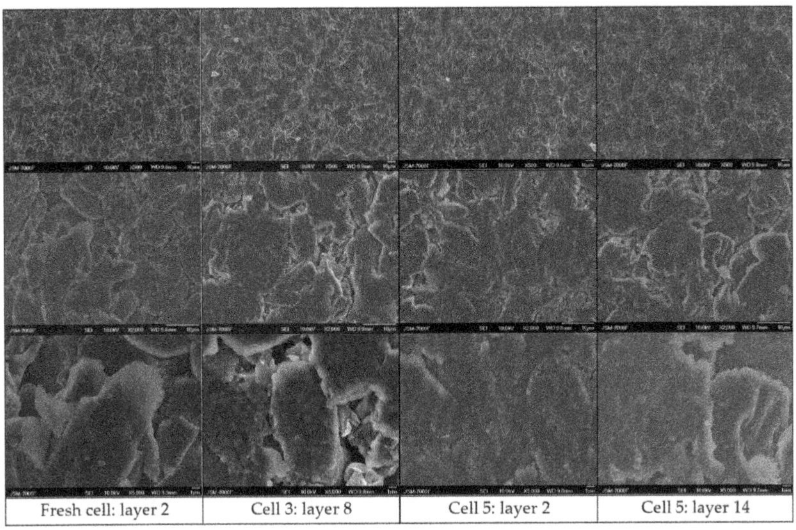

Figure 13. SEM images of the front side of the anode in areas without deposition from a fresh cell, cell 3, and cell 5 at 500×, 2000×, and 5000× magnification.

Figure 14 shows SEM images of the front side of the anode near the top edge of layer 8 in cell 3 and layer 14 in cell 5, which had nearly the highest and the lowest temperature in the stack during cycling, respectively. Depositions can be clearly seen in both cells, indicating the occurrence of lithium plating at the edge [35]. The mossy deposition in cell 3 looks similar to the deposition on the anode of the middle cell in the previous air-cooled stack [21], which was attributed to lithium plating. This is consistent with the earlier study showing that lithium plating can occur near the cell edge due to anode overhang effects [33]. The morphology difference in the depositions between cell 3 and cell 5 could be attributed to the non-uniform temperature distribution and current distribution across the stack, but further investigation is needed.

Figure 15 shows EDS images of the deposition in the lower middle region of layer 2 in cell 5. Elemental oxygen is rich in the deposition area, while element carbon is rich in the area without deposition. The distribution of elemental oxygen supports the earlier discussion that the deposition in cell 5 was from lithium plating, which can react with components in air to form products enriched in oxygen (Li_2O, LiOH, Li_2CO_3, etc.) [36,37]. There is a uniform distribution of the elements fluorine and phosphorus, which are likely from the residue of the electrolyte salt in the pores of the anode that could not be easily washed off. The elements cobalt and copper were also detected but in very low amounts.

Supplementary Figure S9 shows EDS images of a region with deposition and delamination in the upper middle area of layer 8 in cell 3. The deposition is also rich in elemental oxygen but lower in elemental fluorine compared with the deposition in Figure 14. Further investigation is needed to understand the difference in deposition between cell 3 and cell 5.

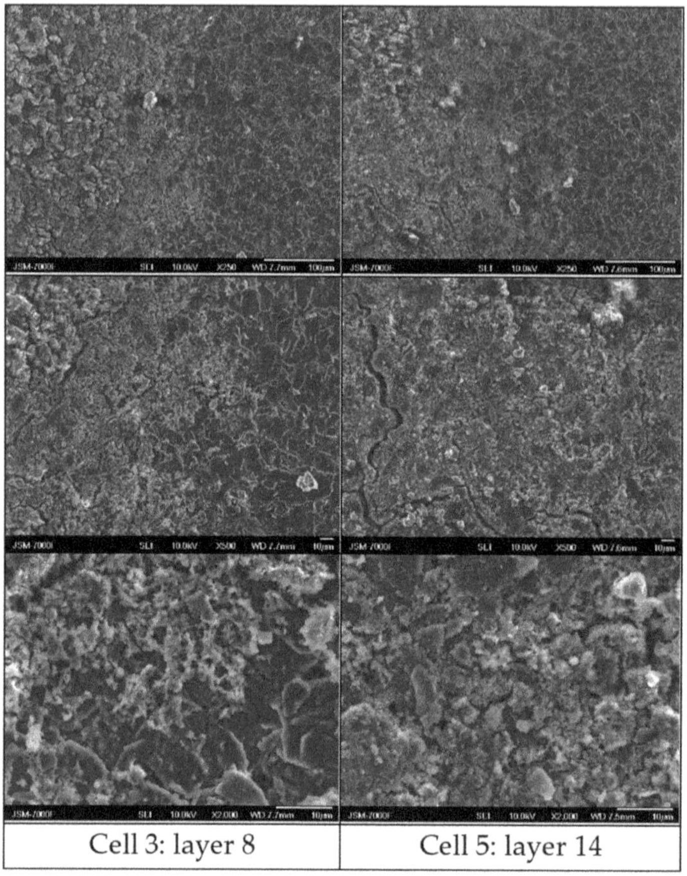

Figure 14. SEM images of the front side of the anode near the top edge of cell 3 and cell 5 at 250×, 500×, and 2000× magnification.

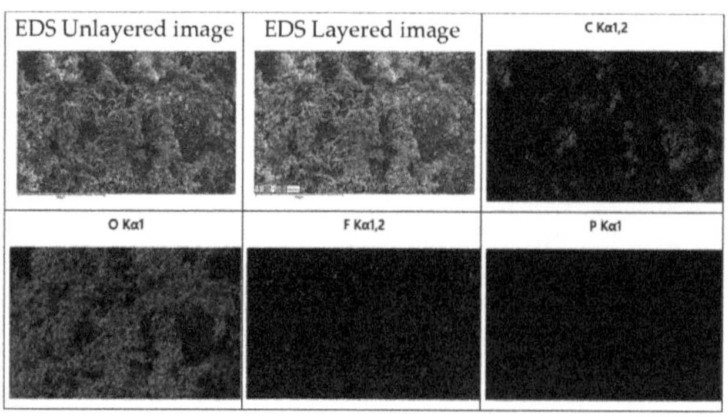

Figure 15. EDS images of deposition in the lower middle region of layer 2 in cell 5.

3.7. Future Work

This study has shown that the degradation behaviors of parallel-connected Li-ion cells in a liquid-cooled stack behave differently from those cells in an air-cooled stack. Although efforts have been made to understand the different degradation behaviors and the phenomenon of capacity drop and recovery due to RPT, further investigation is still needed. In particular, it is noticed that anode delamination only occurred in the upper part (close to tabs) of the anode layers in cell 3 and cell 4, and not in the lower part. In comparison, the lithium plating in cell 5 occurred more often in the lower part of anode. Such non-uniformity could be related to the non-uniform distribution of the local current density and local temperature in the in-plane direction, which is higher closer to the tabs. These observations, as well as the lithium plating at the edge of the anodes, implies that non-uniform degradation occurs in both the through-plane direction and the in-plane direction for large-format Li-ion cells. A more systematic post-mortem analysis of anode samples in the cycled cells is needed. Due to the interactions of multiple parameters, such as local current density, local temperature, and local compression, numerical modeling would be useful to understand the coupled effects.

4. Conclusions

A water-cooled stack of five pouch-format Li-ion cells was assembled and cycled for comparison with a previous air-cooled stack. Two unexpected phenomena were observed. First, it was observed that the middle cell in the water-cooled stack degraded slightly more slowly than the side cells. This trend is opposite to that of cells in the air-cooled stack. The opposite degradation trends of the water-cooled stack and the air-cooled stack could be attributed to the combined effects of the temperature rise and temperature gradient. Second, it was observed that all the cells in the water-cooled stack experienced the phenomenon of capacity drop and recovery after a low C-rate reference performance test and extended rest. This phenomenon was attributed to lithium diffusion between the anode active area and the anode overhang area. In addition, post-mortem analysis showed the occurrence of lithium plating at the edge of the cell, which can also be attributed to the anode overhang effects. Further investigation is needed to confirm the mechanisms.

Supplementary Materials: The following supporting information can be downloaded at: https://www.mdpi.com/article/10.3390/batteries10080274/s1, Figure S1: Fresh cell anodes front side; Figure S2: Fresh cell anodes back side; Figure S3: Cell 3 anodes front sides; Figure S4: Cell 3 anodes back sides; Figure S5: Cell 4 anodes front sides; Figure S6: Cell 4 anodes back sides; Figure S7: Cell 5 anodes front sides; Figure S8: Cell 5 anodes back sides; Figure S9: EDS images of deposition in the upper middle region of layer 8 in cell 3 (at 100× magnification).

Author Contributions: Conceptualization, G.Z.; Methodology, T.I., G.M.C., G.J.N. and G.Z.; Formal analysis, T.I., M.C. and G.Z.; Investigation, T.I., M.C., G.M.C., P.P. and G.Z.; Resources, G.Z.; Data curation, T.I. and M.C.; Writing—original draft, T.I.; Writing—review & editing, M.C., G.M.C., P.P., G.J.N. and G.Z.; Supervision, G.Z.; Project administration, G.Z. All authors have read and agreed to the published version of the manuscript.

Funding: This research received no external funding.

Data Availability Statement: Data are contained within the article.

Acknowledgments: G.Z. acknowledges the financial support from The University of Alabama in Huntsville (UAH) Faculty Startup Fund. T.I. and G.Z. acknowledge helpful discussions with Siyi Liu at UAH.

Conflicts of Interest: The authors declare no conflict of interest.

References

1. Birkl, C.R.; Roberts, M.R.; McTurk, E.; Bruce, P.G.; Howey, D.A. Degradation diagnostics for lithium ion cells. *J. Power Sources* **2017**, *341*, 373–386. [CrossRef]
2. Edge, J.S.; O'Kane, S.; Prosser, R.; Kirkaldy, N.D.; Patel, A.N.; Hales, A.; Ghosh, A.; Ai, W.; Chen, J.; Yang, J.; et al. Lithium ion battery degradation: What you need to know. *Phys. Chem. Chem. Phys.* **2021**, *23*, 8200–8221. [CrossRef] [PubMed]
3. Waldmann, T.; Wilka, M.; Kasper, M.; Fleischhammer, M.; Wohlfahrt-Mehrens, M. Temperature dependent ageing mechanisms in Lithium-ion batteries—A Post-Mortem study. *J. Power Sources* **2014**, *262*, 129–135. [CrossRef]
4. Longchamps, R.S.; Yang, X.-G.; Wang, C.-Y. Fundamental Insights into Battery Thermal Management and Safety. *ACS Energy Lett.* **2022**, *7*, 1103–1111. [CrossRef]
5. Bandhauer, T.M.; Garimella, S.; Fuller, T.F. A Critical Review of Thermal Issues in Lithium-Ion Batteries. *J. Electrochem. Soc.* **2011**, *158*, R1–R25. [CrossRef]
6. Maleki, H.; Hallaj, S.A.; Selman, J.R.; Dinwiddie, R.B.; Wang, H. Thermal Properties of Lithium-Ion Battery and Components. *J. Electrochem. Soc.* **1999**, *146*, 947–954. [CrossRef]
7. Richter, F.; Kjelstrup, S.; Vie, P.J.S.; Burheim, O.S. Thermal conductivity and internal temperature profiles of Li-ion secondary batteries. *J. Power Sources* **2017**, *359*, 592–600. [CrossRef]
8. Steinhardt, M.; Barreras, J.V.; Ruan, H.; Wu, B.; Offer, G.J.; Jossen, A. Meta-analysis of experimental results for heat capacity and thermal conductivity in lithium-ion batteries: A critical review. *J. Power Sources* **2022**, *522*, 230829. [CrossRef]
9. Forgez, C.; Vinh Do, D.; Friedrich, G.; Morcrette, M.; Delacourt, C. Thermal modeling of a cylindrical $LiFePO_4$/graphite lithium-ion battery. *J. Power Sources* **2010**, *195*, 2961–2968. [CrossRef]
10. Li, Z.; Zhang, J.; Wu, B.; Huang, J.; Nie, Z.; Sun, Y.; An, F.; Wu, N. Examining temporal and spatial variations of internal temperature in large-format laminated battery with embedded thermocouples. *J. Power Sources* **2013**, *241*, 536–553. [CrossRef]
11. Zhang, G.; Cao, L.; Ge, S.; Wang, C.-Y.; Shaffer, C.E.; Rahn, C.D. In Situ Measurement of Radial Temperature Distributions in Cylindrical Li-Ion Cells. *J. Electrochem. Soc.* **2014**, *161*, A1499–A1507. [CrossRef]
12. Amietszajew, T.; McTurk, E.; Fleming, J.; Bhagat, R. Understanding the limits of rapid charging using instrumented commercial 18,650 high-energy Li-ion cells. *Electrochim. Acta* **2018**, *263*, 346–352. [CrossRef]
13. Huang, S.; Wu, X.; Cavalheiro, G.M.; Du, X.; Liu, B.; Du, Z.; Zhang, G. In Situ Measurement of Lithium-Ion Cell Internal Temperatures during Extreme Fast Charging. *J. Electrochem. Soc.* **2019**, *166*, A3254–A3259. [CrossRef]
14. Du, X.; Wu, Q.; Wang, Y.-N.; Pan, T.-S.; Wei, Y.-M.; Chen, H.-S.; Song, W.-L.; Fang, D.-N. Visualizing two-dimensional internal temperature distribution in cylindrical Li-ion cells. *J. Power Sources* **2020**, *446*, 227343. [CrossRef]
15. Hunt, I.A.; Zhao, Y.; Patel, Y.; Offer, G.J. Surface Cooling Causes Accelerated Degradation Compared to Tab Cooling for Lithium-Ion Pouch Cells. *J. Electrochem. Soc.* **2016**, *163*, A1846–A1852. [CrossRef]
16. Carter, R.; Kingston, T.A.; Atkinson, R.W.; Parmananda, M.; Dubarry, M.; Fear, C.; Mukherjee, P.P.; Love, C.T. Directionality of thermal gradients in lithium-ion batteries dictates diverging degradation modes. *Cell Rep. Phys. Sci.* **2021**, *2*, 100351. [CrossRef]
17. Klein, M.P.; Park, J.W. Current Distribution Measurements in Parallel-Connected Lithium-Ion Cylindrical Cells under Non-Uniform Temperature Conditions. *J. Electrochem. Soc.* **2017**, *164*, A1893–A1906. [CrossRef]
18. Paarmann, S.; Cloos, L.; Technau, J.; Wetzel, T. Measurement of the Temperature Influence on the Current Distribution in Lithium-Ion Batteries. *Energy Technol.* **2021**, *9*, 2000862. [CrossRef]
19. Werner, D.; Paarmann, S.; Wiebelt, A.; Wetzel, T. Inhomogeneous Temperature Distribution Affecting the Cyclic Aging of Li-Ion Cells. Part I: Experimental Investigation. *Batteries* **2020**, *6*, 13. [CrossRef]
20. Cavalheiro, G.M.; Iriyama, T.; Nelson, G.J.; Huang, S.; Zhang, G. Effects of Nonuniform Temperature Distribution on Degradation of Lithium-Ion Batteries. *J. Electrochem. Energy Convers. Storage* **2020**, *17*, 021101. [CrossRef]
21. Malabet, H.J.G.; Cavalheiro, G.M.; Iriyama, T.; Gabhart, A.; Nelson, G.J.; Zhang, G. Electrochemical and Post-Mortem Degradation Analysis of Lithium-Ion Cells Subjected to Long Term Cycling under Non-Uniform Thermal Conditions. *J. Electrochem. Soc.* **2021**, *168*, 100507. [CrossRef]
22. Yang, X.-G.; Leng, Y.; Zhang, G.; Ge, S.; Wang, C.-Y. Modeling of lithium plating induced aging of lithium-ion batteries: Transition from linear to nonlinear aging. *J. Power Sources* **2017**, *360*, 28–40. [CrossRef]
23. Liu, S.; Zhang, G.; Wang, C.-Y. Challenges and Innovations of Lithium-Ion Battery Thermal Management Under Extreme Conditions: A Review. *ASME J. Heat. Mass. Transf.* **2023**, *145*, 080801. [CrossRef]
24. Iriyama, T.; Cavalheiro, G.M.; Liu, S.; Poudel, P.; Zhang, G. (Digital Presentation) Effects of Non-Uniform Temperature Distribution on the Degradation of Liquid-Cooled Lithium-Ion Cells. In *ECS Meeting Abstracts*; The Electrochemical Society, Inc.: Pennington, NJ, USA, 2022; Volume MA2022-01, p. 442.
25. Schweiger, H.G.; Obeidi, O.; Komesker, O.; Raschke, A.; Schiemann, M.; Zehner, C.; Gehnen, M.; Keller, M.; Birke, P. Comparison of Several Methods for Determining the Internal Resistance of Lithium Ion Cells. *Sensors* **2010**, *10*, 5604–5625. [CrossRef] [PubMed]
26. Gu, W.B.; Wang, C.Y. Thermal-Electrochemical Modeling of Battery Systems. *J. Electrochem. Soc.* **2000**, *147*, 2910–2922. [CrossRef]
27. Thomas, K.E.; Newman, J. Heats of mixing and of entropy in porous insertion electrodes. *J. Power Sources* **2003**, *119–121*, 844–849. [CrossRef]
28. Lin, N.; Röder, F.; Krewer, U. Multiphysics Modeling for Detailed Analysis of Multi-Layer Lithium-Ion Pouch Cells. *Energies* **2018**, *11*, 2998. [CrossRef]

29. Gyenes, B.; Stevens, D.A.; Chevrier, V.L.; Dahn, J.R. Understanding Anomalous Behavior in Coulombic Efficiency Measurements on Li-Ion Batteries. *J. Electrochem. Soc.* **2014**, *162*, A278–A283. [CrossRef]
30. Baghdadi, I.; Briat, O.; Gyan, P.; Vinassa, J.M. State of health assessment for lithium batteries based on voltage–time relaxation measure. *Electrochim. Acta* **2016**, *194*, 461–472. [CrossRef]
31. Somerville, L.; Bareño, J.; Trask, S.; Jennings, P.; McGordon, A.; Lyness, C.; Bloom, I. The effect of charging rate on the graphite electrode of commercial lithium-ion cells: A post-mortem study. *J. Power Sources* **2016**, *335*, 189–196. [CrossRef]
32. Lu, D.; Xu, M.; Zhou, L.; Garsuch, A.; Lucht, B.L. Failure Mechanism of Graphite/LiNi0.5Mn1.5O4Cells at High Voltage and Elevated Temperature. *J. Electrochem. Soc.* **2013**, *160*, A3138–A3143. [CrossRef]
33. Grimsmann, F.; Gerbert, T.; Brauchle, F.; Gruhle, A.; Parisi, J.; Knipper, M. Hysteresis and current dependence of the graphite anode color in a lithium-ion cell and analysis of lithium plating at the cell edge. *J. Energy Storage* **2018**, *15*, 17–22. [CrossRef]
34. Spingler, F.B.; Friedrich, S.; Kücher, S.; Schmid, S.; López-Cruz, D.; Jossen, A. The Effects of Non-Uniform Mechanical Compression of Lithium-Ion Cells on Local Current Densities and Lithium Plating. *J. Electrochem. Soc.* **2021**, *168*, 110515. [CrossRef]
35. Birkenmaier, C.; Bitzer, B.; Harzheim, M.; Hintennach, A.; Schleid, T. Lithium Plating on Graphite Negative Electrodes: Innovative Qualitative and Quantitative Investigation Methods. *J. Electrochem. Soc.* **2015**, *162*, A2646–A2650. [CrossRef]
36. Waldmann, T.; Hogrefe, C.; Flügel, M.; Pivarníková, I.; Weisenberger, C.; Delz, E.; Bolsinger, M.; Boveleth, L.; Paul, N.; Kasper, M.; et al. Efficient Workflows for Detecting Li Depositions in Lithium-Ion Batteries. *J. Electrochem. Soc.* **2024**, *171*, 070526. [CrossRef]
37. Otto, S.-K.; Moryson, Y.; Krauskopf, T.; Peppler, K.; Sann, J.; Janek, J.; Henss, A. In-Depth Characterization of Lithium-Metal Surfaces with XPS and ToF-SIMS: Toward Better Understanding of the Passivation Layer. *Chem. Mater.* **2021**, *33*, 859–867. [CrossRef]

batteries

MDPI

Article

Hybrid Heat Pipe-PCM-Assisted Thermal Management for Lithium-Ion Batteries

Nourouddin Sharifi [1,*], Hamidreza Shabgard [2], Christian Millard [3] and Ugochukwu Etufugh [3]

1 Department of Engineering Technology, Tarleton State University, 1333 W. Washington, Stephenville, TX 76402, USA
2 School of Aerospace and Mechanical Engineering, University of Oklahoma, 865 Asp Ave., Norman, OK 73019, USA; shabgard@ou.edu
3 Department of Mechanical, Environmental, and Civil Engineering, Tarleton State University, 1333 W. Washington, Stephenville, TX 76402, USA; christian.millard@go.tarleton.edu (C.M.); ugochukwu.etufugh@go.tarleton.edu (U.E.)
* Correspondence: sharifi@tarleton.edu

Abstract: A hybrid cooling method for 18650 lithium-ion batteries has been investigated using both experimental and numerical approaches for electric vehicle applications. The experimental setup includes a heater section, a phase change material (PCM) reservoir, and a cooling section. The heater section simulates battery heat generation with two cylindrical aluminum housings, each sized to match an 18650 battery, two cartridge heaters, and an aluminum heat sink. An airflow channel is incorporated into the cooling section. Heat transfers sequentially from the heaters to aluminum housings, the heat sink, through three copper-water heat pipes (HPs), to/from the PCM, and finally to the cooled air in the airflow channel. This innovative design eliminates direct contact between the PCM and the batteries, unlike recent studies where the PCM has been in direct contact with the batteries. Decoupling the PCM reduces system design complexity while maintaining effective thermal management. Temperature measurements at various locations are analyzed under different heater powers, air velocities, and scenarios with and without PCM. Results show that the experimental design effectively maintains battery temperatures within acceptable limits. For a power input of 16 W, steady-state temperatures are reduced by approximately 14%, 10%, and 4% with PCM compared to without PCM for air velocities of 2 m/s, 3 m/s, and 4 m/s, respectively. A transient three-dimensional numerical model was developed in ANSYS-FLUENT to provide insights into the underlying physics. The phase change was simulated using the enthalpy-porosity approach, with computational results showing reasonable agreement with experimental data.

Keywords: thermal management; battery; heat pipes; PCM; EV

1. Introduction

Automakers are creating more electric vehicles (EVs) and accelerating the development of EV technology in response to the growing desire to reduce the carbon footprint of transportation. Despite the growing demand, the broad integration of EVs in the general automotive market continues to be hindered by several challenges. Consumer complaints include the EV's short vehicle lifespan, especially in relation to the battery pack's longevity, and its unreliability for long-distance driving. An EV battery pack's performance is influenced by multiple factors, the primary one being its vulnerability to temperature impacts. Ma et al. [1] reviewed the effects of temperature on lithium-ion batteries.

Typically, the safe temperature range for lithium-ion batteries spans from −20 °C to 60 °C, with the optimal operating range being between 15 °C and 35 °C [1]. Temperatures exceeding the operating limit may cause an irreversible reaction to occur across the lithium-ion battery pack and raise the possibility of thermal runaway. Thermal runaway is characterized by an increase in temperature that releases energy and then causes another increase in temperature [1]. The discharge of thermal energy into the surrounding environment raises the temperature and internal energy of all adjacent batteries. Thermal runaway causes degradation over time and shortens the battery's life. Battery thermal management (BTM), a term for a variety of temperature control techniques, can be used to control this extremely harmful process. Therefore, ongoing developments in effective cooling technologies will increase the overall longevity and safety of the battery packs in electric vehicles. Three primary approaches to battery thermal management have been thoroughly examined by earlier researchers [2–5]: active, passive, and hybrid cooling methods. Passive cooling techniques use mainly heat pipes (HPs) and phase change materials (PCMs), while active cooling techniques use an actively induced flow. The advantages of both active and passive cooling techniques are combined in the hybrid approach. The ability of PCMs to absorb significant amounts of latent heat within a narrow temperature range [6] makes them appealing for passive BTM. Nevertheless, a drawback of most PCMs is their limited heat conductivity, which restricts their range of use.

One approach to improving the heat transfer rate of PCMs is by incorporating HPs. Through the evaporation and condensation of their working fluid, HPs serve as efficient thermal devices that can passively transfer substantial amounts of heat across small cross-sectional areas [7]. The background and history of research on the benefits of integrating HPs with fins and PCM to enhance heat transfer have been investigated extensively by previous researchers [8–11].

In recent advancements, researchers have explored thermal management systems for 18650 lithium-ion batteries utilizing hybrid HP–PCM configurations. These studies commonly consider designs where the PCM and batteries are housed within an enclosure forming a battery pack, with PCM in direct contact with the batteries. Heat pipes are integrated into the PCM such that the evaporator section is positioned within the battery pack while the condenser section is air-cooled. To enhance cooling performance, the evaporator and condenser sections of the HPs are often equipped with fins. Overall, these studies demonstrate that hybrid HP-PCM systems offer notable advantages in reducing maximum battery temperatures and improving temperature uniformity. Below is a summary of recent developments in this field with similar design configurations as explained above, emphasizing their findings and distinguishing features.

Gao et al. [12] conducted a numerical evaluation of a hybrid HP–PCM system where HPs were in direct contact with both the batteries and the PCM. Their results showed a reduction in maximum temperatures by 25.7%, 23.3%, and 34.9% for discharge rates of 2C, 3C, and 4C, respectively, compared to PCM-only systems. Similarly, Sutheesh et al. [13] numerically explored hybrid cooling systems incorporating HPs with nano-enhanced PCM. Their study investigated different cell arrangements, revealing that a horizontal cell arrangement performed better by maintaining the battery pack within the desired temperature range for 65% of the discharge process at a 3C rate. Sharma et al. [14] advanced hybrid systems by integrating fins with flat-plate HPs and PCM through both numerical and experimental analyses. They studied various fin configurations, identifying an optimized design that reduced peak battery temperatures by 2.95 °C and maintained surface temperatures below 60 °C. Experimental validation confirmed the improved thermal performance during charge-discharge cycles, emphasizing the importance of fin optimization in system design.

Similar studies were conducted by Tang et al. [15], with the exception being that they used copper foam-PCM composites. Abbas et al. [16] also investigated hybrid HP–PCM systems and found better battery temperature uniformity when HPs were attached to the batteries compared to when they were not attached.

This study proposes a configuration for a HP-fin-assisted, PCM-based hybrid BTM system designed for 18650-type lithium-ion battery modules. The primary innovation of this work, as compared to the existing literature outlined in the introduction section, lies in the strategic placement of the PCM. Unlike previous designs where the PCM is in direct contact with the batteries, the current configuration decouples the PCM from the battery pack and positions it between the batteries and the air-cooled section. This design offers several practical advantages. By decoupling the PCM, the system removes the PCM size constraints typically imposed by the battery pack dimensions, allowing for greater design flexibility. Moreover, this arrangement simplifies battery replacement, as the PCM is not integrated within the battery pack. As a result, the overall system design is streamlined, and complexity is reduced.

Furthermore, to the authors' knowledge, no prior studies have developed numerical simulations in ANSYS-FLUENT that fully solve the governing equations of HPs, coupled with conjugate heat transfer in the PCM and fins. This aspect strengthens the uniqueness of the present work. In previous studies involving HP–fin–PCM systems, researchers have often treated HPs as materials with high effective thermal conductivity or modeled them as thermal resistance networks, rather than solving their full governing equations. The numerical model developed in this study represents a significant advancement by thoroughly capturing the physical processes within the HPs, PCM, and fins.

2. Experimental Design

The explanation of the experimental setup is divided into four subsections: (*i*) the heating section, (*ii*) the PCM reservoir, (*iii*) the 3D-printed supports and insulation, and (*iv*) the air-cooling section, each of which is described in detail below.

2.1. Heating Section

The experiments in this study required numerous battery charging and discharging cycles, which increased the risk of thermal runaway or potential battery explosion. To avoid the risk of battery explosion and perform the experiment safely in the lab environment, a combination of cartridge heaters and aluminum housings were used to mimic the heat generated by the batteries.

Figure 1a shows the side view of the setup, detailing the various sections, components, dimensions, and thermocouple locations. The heating section was composed of two aluminum housings (AL 6061), each with inner and outer diameters of about 8 mm and 18 mm, and a length of 65 mm. The aluminum housings had the same exterior dimensions as 18650 lithium-ion batteries. Two cartridge heaters (tatoko, 12 V, 100 W, 8 mm diameter and 100 mm length) were inserted concentrically into the aluminum housings. Thermal paste (arctic silver 5, $k \approx 8.7$ W/m·K) was applied to fill the gap between the heaters and the aluminum housings, aiming to reduce thermal contact resistance. The heaters were powered with a DC power supply (B&K Precision 1671A, Yorba Linda, California, Address: 22820 Savi Ranch Pkwy, Yorba Linda, CA 92887, USA).

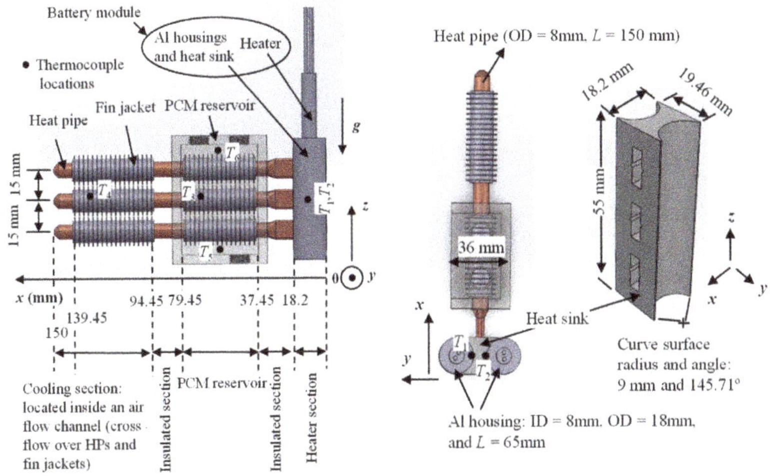

(a) **(b)**

Figure 1. Schematic view of the setup illustrating its components, dimensions, and thermocouple locations: (**a**) Side view; (**b**) Top view.

Three cylindrical copper-water HPs (Advanced Thermal Solutions Inc. (Norwood airport business park, Address: 89 Access Rd STE 27, Norwood, MA 02062, USA): ATS-HP-6L150S71W-129, 8 mm diameter and 150 mm length) were utilized in this setup to transfer heat from the heating section to the PCM and cooling section. To increase compactness on the battery side, a portion of the cylindrical HPs were carefully flattened using a bench vise. Hence, the HPs were referred to as partially flattened cylindrical HPs. The flat portion of each HP had an approximate length of 27.2 mm and measured about 4.2 mm in the y-direction and 10.5 mm in the z-direction. A heat sink was designed to connect the flat portion of the HPs to the aluminum housings, with further details provided later. The heaters, aluminum housings, heat sink, and HPs were all in direct contact. Thermal paste was used at the interface of the aluminum housing–heat sink and the HPs–heat sink. The combination of the heaters, the heat sink, and the Al housings is called the battery module hereafter.

The thermocouples T_1 and T_2 were installed at the mid-height of the aluminum housing–heat sink interface on each side of the heat sink, where they were in contact with the two aluminum housings. The readings of these two thermocouples were very close due to the setup's symmetry with respect to the HP's mid x–z plane. Considering that the temperature difference between T_1 and T_2 was within 0.5 °C at any instant, the temperature T_B, defined as the instantaneous average of T_1 and T_2, $T_B = (T_1 + T_2)/2$, was reported as the battery temperature in all experiments. It should be noted that T_B is not a physical thermocouple installed on the setup, but rather an average of two thermocouples. The thermocouple T_3 recorded the middle HP–fin jacket interface at $x = 69.45$ mm. The purpose of thermocouple T_3 was to capture more detailed phase change phenomena and their impacts in the case with PCM, compared to the scenario without PCM. To securely position this thermocouple, small longitudinal grooves were created on the interior surface of the fin jacket, providing space to insert the thermocouple at the HP–fin jacket interface. Thermocouples T_5 and T_6 were installed within the PCM about 2 mm from the mid-bottom interior surface of the PCM reservoir and the mid-top surface of the PCM, respectively, and were used to record the PCM temperatures to estimate the melting process duration. They were removed for experiments without the PCM.

Figure 1b shows the top view of the setup, including the detailed dimensions of the heat sink. To ensure efficient heat transfer between the aluminum housings and the HPs, a heat sink (AL 6061) was designed and manufactured using a desktop CNC machine. The heat sink has a length of 55 mm and is in contact with the mid 65 mm length of the exterior surface of the aluminum housings. The top and bottom of the exterior surface of the aluminum housings (each 5 mm) are not in contact with the heat sink. These 5 mm lengths will allow extra space between the batteries' bus bar and the batteries' top and bottom caps, mimicking a real battery pack. The heat sink includes three slots, each with an approximate dimension of 11 mm by 4.2 mm. Each partially flattened cylindrical HP was inserted into the slots. Grooves (~0.6 mm in radius) were machined in each slot of the heat sink to accommodate thermocouples.

2.2. PCM Reservoir

In this study, 1-Tetradecane (99% pure $C_{14}H_{30}O$) was selected as the PCM, with a melting temperature of approximately 38 °C [17]. While various PCMs with different melting temperatures have been used in the literature, Jiang et al. [18] reported that PCMs with melting temperatures lower than 45 °C are more effective for heat dissipation when the maximum allowable battery temperature is below 50 °C. Since the maximum allowable battery temperature in this study was below 50 °C, a PCM with a melting temperature approximately 5–7 °C lower than 45 °C, such as 1-Tetradecane, was chosen. Moreover, this PCM is safe for use up to 100 °C, and is readily available.

A 3D-printed enclosure (PCM reservoir) was designed and created to hold a measured amount of PCM (see Figure 1a,b). The interior dimensions of the enclosure are 42 mm in the x-direction, 26 mm in the y-direction, and 59 mm in the z-direction, and the wall thickness is about 5 mm. The reservoir has six holes in the x-direction, each with a diameter of about 8 mm. It was made to properly house the HPs, as well as mate well with the battery module support surface. The holes for the HPs are about 15 mm apart from center to center. There are additional threaded thru holes on the top and bottom for filling and draining the PCM.

To enhance the heat transfer rate within the PCM, three aluminum fin jackets (AL 6061) were designed and machined using a manual lathe. The annular fin jackets in the PCM reservoir each have a length of 40 mm, an inner diameter of about 8 mm, a wall thickness of about 1.5 mm, an approximate fin thickness of 0.5 mm, a fin length of 2 mm, and a fin spacing of 2.5 mm.

The PCM reservoir was open on the side wall, and a lid was used to close the opening. This opening facilitated the inserting of the fin jackets and the installation of thermocouples during assembly. Caps were used to close the top and bottom openings of the PCM reservoir. An aquarium silicone sealant was then used to prevent any PCM leakage from the PCM reservoir.

2.3. 3D-Printed Supports and Insulation

Figure 2 presents an isometric view of the experimental setup, both with and without the insulation box. As shown in Figure 2a, PLA plastic 3D-printed parts called battery module supports were created to hold and support the aluminum housings, heaters, heat sink, and HPs, keeping these components in place. The battery module support was made to be screwed together at four threaded thru holes at the top and bottom. It also accounted for the thermocouples that were attached during experimentation. To achieve this, 2 mm semi-circular slots were incorporated into the design. Furthermore, 3D-printed heater sleeves were also created to insulate and support the heaters. A 3D-printed base plate was created to secure the components in place and allow for better leveling of the setup. As illustrated in Figure 2b, a double-pane (~5 mm air gap) insulation box was also 3D-printed to enclose the setup for insulation purposes. The setup is geometrically symmetric with respect to its (x-z) mid-plane.

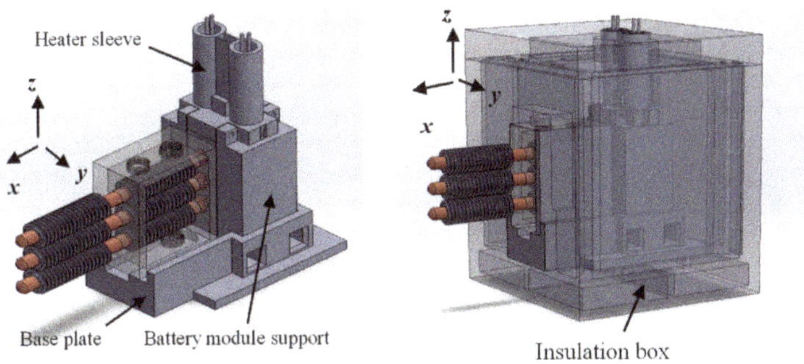

(a) (b)

Figure 2. Isometric view of the experimental setup showing: (**a**) Without insulation box, including the base plate, battery module support, and heater sleeve; (**b**) With all components and the insulation box (the insulation box is transparent in this figure for clarity).

To enhance thermal insulation, a 10 mm-thick aerogel insulation pad (MARSTEP-NASA Tech, $k \approx 0.025$ W/mK [19]) was placed between the 3D-printed battery module support and the PCM reservoir, between the PCM reservoir and the flow channel wall, and around the battery module supports and heaters within the insulation box.

2.4. Cooling Section

Figure 3 illustrates a schematic of the flow channel and the setup. As shown in Figure 3a, an air flow channel was constructed to provide a forced convection thermal boundary condition on the condenser section of the HPs (portion of the HPs within the flow channel). The internal measurements of the flow channel are approximately 1 m in length (y-direction), 160 mm in width (x-direction), and 180 mm in height (z-direction). The centerline of the setup, along the x-direction, is approximately 620 mm from the inlet of the channel. The flow channel walls were constructed from a 5 mm thick acrylic sheet. A fan (VEVOR inline duct fan, 4-Inch 200 CFM with variable speed controller, Model: BT-FL14A, 120 VAC/60 Hz 31 W) was used to create a steady flow of air through the channel by creating suction at the outlet. A bank of 100 mm long, thin-walled, 5 mm-diameter plastic tubes (drinking straws) was used as a honeycomb-type flow straightener. The air speed was measured using a hot wire anemometer (Koselig Instruments, Model: KI-10001, ASIN: B08QRC756F, 0.1–25 m/s, accuracy: 1% full range, resolution: 0.01).

Fin jackets were also employed to enhance the heat transfer rate from the HPs to the cooling air. The fin jackets have the same dimensions as those used within the PCM reservoir, with the exception that their length is about 45 mm. Thermal paste was used between the fin jackets and the HPs.

The component dimensions in the experimental setup were estimated based on the minimum heat transfer area required to achieve a cooling capacity of 8W per battery, within a fan speed range of 0 to 4 m/s. The 8W power corresponds to a high discharge rate of 5C for an individual battery [20]. Specifically, the number of HPs, their dimensions, and the fin dimensions were calculated by determining the necessary heat transfer area through convective heat transfer analysis. The fin lengths were constrained by the heat sink length, the number of HPs, and their diameters. The dimensions of the aluminum housings and heat sink were constrained by the size of the 18650 lithium-ion battery. The

PCM reservoir dimensions were selected based on the minimum volume of PCM required to encapsulate all three HPs with fin jackets within the reservoir. Additionally, sufficient space was allocated to facilitate the installation of the fin jackets on the HPs within the PCM reservoir, as well as the necessary placement of thermocouples.

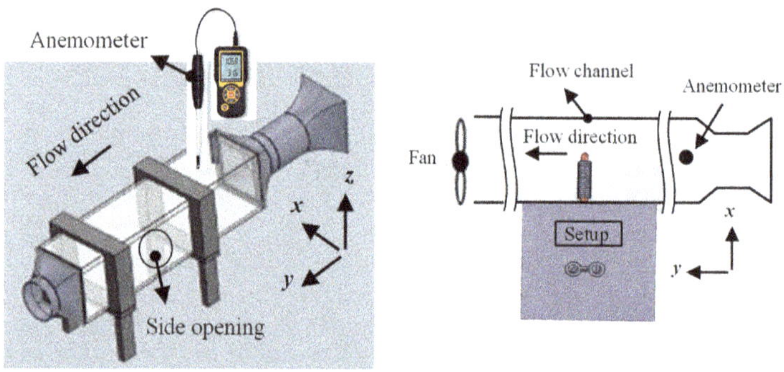

(**a**) (**b**)

Figure 3. Isometric of the experimental setup within its insulation box with flow channel: (**a**) Isometric view showing the flow channel for the air-cooling section of the setup; (**b**) Top view showing the air-cooling section of the setup positioned with the flow channel through its side opening, with air flowing crosswise over the HPs and fin jackets.

Temperatures at specific locations were recorded with Teflon-coated, 254-μm-diameter chrome-alumel (K-type) thermocouples. The calibration of the thermocouples was verified using the boiling and freezing points of distilled water. The thermocouples were positioned with an accuracy of approximately ±1 mm. Aluminum tape was used to secure the thermocouples in place. Thermocouple voltage readings were recorded at 1s intervals using four OMEGA TC-08 USB thermocouple data acquisition modules, managed through Picolog 6 software. The measurement error for each point was ±0.5 °C, attributed to system uncertainties involving the data acquisition device and thermocouples. Additionally, the uncertainty in the heating power measurement was less than 5%. Table 1 displays the thermophysical properties of 1-Tedradecanol, the HP, and the fin jacket.

Table 1. Thermophysical properties of HP and fin jacket.

	PCM	HP Working Fluid	HP Wall (≈0.5 mm) and Wick (≈1 mm)	Fin Jacket
Material	1-Tedradecanol [10]	Water [10]	Copper [8]	Aluminum 6061 [21]
Density, ρ (kg/m^3)	823 (liquid)	996.6 (liquid)	8933	2700
Thermal conductivity, k (W/m K)	0.481 (solid) 0.22 (liquid)	0.6132 (liquid) 0.01932 (vapor)	401	167
Specific heat, c_p (J/kg K)	2407	4181 (liquid) 1882 (vapor)	385	896
Viscosity, μ (Pa.s)	0.0064	8614×10^{-7} (liquid) 91.74×10^{-7} (vapor)		
Latent heat of fusion, h_{sl} (kJ/kg)	235.41			
Latent heat of vaporization, h_{fg} (kJ/kg)		2434.9		
Melting temperature, T_m (°C)	38			

Table 1. *Cont.*

	PCM	HP Working Fluid	HP Wall (≈0.5 mm) and Wick (≈1 mm)	Fin Jacket
Thermal expansion coefficient β (1/K)	0.00084			
Vapor pressure, P_{sat} (Pa)		4352		

3. Experimental Procedure

Two cooling scenarios were investigated in this study: (*i*) without the PCM, and (*ii*) with the PCM, each at various input powers and air velocities. For the case without PCM, the aerogel insulation pad was cut into small pieces and inserted into the PCM reservoir to insulate the portion of the HPs within the PCM reservoir. For the PCM experiments, testing commenced with the PCM in a solid state and at a uniform temperature. Precautions were taken to ensure the PCM was void-free. To do so, the PCM was slowly melted in a beaker over a plate heater. The warm, molten PCM was carefully poured into the PCM reservoir (refer to the setup in Figure 2a with the base plate), forming an initial PCM layer approximately 5 mm thick. The test cell was then allowed to cool naturally in ambient air ($T_{air} < T_m$) to solidify the PCM. Once this layer was hardened, the pouring and solidifying process was repeated until the enclosure was filled with solid PCM. To account for the expansion of the PCM upon melting, a 3 mm air gap was maintained in the enclosure.

The setup was left to stabilize at room temperature (~21.5 °C) for approximately 24 h. Afterwards, it was insulated with aerogel pads and placed inside the insulation box (see Figure 2b). The insulation box with the setup was connected to the flow channel by placing the cooling section of the HPs into the air channel through its side opening, as shown in Figure 3b. Four heater power levels were tested, with each heater individually powered at 4 W, 6 W, 8 W, and 10 W, yielding total powers of 8 W, 12 W, 16 W, and 20 W per experiment. The fan speed was adjusted to achieve air velocities of 0 m/s, 2 m/s, 3 m/s, and 4 m/s, corresponding to air flow rates of 0, 0.058 m^3/s, 0.087 m^3/s, and 0.116 m^3/s, respectively, through a flow channel with an interior cross-sectional area of about 0.029 m^2. The different combinations of heating power and air speed produced temperatures comparable to the temperatures within an 18650 lithium-ion battery, specifically for relatively higher discharge rates, i.e., from 2 C to 4 C (see Figure 7 of reference [5] and Figure 17a of reference [20]). The setup was assumed to be thermally symmetric with respect to the HPs mid x–z plane.

4. Physical Model and Mathematical Formulation

A three-dimensional numerical model was developed using ANSYS-FLUENT 2023 software to explore the underlying physics of the problem, offering insights that could not be obtained solely from experimental studies. This model also facilitated optimization studies of the setup. The numerical approach employed the transient enthalpy-porosity technique. The conservation equations for mass, energy, and momentum were fully solved across the entire computational domain. The heat transfer phenomena were driven by transient and coupled effects, including: (*i*) conduction through solid walls (such as Al housings, heat sinks, fin jackets, and HPs' walls and wicks), (*ii*) vaporization, condensation, and incompressible flow of the HPs' working fluid, (*iii*) natural convection occurring within the PCM during phase transitions, and (*iv*) forced convection in the cooling section of the experimental setup. A User-Defined Function (UDF) was created and incorporated into the FLUENT software to improve the detailed modeling of the HPs. To reduce the complexity of the numerical simulation, it was divided into two different parts. In the first part, air flow over the HPs and fin jackets within the flow channel was simulated to determine the corresponding average convective heat transfer coefficients. The computational domain for this part is depicted in Figure 4a. The interior dimensions of the computational domain and the positioning of the HPs relative to the airflow channel align with the details provided in the cooling section of the experimental design. Due to the model's symmetry along the x–y plane at the center of the middle HP in this configuration, only the top half of the physical domain was considered for simulation. In the second part, the setup comprising the heat sink, HPs, PCM, and fin jackets was simulated. The computational domain for this part is illustrated in Figure 4b, shown in two isometric views of the same domain. Due to the symmetry of the model along the x–z plane, only half of the physical domain is represented. The average convective heat transfer coefficients obtained from the first part, with the computational domain shown in Figure 4a, were uniformly applied as convective thermal boundary conditions on the exterior surfaces of the HPs and fin jackets exposed to the air.

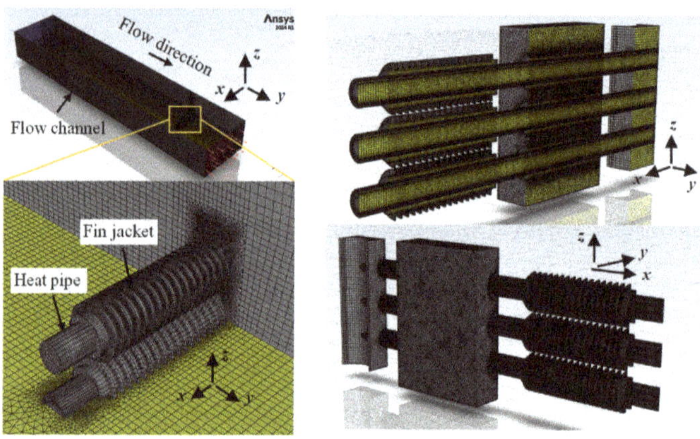

(a) (b)

Figure 4. Computational domains: (**a**) Air-cooling section of the setup within the flow channel; (**b**) Half of the HP-fin-PCM physical domain.

The model was based on several assumptions: all thermophysical properties were considered constant, except for the vapor phase of the HPs' working fluid, which was treated as an ideal gas, resulting in a density that varies with temperature and pressure. The solid PCM was assumed to be pure, and the molten PCM was assumed to have the same density as the solid PCM. Natural convection in the molten PCM was modeled using the Boussinesq approximation. The phase change process was represented through the enthalpy-porosity method [22]. Thermal radiation was considered negligible due to the relatively low temperatures involved. The vapor phase of the HPs' working fluid was modeled as undergoing three-dimensional laminar incompressible flow, with the wick fully saturated. At the wick interface, both the vapor and liquid phases of the working fluid were in a saturated state. It was assumed that the porous wick generated sufficient capillary pressure to drive the liquid flow through the wick under the operating conditions considered in this study.

Multiple physical boundaries divided the different materials, material phases, and components within the conjugate system. Thermal contact resistances were ignored. The computational domain depicted in Figure 4b is governed by the following boundary and initial conditions. Symmetry conditions were applied along the mid x–z plane of the setup. A constant, uniform heat flux was imposed on the external surface of the heat sink in contact with the aluminum housing. The remaining external boundaries were set as adiabatic and impermeable, with no-slip conditions applied. A convective heat transfer condition was applied to the outer surfaces of the HPs and fin jackets within the flow channel. The initial temperature of the entire domain was initially set to the laboratory temperature of approximately 21.5 °C.

Given the aforementioned assumptions, the governing equations for the HPs, PCM, and solid walls are expressed as follows:

4.1. Descriptive Equations for the Heat Pipe

The vapor flow within the HP is described by the following Equations [23]:

$$\frac{\partial}{\partial t}(\rho_v) + \nabla \cdot (\rho_v \mathbf{V}) = 0 \tag{1}$$

$$\frac{\partial}{\partial t}(\rho_v \mathbf{V}) + \nabla \cdot (\rho_v \mathbf{V}\mathbf{V}) = -\nabla P + \mu_v \nabla \cdot \left(\nabla \mathbf{V} - \nabla \mathbf{V}^{\mathrm{T}}\right) \tag{2}$$

$$\frac{\partial}{\partial t}(\rho_v \mathbf{V}) + \nabla \cdot (\rho_v \mathbf{V}\mathbf{V}) = -\nabla P + \mu_v \nabla \cdot \left(\nabla \mathbf{V} - \nabla \mathbf{V}^{\mathrm{T}}\right) \tag{3}$$

The ideal gas law is used to determine the vapor density as a function of its pressure and temperature.

$$\rho = \frac{P}{R_g T} \tag{4}$$

where R_g is the gas constant and is calculated by dividing the universal gas by atomic mass of the vapor. The compressibility factor was calculated to be approximately 0.99 and remained nearly constant under the operating conditions of this study, confirming the validity of the ideal gas assumption for water vapor within the HP.

The heat flux at the wick-vapor interface was used to calculate the evaporation/condensation rate at the interface. This was accomplished by using the following equation:

$$q''_{\text{interface}} = \left(\dot{m}'' h_{fg}\right)_{\text{interface}} = \left(\rho_v v_v h_{fg}\right)_{\text{interface}} \tag{5}$$

Equation (5) was also used to calculate the vapor velocity at the interface.

The local wick-vapor interfacial temperatures were obtained from the local vapor pressures adjacent to the interface. During evaporation, the temperature of the vapor is equal to the saturation temperature corresponding to the local pressure. Here, the Antoine equation was used to establish the relationship between the saturation temperature and pressure [24].

$$T_{\text{sat}} = 39.72\ \text{K} + \frac{1730.63}{8.07 - \log_{10} P} \tag{6}$$

where T_{sat} is in K, and P is the absolute pressure of the vapor in the computational cell adjacent to the interfacial node in mmhg (1 Pa = 0.0075 mmhg). The interfacial vapor velocity and saturation temperature, obtained from Equation (5) and Equation (6), respectively, were implemented by a UDF written in C++ and integrated into the ANSYS-FLUENT solver.

In accordance with Cao and Faghri [25], it was assumed that the porous wick of the HP was fully saturated with liquid, and liquid flow within the wick was neglected. The heat conduction equation for all solid materials, including the aluminum housings, heat sink, HP walls and wicks, and fin jackets in the setup, is governed by

$$\frac{\partial \left(\rho c_p\right)_w}{\partial t} = k_w \nabla^2 T \tag{7}$$

4.2. Descriptive Equations for the PCM

The enthalpy-porosity method, implemented in ANSYS-Fluent, was used to model the solid-liquid phase change of the PCM. The continuity, momentum, and energy conservation equations for the PCM are formulated as follows [22]:

$$\frac{\partial}{\partial t}(\rho) + \nabla\cdot(\rho\mathbf{V}) = 0 \tag{8}$$

$$\frac{\partial}{\partial t}(\rho\mathbf{V}) + \nabla\cdot(\rho\mathbf{V}\mathbf{V}) = -\nabla P + \mu_v\nabla^2\mathbf{V} + \rho g[\beta(T - T_m) - 1] + S_m \tag{9}$$

$$\frac{\partial}{\partial t}(\rho H) + \nabla\cdot(\rho\mathbf{V}H) = k\nabla^2 T \tag{10}$$

The specific enthalpy H, as used in Equation (10), is defined as the sum of the sensible enthalpy (h) and the latent enthalpy (ΔH) [22]:

$$H = h + \Delta H = h_{\text{ref}} + \int_{T_{\text{ref}}}^{T} C_p dT + \lambda(T)L_f \tag{11}$$

Here, the enthalpy h_{ref} is calculated at reference temperature T_{ref}, where L_f represents the latent heat of fusion, and T_{ref} is set to 21.5 °C. The liquid fraction $\lambda(T)$, by definition, is temperature-dependent and is mathematically expressed as follows [22]:

$$\lambda(T) = \frac{\Delta H}{L_f} = \frac{T - T_s}{T_l - T_s} = \begin{cases} 0 \text{ if } T < (T_m - \Delta T) \\ \frac{T - T_m + \Delta T}{2\Delta T} \text{ if } (T_m - \Delta T) \leq T \leq (T_m + \Delta T) \\ 1 \text{ if } T > (T_m + \Delta T) \end{cases} \tag{12}$$

From Equation (12), the liquid fraction is zero for the solid PCM and equal to 1 for the liquid PCM. The source term in the momentum equation, S_m in Equation (9), accounts for the flow through porous media in the mushy zone, where the phase transition takes place. This is modeled using a modified version of the Carman–Kozeny equation [22]. S_m can be expressed as:

$$S_m = A\mathbf{V} = -\frac{A_{\text{mush}}(1-\lambda)^2}{\lambda^3 + \varepsilon}V \tag{13}$$

As outlined in [26], the purpose of the source term $A\mathbf{V}$ in Equation (13) is to suppress flow within the mushy zone and gradually halt it as the PCM approaches complete solidification. Specifically, the value of A is zero when the PCM is fully in the liquid phase (λ = 1), while it becomes significantly large when the PCM is fully solidified (λ = 0). This source term in the momentum equation is active only during the phase transition, specifically within the melting regime. Further details on the importance and implementation of this source term can be found in the literature [22]. Consequently, the value of A_{mush} must be sufficiently high, typically $\geq 10^3$ kg/(m^3·s), to effectively inhibit flow in the solid phase of the PCM. Larger values of A_{mush} result in a relatively slow-moving melting front, as indicated by the reduced velocity magnitudes within the PCM, in accordance with Equation (13). Conversely, smaller values of A_{mush} may lead to an overestimation of the melting rate. Additionally, the value of A_{mush} is influenced by the temperature range of the phase transition, as noted in [22]. In this study, the default value of A_{mush} value of 100,000 provided a satisfactory agreement with the experimental data. A small constant ε (= 0.001) was employed to maintain the validity of Equation (13) when $\lambda \longrightarrow 0$.

5. Results and Discussion

Following the detailed explanation of the experimental design, procedure, and the underlying physical model, the results obtained from both experimental and numerical analyses are now presented and discussed. This section is divided into two subsections: experimental results, where the findings from the physical tests are analyzed, and numerical results, which focuses on the computational simulations. These results are compared and contrasted to assess the performance of the hybrid HP–PCM system, providing a comprehensive understanding of its thermal management capabilities.

5.1. Experimental Results

The following subsections present the experimental results for two scenarios: with and without PCM, along with the related discussions. In these experiments, the total power associated with two cartridge heaters was denoted as P_t. The power of each battery was $P_B = P_t/2$. The total power of P_t was transferred from the three HPs toward the PCM (if the PCM is used) and then to the cooling air in flow channel. The initial temperature for all experiments was approximately 21.5 °C, with a variation of approximately ±0.5 °C. Each experiment was performed twice to evaluate the repeatability of the results and to analyze the influence of initial temperature variations on the outcomes. Slight discrepancies were noted between the two sets of results, with instantaneous temperature measurements differing by no more than 1 °C. The experimental results presented in this study represent the averaged values obtained from both experiments.

5.1.1. Without PCM

Figure 5 shows the battery temperature for different total powers and air velocities when no PCM was used (referred to as W/O PCM in figures). Heat pipes were cooled by natural convection when V_{air} was zero, and were cooled by forced convection when V_{air} was non-zero. As expected, the battery temperature was increased (decreased) by increasing the total input power (air velocity). Temperatures reached a steady-state condition within approximately 2 h of operation when $V_{air} \geq 2$ m/s. Steady-state conditions are defined when temperature variations over 5 min are less than 1 °C. For natural convection cooling experiments, the testing was halted once the battery temperature (T_B) reached 80 °C. This precaution was taken to prevent potential thermocouple failure and ensure that temperatures remained within the calibration range of 0 °C to 100 °C. As illustrated in Figure 5, forced convection airflow exerted a substantial influence on reducing battery temperatures across all total power levels. The impacts of air velocities were relatively more significant at higher power.

The experiment deliberately included various power levels and air velocities, irrespective of whether the battery temperature exceeded its safe operational range (below 50 °C). This approach ensured sufficient data to investigate: (*i*) the correlation between air velocity and battery temperature at each power level within the designed setup, (*ii*) battery temperature under unexpected conditions such as fan malfunction, and (*iii*) the identification of combinations of power levels and air velocities that maintained safe battery temperatures for the designed setup.

Inspection of Figure 5 shows that the air velocity had an approximately linear impact on the battery temperatures at a fixed power. Furthermore, it is observed that battery temperatures remained within the safe operational range (below 50 °C) throughout the duration of applied power when the total power levels were below 12 W (corresponding to a battery discharge rate of approximately 4-C [20]) or when air velocities exceeded 4 m/s. However, for other combinations of power levels and air velocities, the duration of input power was constrained to the period before the battery temperature reached 50 °C. For instance, at an air velocity of 2 m/s, the permissible duration of input power was limited to approximately 39 min, 14 min, and 10 min for input powers of 12 W, 16 W, and 20 W, respectively. These durations exhibit an approximate linear correlation with the input power levels. To ensure the battery temperature remained under 50 °C for the entire duration of application of the above power levels, it was necessary to increase the air velocity to a minimum of 3 m/s (50% increment), 4 m/s (100% increment), and 6 m/s (200% increment). These data suggest an approximate linear relationship between the input power and the required minimum velocity increment. Consequently, this observation supports a practical recommendation for designs utilizing forced convection cooling, emphasizing the necessity of smart fan operation to dynamically control the fan speed based on temperature feedback from the battery. Such an investigation was performed by Zhuang et al. [27] in a battery pack to determine the optimal duration of air circulation for effective cooling and to implement an adaptive cooling strategy that adjusts the cooling intensity based on fluctuations in heat dissipation and energy consumption.

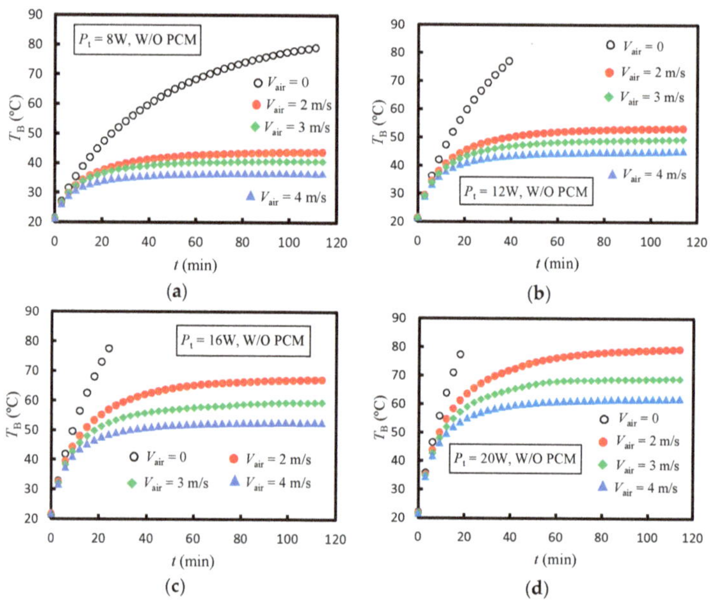

Figure 5. Battery temperature for different total powers and air velocities (without PCM): (**a**) 8 W; (**b**) 12 W; (**c**) 16 W; and (**d**) 20 W.

5.1.2. With PCM

Figure 6 illustrates the measured temperatures primarily for the "with PCM" scenario at a total power of 16 W and varying air velocities. For comparison, some results from the "without PCM" scenario are also included to highlight the thermal performance improvement achieved with PCM. The 16W total input power selected in this study (equivalent to 8W per individual battery) corresponds to a relatively high discharge rate of 5C [20]. The positions of thermocouples T_3, T_5, and T_6, as well as the definition of T_B derived from T_1 and T_2, are detailed earlier in the experimental design section. The melting temperature of the PCM ($T_m \approx 38$ °C) is indicated by a solid dashed line.

As shown in Figure 6, the battery temperature (T_B) in the configuration with PCM was significantly lower than in the configuration without PCM. This reduction is attributed to the PCM's capacity to absorb substantial heat as latent heat during its melting process. The impact of the PCM was more pronounced for relatively lower air velocities because more heat was transferred into the PCM rather than air. For 16 W total power, the battery temperatures (T_B) at $t \approx 110$ min were reduced by about 14%, 10%, and 4% when using the PCM compared to the case without the PCM for air velocities of 2 m/s, 3 m/s, and

4 m/s, respectively. The battery temperature was around or less than 50 °C specifically for cooling air velocities of 3 m/s and 4 m/s. This suggests that, for this setup, air velocities below 3 m/s are not recommended when the total input power is 16 W, unless a smart fan operation increases the fan speed once the battery temperature reaches 50 °C, as discussed in relation to Figure 5.

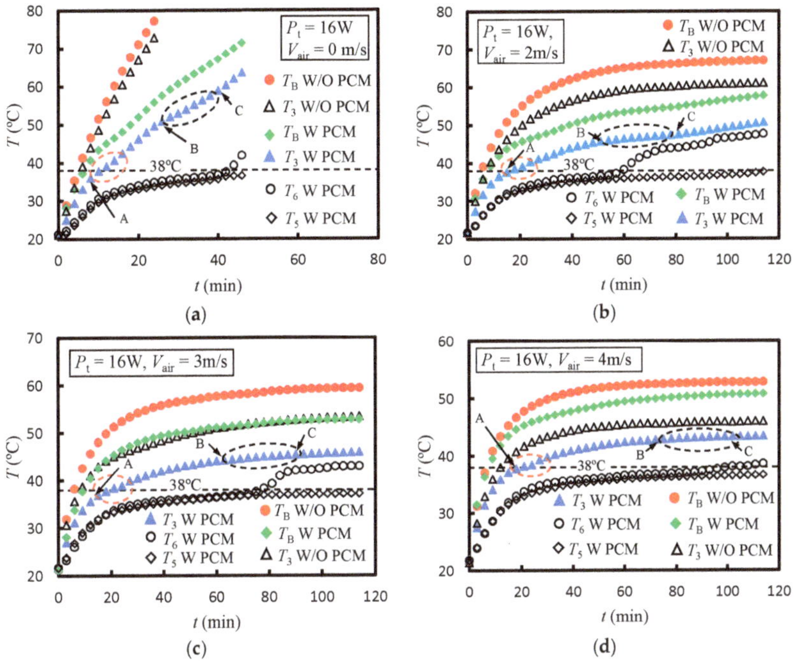

Figure 6. Local temperature histories for a total power of 16W and different air velocities with and without the PCM: (**a**) 0 m/s; (**b**) 2 m/s; (**c**) 3 m/s; and (**d**) 4 m/s.

A comparison between the time required for T_6 and T_5 to reach 38 °C reveals that the melting process predominantly advanced towards the upper region of the PCM reservoir rather than the lower region. This is because, as melting progressed outward from each fin jacket in the PCM reservoir, the warm molten PCM tended to accumulate towards the top portion of the reservoir.

A comparative analysis between T_3 with and without PCM further emphasized the effect of the PCM's latent heat in absorbing thermal energy from the HPs within the PCM reservoir. For each tested air velocity, T_3 (with PCM) exhibited significantly lower temperatures compared to T_3 (without PCM). The difference between T_3 (with PCM) and T_3 (without PCM) was more pronounced at lower air velocities. At lower air velocities, a larger portion of the input heat was absorbed by the PCM rather than being transferred to the air in the flow channel through forced convection.

A detailed inspection of the temperature profile T_3 (with PCM) revealed an initial linear increase from $t = 0$ until the onset of PCM melting around the fin jacket within the PCM, marked as point A in Figure 6. This initial heat absorption was characterized by the PCM's sensible heat, associated with its heat capacity and thermal conduction. The onset of PCM melting occurred when the temperature T_3 reached approximately 38 °C, leading to a reduced rate of temperature increase as the PCM began to absorb heat primarily in the form of latent heat, which was significantly higher than its sensible heat. For example, with an air velocity of V_{air} = 2 m/s, the onset of PCM melting began at about t = 15 min and lasted for approximately 6 min, during which the PCM adjacent to the fin jacket transitioned from solid to mushy and eventually to liquid phase (the region enclosed by the red dashed oval in Figure 6, corresponding to a high heat absorption period). The temperature increase in T_3 (with PCM) slowed down until point B ($t \approx 48$ min for V_{air} = 2 m/s). The melting process in segment AB was governed by mixed conduction and natural convection within the molten PCM, resulting in relatively constant heat absorption.

In segment BC, natural convection within the molten PCM surrounding the fin jacket accelerated, enhancing the heat transfer rate from the fin jacket to the PCM and causing a second decline in the temperature rise, shown as the region encircled by the black dashed oval in Figure 6 ($t \approx 78$ min for V_{air} = 2 m/s at point C). At some point during segment BC, all the molten PCM around the three fin jackets became interconnected. Beyond point C, the PCM reservoir was predominantly occupied by molten PCM, and the subsequent temperature increase in T_3 (with PCM) was primarily due to the absorption of sensible heat rather than latent heat, corresponding to a period of decreasing heat absorption.

Despite the setup being well-insulated to minimize thermal losses to the environment, an energy balance was conducted between the input and output heat using measured data from various locations on the setup. This involved applying conduction analysis through the thickness of the 3D-printed parts and the aerogel insulation pad on the battery module and PCM reservoir, as well as convection analysis on the flow channel side. For a total power of P_t = 16W, it was determined that heat loss constituted approximately 4–6% of P_t at air velocities ranging from 4 m/s to 2 m/s.

5.1.3. Cyclic Mode

Experimental results indicate that the PCM effectively reduces and regulates the battery temperature primarily through latent heat absorption during its melting phase. This phenomenon is particularly advantageous when the PCM is subjected to fluctuating thermal loads. Fluctuating thermal loads are commonly encountered in electric vehicles when the batteries experience repeated charging and discharging cycles. To study the performance of the developed BTM under cyclic thermal load, an experiment was run for a cyclic power condition with a maximum total power of 16W and an air velocity of 2 m/s. The heaters (P_t = 16 W) were turned on for 10 min and turned off for the following 5 min, and the cyclic power condition was repeated. The cyclic periods were consistent with the electric car cyclic mode operation presented in [28]. Figure 7 shows the battery temperature under the cyclic condition. As shown in this figure, the PCM dampened the battery temperature fluctuations compared to those without the PCM. Temperature T_3 in this figure shows that the PCM experienced both melting and solidification as T_3 moved to about 38 °C. During the melting process, the PCM absorbs thermal energy, while during solidification, it releases stored heat.

A comparison between Figures 6b and 7 demonstrates an additional advantage of incorporating PCM in the design, particularly during cyclic operations. As shown in Figure 6b, the battery temperature surpasses 50 °C after approximately 38 min, whereas in Figure 7, it remains below 50 °C for up to 2 h and is expected to stay below 50 °C until the average battery temperature stabilizes at a steady-state condition, as nearly depicted in Figure 7. The average battery temperature in the steady-state condition for cyclic operation is around 45 °C in Figure 7. This indicates that, for the same air velocity, the PCM configuration can potentially accommodate relatively higher fluctuating thermal loads in cyclic operation, keeping the battery temperature below 50 °C for a longer duration compared to non-cyclic operation.

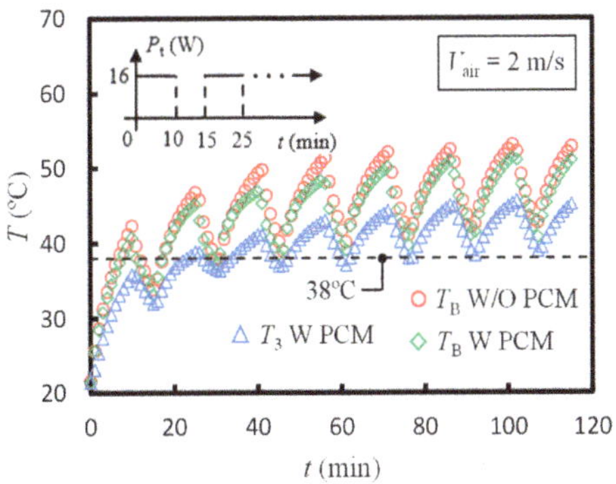

Figure 7. Battery temperature in cyclic operation at an air velocity of 2 m/s, with and without the PCM.

The battery temperature in a non-cyclic input power operation represents an extreme case of cyclic input power operation, where the power time ratio (defined as the duration when the power is on to the duration when the power is off, such as 10 min/5 min in Figure 7) approaches infinity. For any power time ratio between 0 and infinity during cyclic operation, the average battery temperature is expected to be lower than in non-cyclic operation for the same power and air velocities. This demonstrates the robustness and applicability of the cooling system incorporating PCM under varying operational conditions.

5.2. Numerical Results

Numerical simulations were conducted to complement the experimental results and provide deeper insight into the physics of the problem, which are explained here. The heat transfer coefficients on the air-cooled section of the HPs were determined numerically using the computational domain shown in Figure 4a. Air entered the test section at uniform velocities ranging from 2 to 4 m/s, and a boundary layer began to develop along the walls. However, the distance from the inlet to the HPs was shorter than the length required for the boundary layer to transition into a turbulent regime, ensuring that the airflow over the HPs remained laminar. Additionally, the Richardson number was calculated to be approximately 0.003 at the lowest airspeed, indicating that the flow was sufficiently fast to suppress the effects of natural convection at the given velocity. A constant power of 8 W (corresponding to a total power, P_t, of 16 W) was applied at the outer surface of the heat sink in contact with the aluminum housing. Once steady-state conditions were achieved, the average heat flux and temperature at the finned and unfinned surfaces of the HPs exposed to the airflow were calculated. The heat flux was subsequently divided by the temperature difference between the exposed surface and the inlet air to calculate the average heat transfer coefficient. Convective heat transfer coefficients of 38, 51, and 64 W/m^2·K were obtained for inlet air velocities of 2, 3, and 4 m/s, respectively. The convective heat transfer coefficients were found to be nearly linear within the range of velocities studied. The simulation was not conducted for the case where V_{air} = 0, which corresponds to natural convection, as the time required for the battery to reach 50 °C is relatively short compared to the forced convection cases and is not of practical interest. The following Fluent settings outlined in Table 2 were employed in the integrated HP–fin–PCM model:

Table 2. Settings used in the integrated HP–fin–PCM BTM model implemented in ANSYS Fluent.

Velocity-pressure coupling algorithm	SIMPLE
Gradient spatial discretization scheme	Least squares cell based
Pressure spatial discretization scheme	PRESTO!
Density spatial discretization scheme	QUICK
Momentum spatial discretization scheme	QUICK
Energy spatial discretization scheme	QUICK
Transient formulation	First order implicit
Mesh size (Number of control volumes)	1,350,000
Under relaxation factors	Pressure: 0.5 Density: 1 Momentum: 0.5 Liquid fraction: 1 Energy: 1
Residuals	Continuity: 10^{-4} velocity: 10^{-4} Energy: 10^{-9}
Mushy zone constant	10^5
Time step size	0.5 s

5.2.1. Comparison of Numerical Predictions with Experimental Data

The numerical predictions of battery temperatures were validated against experimental results, as illustrated in Figure 8, for a total input power of 16 W and varying air velocities. The predicted battery temperatures exhibited behavior consistent with experimental measurements. The agreement between the predicted values and the measurements was within an acceptable range. The discrepancies observed between the predicted values and experimental data were attributed to uncertainties in the PCM properties and differences between the computationally determined convective heat transfer coefficients and the actual values, as well as the assumptions made in the model.

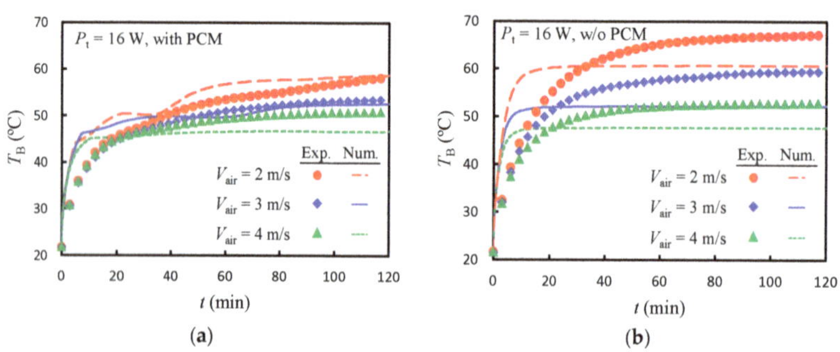

Figure 8. Comparison of predicted and measured battery temperatures for a total input power of 16 W at different air velocities: (**a**) with PCM; (**b**) without PCM.

As can be observed in Figure 8, at t = 120 min the computational results underpredicted the battery temperature. This underprediction is quantified by dividing the difference between the experimental and numerical battery temperatures by the experimentally measured temperature rise of the battery relative to its initial temperature. For cases with (without) PCM at t = 120 min, the numerical results underpredicted the experimental measurements by −1% (20%), 12% (24%), and 20% (22%) for air velocities of 2, 3, and 4 m/s, respectively. At earlier times, the computational results overpredicted the measured temperatures. This discrepancy was primarily attributed to the thermal mass of the acrylic walls, insulation, and the heater element, which were not accounted for in the computational model.

5.2.2. Heat Absorption Predictions by the PCM

Figure 9 shows the computationally predicted evolution of the PCM melt fraction and the normalized power (heat transfer) to the PCM for air velocities of 2 m/s and 4 m/s, with a total power P_t of 16 W. The heat transfer to the PCM is normalized by dividing it by the heat input from the battery (P_t = 16 W). As evident, the heat transfer to the PCM was greater when the air velocity was smaller. This was attributed to the reduced convective heat transfer to the air, which resulted in a higher amount of heat being absorbed by the PCM. For both cases, the heat transfer to the PCM initially increased and then followed a generally decreasing trend. The initial increase was attributed to the rapid rise in the HP temperature. When the PCM temperature reached the melting temperature, melting commenced, and the melt layer thickness continuously increased, corresponding to an increasing thermal resistance. On the other hand, the thermal resistance between the finned surface of the HPs and air remained constant. As such, the heat transfer to the PCM slowly deceased during melting. Once the melting process was complete, the PCM's heat absorption diminished rapidly, as observed for an air velocity of 2 m/s.

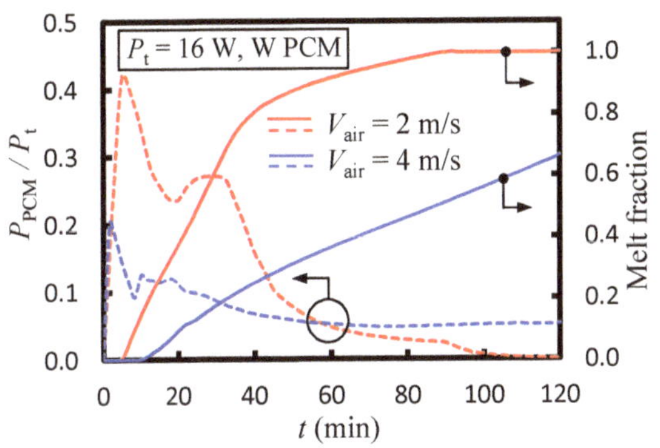

Figure 9. Computationally determined time variations of the melt fraction (solid lines) and normalized heat transfer to the PCM (dashed lines) for air velocities of 2 m/s and 4 m/s, with P_t = 16 W.

Regarding the PCM melt fraction, as shown in Figure 9, increasing the air velocity decreased the melting rate due to more effective convective heat transfer to the air, which reduced the heat transfer to the PCM and slowed down the melting process.

5.2.3. Representative Contours of Temperature

Figure 10 illustrates the temperature contours on the symmetry plane of the computational domain (*x-z* plane, as shown in Figure 4b) at 15 min intervals for a total power input of 16 W and an air velocity of 2 m/s. This figure illustrates the temporal evolution of natural convection within the molten PCM over time, as well as the nearly isothermal nature of the HPs throughout the entire duration of the experiment. As anticipated, the lowest temperatures were observed at the bottom of the PCM reservoir, while the highest temperatures occurred at the aluminum heat sink.

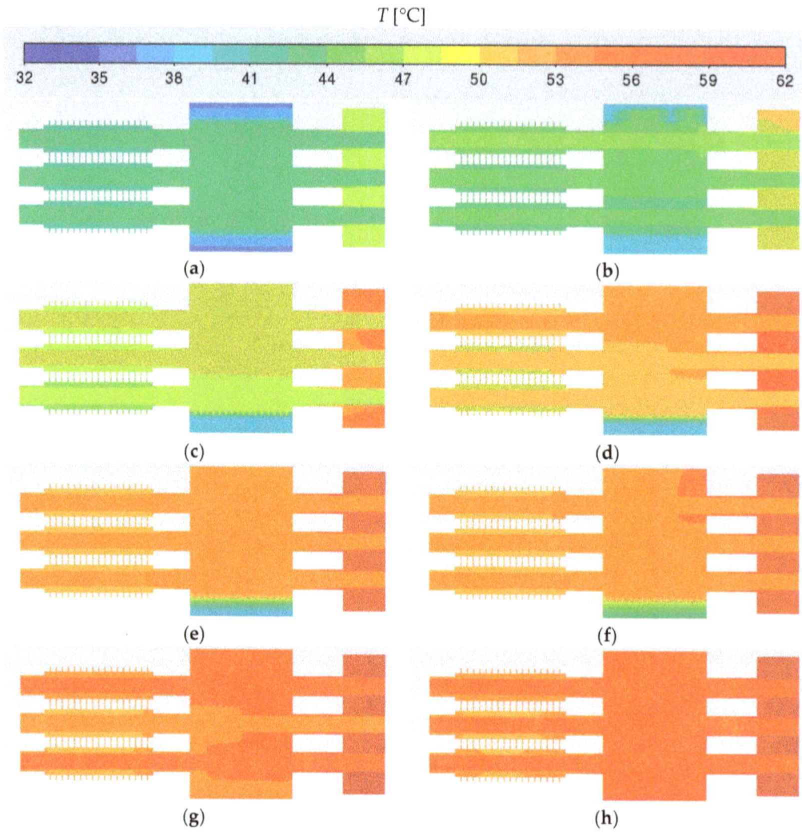

Figure 10. Temperature contours on the symmetry plane (x-z plane in Figure 4b) at 15-*min* intervals for a total power of 16 W and an air velocity of 2 m/s: (**a**) t = 15 min, (**b**) t = 30 min, (**c**) t = 45 min, (**d**) t = 60 min, (**e**) t = 75 min, (**f**) t = 90 min, (**g**) t = 105 min, (**h**) t = 120 min.

The temperature contours within the PCM reservoir, depicted in Figure 10b, reveal that natural convection within the molten PCM intensified as the warm molten PCM migrated toward the top of the PCM reservoir. Consequently, the lower portion of the PCM remained relatively cooler compared to the upper region. Examination of the temperature distribution in the lower PCM reveals a gradual warming trend toward its melting temperature over time, eventually reaching this state after around 90 min, as shown in Figure 10f. This observation aligns with the experimental results presented in Figure 6, particularly with the discussion regarding the temperature comparison between T_5 and T_6.

After about 90 min, the entire PCM transitioned to a liquid phase, as evidenced by Figure 10g,h. The HP cores (vapor space) exhibit a uniform temperature distribution throughout at each time interval shown in Figure 10, highlighting the near-isothermal behavior of the HPs and their efficiency in transferring heat.

To further analyze and investigate the internal heat transfer mechanisms of the HP, particularly the interaction between the HP and PCM, the heat flux distribution at the wick-vapor interface of the middle HP was examined at representative time intervals: t = 1 min, during the initial stages; t = 30 min, when natural convection within the molten PCM was accelerating; and t = 120 min, when the system was approaching thermal equilibrium. The heat flux distributions for these time intervals are presented in Figure 11.

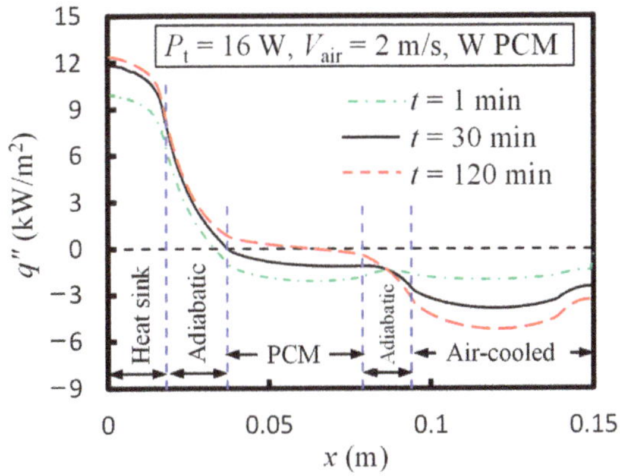

Figure 11. Heat flux distribution at the wick-vapor interface of the middle HP at t = 1, t = 30 and 120 min for a cooling air velocity of 2 m/s and a total power of 16 W.

In the figure, regions with positive heat flux correspond to evaporation zones within the HP, where heat is transferred into the HP, while regions with negative heat flux indicate condensation zones, where heat is released from the HP. As time progressed, the highest heat flux was observed in the heat sink section of the HP, where heat is absorbed, while the lowest heat flux was seen in the air-cooled section, where heat is dissipated to the ambient air.

A notable heat transfer phenomenon was observed within the PCM section. At the early stage (t = 1 min), the PCM absorbed heat from the HP at its highest rate, as indicated by the relatively large negative heat flux. This occurred because the HP was significantly warmer than the PCM. By t = 30 min, at the intermediate stage, heat absorption from the HP continued, but at a reduced rate. Finally, at t = 120 min, as the system neared thermal equilibrium, the PCM simultaneously absorbed and released heat to the HP at its lowest rate. This is indicated by the coexistence of positive and negative heat flux regions within the PCM section.

In the air-cooled section of the HP, the magnitude of localized heat flux decreased toward the tip of the HP compared to its midsection. This decrease was attributed to the portion of the HP within the flow channel that was not covered by the fin jacket (approximately 10.55 mm). Overall, these findings highlight the complex heat transfer dynamics within the HP–fin–PCM system and the effectiveness of the numerical model in capturing these phenomena.

More detailed predictions, including contours of PCM melt fraction and velocity fields, are provided in Appendix A for further reference.

6. Conclusions

An experimental and numerical investigation was carried out on a new setup comprising heat pipes, fins, and PCM to evaluate the thermal performance of a hybrid cooling method for lithium-ion battery thermal management. Unlike recent studies, where the PCM is typically in direct contact with the batteries, this design innovatively decouples the PCM from the batteries and positions it between the batteries and the air-cooled section. This decoupling not only reduces the complexity of the system but also enhances design flexibility and facilitates easier battery maintenance or replacement.

The study encompassed various operating conditions, including different input powers, air velocities, and operating modes (with and without PCM, and cyclic mode). In general, incorporating PCM led to a decrease in the battery surface temperature when compared to scenarios without PCM. Specifically, for a total input power of 16W, steady-state battery temperatures were reduced by approximately 14%, 10%, and 4% with PCM compared to without PCM for air velocities of 2 m/s, 3 m/s, and 4 m/s, respectively. The battery temperature was maintained at about and below 50 °C for cooling air velocities of 3 m/s and 4 m/s. The advantages of PCM were observed to be more significant at lower air velocities and in cyclic battery operation. The predictions exhibited satisfactory agreement with the experimental data, validating the numerical model's effectiveness in accurately simulating the underlying physics. The proposed model provides a robust foundation for future optimization studies. The new HP–fin–PCM design emerges as a promising candidate for BTM systems in electric vehicles. While this study focused on a single module of the HP–fin–PCM system, future research should extend this investigation to multiple modules within a battery pack. Such studies could reveal the full extent of the benefits offered by the HP–fin–PCM configuration. Additional investigations are needed to explore the effects of various PCMs with different melting temperatures, air inlet temperatures, the impact of the number of fins, and the use of smart fans on the battery surface temperature.

Author Contributions: Conceptualization and methodology, N.S. and H.S.; SolidWorks Design and construction, C.M., U.E. and N.S.; data collection, N.S.; numerical simulations, H.S.; writing, N.S. and H.S.; proof read, C.M. and U.E.; supervision, N.S.; project administration, N.S.; funding acquisition, N.S. All authors have read and agreed to the published version of the manuscript.

Funding: This research was funded by Tarleton State University President's Excellence in Research Scholars (PERS) grant and Mayfield College of Engineering.

Data Availability Statement: Data are available on request.

Acknowledgments: This study was supported by the Tarleton State University President's Excellence in Research Scholars (PERS) grant and Mayfield College of Engineering. The faculty also wish to thank undergraduate students Christian Millard, Ugochukwu Etufugh, and Griffin DaRosa for their assistance inconstructing the experimental setup.

Conflicts of Interest: The authors declare no conflicts of interest.

Nomenclature

A	mushy-zone constant function
A_{mushy}	mushy-zone constant
BTM	battery thermal management
CNC	computer numerical control
c_p, C_p	specific heat (J/kg K or J/kg °C)
EV	electric vehicle
g	gravitational acceleration (m/s^2)
h, H	enthalpy (J/kg)
h_{fg}	latent heat of vaporization of the HP working fluid (kJ/kg)
h_{sl}	latent heat of fusion of the PCM (kJ/kg)
HP	heat pipe
ID	inner diameter (m or mm)
k	thermal conductivity (W/m·K)
L	length (m or mm)
L_{f}	latent heat of fusion (kJ/kg)
$\dot{m}''$	mass flux (kg/m^2 s)
OD	outer diameter (m or mm)
P	pressure (Pa)
PLA	polylactic acid (3D printer filament material)
PCM	phase change material

P_B	individual battery power (W) or individual heater power (W)
P_{PCM}	heat transfer or power to the PCM (W)
P_t	total heater power (W); $P_t = 2P_B$
q''	heat flux (W/m^2 or kW/m^2)
R_g	gas constant (J/kg K)
S_m	source term (N/m^3)
t	time (s or min)
T	temperature (K or °C)
T_{air}	air temperature (K or °C)
TC	thermocouple
T_m	melting temperature (K or °C)
V	velocity vector (m/s)
$\mathbf{V}^T$	transposed velocity vector (m/s)
x, y, z	coordinates (m or mm)
Greek	
ΔT	half width of temperature range (K)
β	thermal expansion coefficient (1/K)
ε	small value
λ	liquid fraction
μ	viscosity (Pa.s)
ρ	density (kg/m^3)
Subscript	
B	battery
l	liquid
ref	reference
s	solid
sat	saturation
v	vapor
w	wall

Appendix A

The following figures provide additional insights into the thermal and fluid dynamics observed during the simulation. These figures aim to complement the main results by illustrating specific phenomena in greater detail.

Figure A1 shows the PCM melt fraction and *z*-velocity vectors on the *y*–*z* plane crossing through the middle of the PCM reservoir at $t = 25$ min, when natural convection is in the process of accelerating. The melted region is relatively greater around the top HP due to the influence of natural convection, which drives the warmer PCM upward. Accumulation of warmer PCM at the upper region creates more melting relative to the lower region as depicted in Figure A1a. Within the melted region, the liquid PCM near the HPs is relatively warmer and moves upward due to relatively lower density. Near the interface with the solid PCM, the liquid PCM is relatively colder and descends due to higher density as seen in Figure A1b.

Figure A2 shows the axial water vapor velocity contours within the HPs vapor spaces at time intervals: $t = 1$ min, during the initial stages; $t = 30$ min, when natural convection within the molten PCM is accelerating; and $t = 120$ min, when the system approaches thermal equilibrium. The right end is in contact with the aluminum heat sink, which serves as the heating section. To enhance clarity and avoid visual clutter, the vapor velocity vectors are displayed for only one HP, providing a clearer representation of the flow direction during acceleration and deceleration. As evident, the maximum vapor velocities occur somewhere between the heating section and the PCM reservoir. In this region, thermal energy enters the HPs and evaporates the HPs working fluid. The vapor flows toward the relatively colder sections of the HP, while more vapor is added to the vapor space due to ongoing evaporation. When the vapor reaches the cooled sections of the HPs, condensation removes mass from the vapor space, resulting in decreasing axial velocity. It is observed that the vapor velocities decrease as time increases. The smaller velocities are attributed to the greater vapor temperatures, which correspondingly create greater pressures at later times due to the coupled relationship between vapor temperature and pressure under saturation conditions (see Figure 10). As the vapor pressure increases, its density increases and the same amount of heat can be transported by a smaller vapor volumetric flowrate. The average absolute vapor pressures at t = 1, 30, and 120 min were about 3.6 kPa, 9.5 kPa and 14.4 kPa, respectively. Meanwhile, the average vapor densities were 0.026 kg/m^3, 0.065 kg/m^3 and 0.096 kg/m^3, for t = 1, 30, and 120 min, respectively.

Figure A1. (**a**) PCM melt fraction, and (**b**) z-velocity -vectors on y-z plane crossing through the middle of the PCM reservoir at time t = 25 min, with a cooling air velocity of 2 m/s and a total power of 16 W.

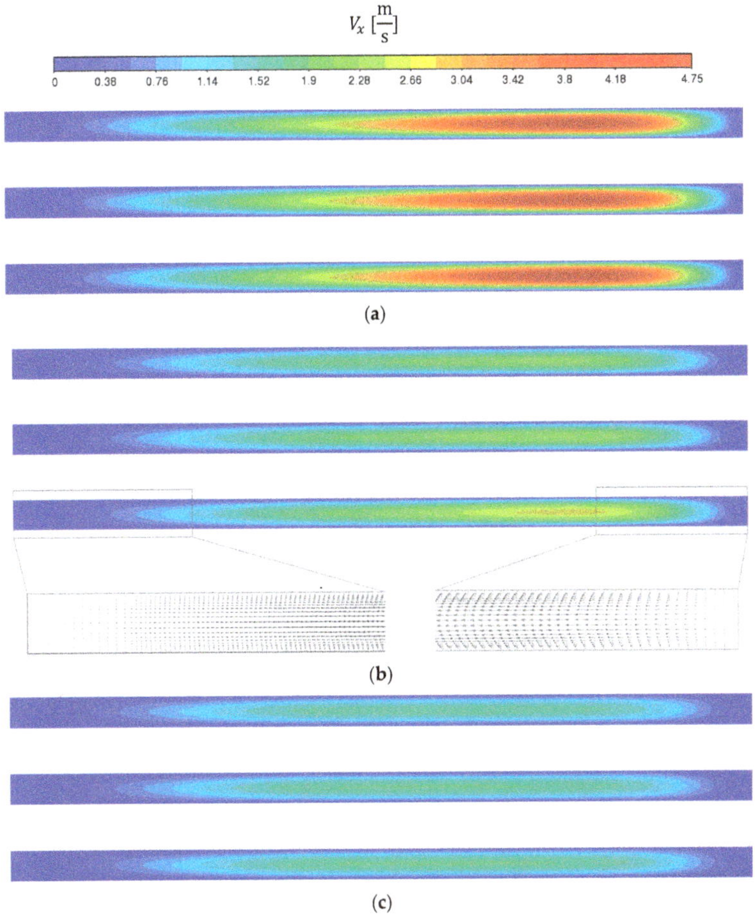

Figure A2. Axial vapor velocity contours within the HPs for a cooling air velocity of 2 m/s and a total power of 16 W: (**a**) t = 1 min; (**b**) t = 30 min; (**c**) t = 120 min (the right end is in contact with the hot aluminum heat sink).

References

1. Ma, S.; Jiang, M.; Tao, P.; Song, C.; Wu, J.; Wang, J.; Deng, T.; Shang, W. Temperature effect and thermal impact in lithium-ion batteries: A review. *Prog. Nat. Sci. Mater. Int.* **2018**, *28*, 653–666. [CrossRef]
2. Rahmani, A.; Dibaj, M.; Akrami, M. Recent Advancements in battery thermal management systems for enhanced performance of Li-Ion batteries: A comprehensive review. *Batteries* **2024**, *10*, 265. [CrossRef]
3. Ahmadian-Elmi, M.; Zhao, P. Review of thermal management strategies for cylindrical lithium-ion battery packs. *Batteries* **2024**, *10*, 50. [CrossRef]
4. Weragoda, D.M.; Tian, G.; Burkitbayev, A.; Lo, K.H.; Zhang, T. A comprehensive review on heat pipe based battery thermal management systems. *Appl. Therm. Eng.* **2023**, *224*, 120070. [CrossRef]
5. Bernagozzi, M.; Georgoulas, A.; Miché, N.; Marengo, M. Heat pipes in battery thermal management systems for electric vehicles: A critical review. *Appl. Therm. Eng.* **2023**, *219*, 119495. [CrossRef]
6. Faghri, A.; Zhang, Y. *Transport Phenomena in Multiphase Systems*; Academic Press, Elsevier: New York, NY, USA, 2006.
7. Faghri, A. *Heat Pipe Science and Technology*, 2nd ed.; Global Digital Press, 2016.
8. Sharifi, N.; Bergman, T.L.; Allen, M.J.; Faghri, A. Melting and solidification enhancement using a combined heat pipe, foil approach. *Int. J. Heat Mass Transf.* **2014**, *78*, 930–941. [CrossRef]
9. Shabgard, H.; Faghri, A.; Bergman, T.L.; Andraka, C.E. Numerical simulation of heat pipe-assisted latent heat thermal energy storage unit for dish-stirling systems. *J. Sol. Energy Eng.* **2014**, *136*, 021025. [CrossRef]

10. Sharifi, N.; Roesler, D.; Gold, A.; Shabgard, H. Thermal management system for lithium-ion batteries using phase change material, heat pipes and fins. *ASME Int. Mech. Eng. Congr. Expo.* **2023**, *87677*, V010T11A049.
11. Zhang, C.; Yu, M.; Fan, Y.; Zhang, X.; Zhao, Y.; Qiu, L. Numerical study on heat transfer enhancement of PCM using three combined methods based on heat pipe. *Energy* **2020**, *195*, 116809. [CrossRef]
12. Gao, C.; Sun, K.; Song, K.; Zhang, K.; Hou, Q. Performance improvement of a thermal management system for Lithium-ion power battery pack by the combination of phase change material and heat pipe. *J. Energy Storage* **2024**, *82*, 110512. [CrossRef]
13. Sutheesh, P.M.; Nichit, R.B.; Rohinikumar, B. Numerical investigation of thermal management of lithium ion battery pack with nano-enhanced phase change material and heat pipe. *J. Energy Storage* **2024**, *77*, 109972. [CrossRef]
14. Sharma, D.K.; Agarwal, P.; Prabhakar, A. Effect of fin design and continuous cycling on thermal performance of PCM-HP hybrid BTMS for high ambient temperature applications. *J. Energy Storage* **2023**, *74*, 109360. [CrossRef]
15. Tang, Z.; Feng, R.; Huang, P.; Bai, Z.; Wang, Q. Modeling analysis on the cooling efficiency of composite phase change material-heat pipe coupling system in battery pack. *J. Loss Prev. Process Ind.* **2022**, *78*, 104829. [CrossRef]
16. Abbas, S.; Ramadan, Z.; Park, C.W. Thermal performance analysis of compact-type simulative battery module with paraffin as phase-change material and flat plate heat pipe. *Int. J. Heat Mass Transf.* **2021**, *173*, 121269. [CrossRef]
17. 1-tetradecanol: 112-72-1. ChemicalBook. Available online: https://www.chemicalbook.com/ChemicalProductProperty_EN_CB8689436.htm (accessed on 3 January 2025).
18. Jiang, Z.Y.; Qu, Z.G. Lithium–ion battery thermal management using heat pipe and phase change material during discharge–charge cycle: A comprehensive numerical study. *Appl. Energy* **2019**, *242*, 378–392. [CrossRef]
19. Available online: https://us.amazon.com/Aerogel-Insulation-Padding-Manufacturing-Lightweight/dp/B0BWVT9ZSK (accessed on 3 January 2025).
20. Akula, R.; Minnikanti, A.; Balaji, C. Pin fin-PCM composite heat sink solution for thermal management of cylindrical Li-ion battery. *Appl. Therm. Eng.* **2024**, *248*, 123146. [CrossRef]
21. Available online: https://www.graspengineering.com/material-properties-aluminum-6061-t6-6061-t651/ (accessed on 3 January 2025).
22. Mishra, G.; Memon, A.; Gupta, A.K.; Nirmalkar, N. Computational study on effect of enclosure shapes on melting characteristics of phase change material around a heated cylinder. *Case Stud. Therm. Eng.* **2022**, *34*, 102032. [CrossRef]
23. Mahdavi, M.; Faghri, A.; Shabgard, H. Thermal performance of U-shaped and L-shaped heat pipes. *Numer. Heat Transf. Part A Appl.* **2021**, *80*, 411–435. [CrossRef]
24. Calle-Arroyo, C.; Lopez-Fidalgo, J.; Rodríguez-Aragon, L.J. Optimal designs for Antoine Equation. *Chemom. Intell. Lab. Syst.* **2021**, *214*, 10433. [CrossRef]
25. Cao, Y.; Faghri, A. Transient two-dimensional compressible analysis for high temperature heat pipes with pulsed heat input. *Numer. Heat Transf. Part A Appl.* **1990**, *18*, 483–502. [CrossRef]
26. Voller, V.R.; Prakash, C. A fixed grid numerical modelling methodology for convection-diffusion mushy region phase-change problems. *Int. J. Heat Mass Transf.* **1987**, *30*, 1709–1719. [CrossRef]
27. Zhuang, W.; Liu, Z.; Su, H.; Chen, G. An intelligent thermal management system for optimized lithium-ion battery pack. *Appl. Therm. Eng.* **2021**, *189*, 116767. [CrossRef]
28. Liu, H.; Chen, F.; Tong, Y.; Wang, Z.; Yu, X.; Huang, R. Impacts of driving conditions on EV battery pack life cycle. *World Electr. Veh. J.* **2020**, *11*, 17. [CrossRef]

batteries

MDPI

Article

Air-Outlet and Step-Number Effects on a Step-like Plenum Battery's Thermal Management System

Olanrewaju M. Oyewola [1,*], Emmanuel T. Idowu [1], Morakinyo J. Labiran [1], Michael C. Hatfield [2] and Mebougna L. Drabo [3]

1 Department of Mechanical Engineering, University of Alaska Fairbanks, Fairbanks, AK 99775, USA; etidowu@alaska.edu (E.T.I.); mjlabiran@alaska.edu (M.J.L.)
2 Department of Electrical and Computer Engineering, University of Alaska Fairbanks, Fairbanks, AK 99775, USA; mchatfield@alaska.edu
3 Department of Mechanical & Civil Engineering and Construction Management, Alabama A&M University, Huntsville, AL 35811, USA; mebougna.drabo@aamu.edu
* Correspondence: ooyewola@alaska.edu or oooyewola001@gmail.com

Abstract: Optimizing the control of the battery temperature (T_b), while minimizing the pressure drop (ΔP) in air-cooled thermal management systems (TMSs), is an indispensable target for researchers. The Z-type battery thermal management system's (BTMS's) structure is one of the widely investigated air-cooled TMSs. Several designs of air-cooled BTMSs are often associated with the drawback of a rise in ΔP, consequently resulting in an increase in pumping costs. In this study, the investigation of a Step-like plenum design was extended by exploring one and two outlets to determine possible decreases in the maximum battery temperature (T_{max}), maximum battery temperature difference (ΔT_{max}), and pressure drop (ΔP). The computational fluid dynamics (CFD) method was employed to predict the performances of different designs. The designs combine Step-like plenum and two outlets, with the outlets located at different points on the BTMS. The results from the study revealed that using a one-outlet design, combined with a Step-like plenum design, reduced T_{max} by 3.52 K when compared with that of the original Z-type system. For another design with two outlets and the same Step-like plenum design, a reduction in T_{max} by 3.45 K was achieved. For ΔT_{max}, the use of a two-outlet design and a Step-like plenum design achieved a reduction of 6.34 K. Considering the ΔP performance, the best- and poorest-performing designs with two outlets reduced ΔP by 5.91 Pa and 3.66 Pa, respectively, when compared with that of the original Z-type design. The performances of the designs in this study clearly show the potential of two-outlet designs in reducing ΔP in systems. This study, therefore, concludes that the operational cost of the Step-like plenum Z-type BTMS can be reduced through the careful positioning of the two-outlet section, which will promote the design and development of current and future electric vehicle (EV) technologies.

Keywords: air outlet; step number; pressure drop; temperature; BTMS

1. Introduction

The performance enhancement of battery thermal management systems (BTMSs) has been a vital research focus in recent times because of their roles in advancing the design and development of novel highly efficient electric vehicles (EVs). Over the years, EVs have proven to be reliable in the bid to minimize environmental pollution and fossil fuel consumption [1] because of the choice of batteries as an energy source as an alternative to conventional fossil fuels. Different types of batteries have been used to power EVs; however, the most common type is the lithium-ion battery. The power consumption, mileage, and cost of electric vehicles generally depend on the health and capacity of batteries [2]. Interestingly, despite their wide acceptance, lithium-ion batteries generate too much heat during charging and discharging, which poses severe threats to the health and service life of the batteries. In order to overcome this challenge, novel BTMSs that are highly efficient and effective must be developed. Generally, BTMSs are required to maintain the working temperature of battery packs between 25 °C and 40 °C, while the maximum temperature difference (ΔT_{max}) between any two batteries in a pack remains below 5 °C [3,4]. This is necessary because a ΔT_{max} value higher than the recommended value and temperatures outside the recommended range will negatively impact the performance of batteries [5]. Several cooling methods for BTMSs have been developed over the years and can be categorized as passive and active cooling methods. For active cooling, batteries are cooled by passing air across the batteries [6,7], by the direct contact of liquid on the batteries, by passing the liquid through cooling plates or mini channels [8,9], or by placing heat pipes between and around batteries [10,11]. Passive cooling utilizes phase-change materials (PCMs), which will be in direct contact with batteries to maintain the batteries' operational temperature [12,13]. The combination of two methods has also been explored, such as passive/passive, active/active, or passive/active methods [14–16]. Despite the progress recorded in the design of BTMSs, the air-cooling BTMS still remains one of the most preferred choices because of its numerous benefits of design affordability, simplicity, stability, light weight, and longevity [17]. It has been employed by many companies, such as Nissan, Toyota, Volkswagen, and Chevrolet [17,18].

The research on the air-cooling technique is being explored, with new designs supplementing the air-cooling techniques by adding other enhancement techniques, to form hybrid thermal management systems (HTMSs). Focusing HTMSs, Mo et al. [1] combined phase-change materials (PCMs) and air-cooling techniques for managing the performance of batteries under all climate conditions. In a study by Qin et al. [19], PCMs and air-cooling techniques were integrated together to form a hybrid BTMS, which was investigated experimentally and numerically. Behi et al. [20] proposed a novel concept that combines heat pipes and air-cooling for managing battery temperatures in electric vehicles. Zhang et al. [21] carried out a performance analysis of another hybrid BTMS, which combines PCMs and air-cooling, by leveraging the potential of machine-learning algorithms to optimize the choice of the operational and structural parameters of the design. Zhou et al. [22] utilized liquid cooling while designing the channel for the liquid flow in the form of steps. A new concept has also emerged for PCMs, as reported by Song et al. [23], where a dual-layer PCM was designed, while air provided additional cooling. A recent concept proposed by Zhu et al. [24] combined two different outlet positions employed in the Z-type BTMS. The PCM-based hybrid design has been further investigated using nano-PCM, fins, and metal foam [25,26]. The modification of the structural orientations of battery packs to optimize their performance has also been widely employed. Argade and De [27] optimized the design of the Z-type BTMS by adopting a tapered air inlet and secondary vents. Oyewola and Idowu [28] investigated the waved and straight structures of the divergence plenum of the Z-type BTMS to alter the flow of the air to improve the thermal uniformity. Suo et al. [29] also investigated the effects of different plenum shapes, such as slanted, convex, concave, and stepped, and different plenum positions on the heat dissipation performances of batteries. Chen et al. [30] explored a control strategy based on the temperature difference among batteries for a J-type-structured BTMS. Alzwayi and Paul [31] considered the control of flow patterns while introducing vertical and spiral fins to enhance the cooling performance of the system. The installation of baffles, spoilers, and fins along the airflow path has been widely adopted. For instance, Qin et al. [32] carried out experimental and numerical simulation studies for a design with a combination of internal fin structures and air cooling to enhance the thermal performance of a system. Wang et al. [33] installed spoilers along the airflow path to enhance the thermal performance of the Z-type BTMS. Wang et al. [34] introduced parallel plates to the airflow of the same Z-type BTMS to enhance the performance of the system. Oyewola et al. [35] investigated the use of straight and inclined baffles in enhancing the performance of the Z-type BTMS. Alzwayi and Paul [36] examined the heat transfer enhancement of batteries using vertical and spiral cooling fins. Each of these studies provides distinct performances and demonstrates huge potential in enhancing the performance of Z-type BTMSs.

The findings from a few of the aforementioned studies are further presented to demonstrate how they impact the thermal performances of batteries. A study by Wang et al. [34] added parallel plates to the Z-type BTMS to enhance its performance. It was revealed that this technique reduced T_{max} and ΔT_{max} by 3.37 K (6.17%) and 5.5 K (71.9%), respectively, in comparison with those of the original Z-type design. In a study by Oyewola et al. [35], where straight and inclined baffles were installed to understand how they influence the performance of the system, the results showed that the baffle thickness can either decrease or increase the T_{max} values of the batteries, while inclined baffles generally produce better T_{max} and ΔP performances. In a study by Alzwayi and Paul [36], the heat transfer enhancement in a Z-type BTMS with cylindrical lithium-ion batteries having vertical and spiral fins installed on them was investigated. It was observed that spiral fins reduced the battery temperature by 3.2% when compared with longitudinal fins, which also resulted in a 65.6% reduction in the amount of material used. Other studies that explore the hybrid approach have also reported enhancements in performance. For instance, Moaveni et al. [26] investigated passive and hybrid BTMSs by employing fins, nano-PCMs, and metal foam, which served as flow control mechanisms. From the study, using 9% nanoparticles and four fins in the hybrid system, T_{max} reduced by 5.18 K and 10.36 K, when compared with that of the system without fins and nanoparticles, respectively. Song et al. [23] considered a hybrid BTMS having a double-layer PCM and air-cooling techniques. The T_{max} value of the hybrid BTMS achieved a reduction of 21.5% in comparison with that of the PCM BTMS at an ambient temperature of 40 °C. The aforementioned studies reported improvements in the thermal performance of the BTMS, with most of the techniques requiring the addition of materials, which, consequently, implies an increase in the overall weight of the system. This is not favorable in applications where weight reduction is crucial to achieve the optimal operation. Hence, more studies have been carried out to improve thermal performance without necessarily adding more materials. One such approach can be seen in a study by Oyewola et al. [28], where different wave-like and straight configurations were investigated. The study reported that ΔT_{max} and T_{max} reduced by 8 K and 4.61 K, respectively, using a wave-inclined plenum configuration. Targeting improvement in thermal performance without adding materials, Argade and De [27] created outlet vents on the convergence plenum while further redesigning the divergence plenum to have an inclined configuration. The study revealed that T_{max} and ΔT_{max} were reduced by 2.18% and 72.84% when compared to those of the system without vents.

From the literature presented in this study, it is evident that Z-type BTMSs have been extensively investigated through different structural modifications. One of these designs that has shown huge thermal enhancement potential is the Step-like plenum BTMS, with a significant reduction in T_{max}. A lot of investigations have also been conducted on the Z-type BTMS with a Step-like plenum design to gain detailed insight into the performance of the design. Zhou et al. [22] proposed stepped channels to achieve a lightweight design, using liquid as a cooling fluid. Zhu et al. [24] combined the Step-like design with two outlet positions. Oyewola et al. [37] also investigated the Step-like design, focusing on four different numbers of steps, with each design having the same step height. In order to better understand the Step-like design, Oyewola et al. [38] investigated each number of steps with variable step heights. A further study by Oyewola et al. [39] leveraged machine learning's benefits to investigate additional numbers of steps and various inlet air velocities and temperatures. Each of these studies demonstrates different performances in terms of T_{max}, ΔT_{max}, and ΔP. By installing steps on the divergent plenum and changing the positions of the outlets, the performances of the designs varied. However, increases in the pressure drop were recorded for all the Step-like plenum designs. This work differs from the previous works by investigating the performance of Step-like plenum BTMSs through the introduction of two outlets, positioned at different locations on the convergent plenum, with the aim of maintaining a reduction in T_{max} while reasonably reducing ΔP. Considering the potential of increasing the number of steps while adopting and repositioning two outlet channels, more investigations can be carried out for the Step-like plenum configuration. In view of these, the current study explored additional possibilities not covered by Oyewola et al. [37] and Oyewola et al. [39] by investigating seven different numbers of steps, combined with one or two outlets, placed at different flow pattern positions. The numerical simulation method was employed to investigate different structural designs, thereby proposing the best structure in terms of reductions in T_{max}, ΔT_{max}, and ΔP. It should be noted that the performances of these designs will be presented based on each of the aforementioned parameters and not based on a combined effect of the three parameters. A future study is underway to optimize the designs and select the performance metric that balances the tradeoffs between the parameters.

2. Simulation Method

A computational fluid dynamics (CFD) approach was adopted, with the reference Z-type BTMS as the primary design. From the literature, it can be seen that several modifications of Z-type BTMs have been developed and investigated, with the aim of minimizing heat accumulation and pressure loss during cooling. The studies have revealed that parameters such as the maximum temperature (T_{max}), maximum temperature difference (ΔT_{max}), and pressure drop (ΔP) are significant in evaluating the performance of the system [30,31,40]. Air-cooled BTMSs aim to prevent excess temperature accumulation through reductions in T_{max}, promote thermal homogeneity through reductions in ΔT_{max}, and minimize pumping costs through reductions in ΔP. These are general performance metrics extensively reported in air-cooled battery thermal management system studies. Furthermore, a reduction in ΔT_{max} indicates improvement in thermal homogeneity (uniform temperature distribution) among the batteries in a pack. Improvement in thermal homogeneity decreases the possibility of irregular heating, which could cause the overheating of a battery at the expense of the other batteries in the same pack, maintaining significantly lower temperatures. Such overheating affects the lifespan of batteries and could lead to explosions [5,37,40]. The CFD tool used for this study is the well-known ANSYS software 2024 R1 package. This software was used for developing physical geometries, meshing and independence tests, and computing the appropriate governing equations to predict the airflow and temperature profiles of BTMSs.

2.1. Physical Geometries

2.1.1. Z-Type Geometry

The widely investigated Z-type BTMS (denoted as Z-1) structure was considered in this study because of its simplistic shape, configuration, and arrangement of batteries. An experimental study has been carried out for this structure to establish and evaluate its reliability for real-life applications. Additionally, 3D and 2D geometries of the BTMS have been investigated, with the results showing very good comparative performances [40]. A numerical simulation by Chen et al. [40] compared the performances of conventional Z-type BTMSs using 2D and 3D geometries. The results showed a negligible difference between the two geometries, with estimated average errors of 1% and 0.1% for the airflow and temperature of the batteries, respectively. Moreso, simulations with 2D geometries were faster than those with 3D geometries. Based on this assertion, 2D geometries were considered for all the numerical simulations in this study. Figure 1a–d shows a 2D Z-type BTMS, a 2D battery, a 3D Z-type BTMS, and a 3D battery, respectively. The BTMS consists of an inlet section and an outlet section for passing in air and rejecting air, respectively, while eight batteries in a series arrangement are cooled down. The air passes through cooling channels (*CCs*) during the cooling process. A cooling channel has fixed length (CC_l) at 3 mm, while the velocity (V_i) and temperature (T_i) of the air at the inlet are taken as 3 m/s and 299.15 K, respectively, with a heat power of 11.8 W. The heat generation rate of the battery can be estimated using Equation (1), for the 5-current constant discharge rate of the battery (12 Ah $LiFePO_4$) estimated at 69,370 W/m^3 [40–42]. The properties of the battery and air are presented in Table 1 [34,37]. The description of the original Z-type BTMS is presented in Table 2.

$$Q_b = \left(I^2 R_b - I T_b \frac{du}{dT}\right) / V_b \tag{1}$$

$$R_b = 0.00705 - 0.01853 \times SOC + 0.05894 \times SOC^2 - 0.09151 \times SOC^3 + 0.06579 \times SOC^4 - 0.01707 \times SOC^5 \tag{2}$$

where I is the discharge current, R_b is the battery's equivalent resistance, T_b is the battery temperature, V_b is the volume of the battery, and $\frac{du}{dT} = -0.22$ mV/K is a coefficient obtained from electrochemical–calorimetric techniques.

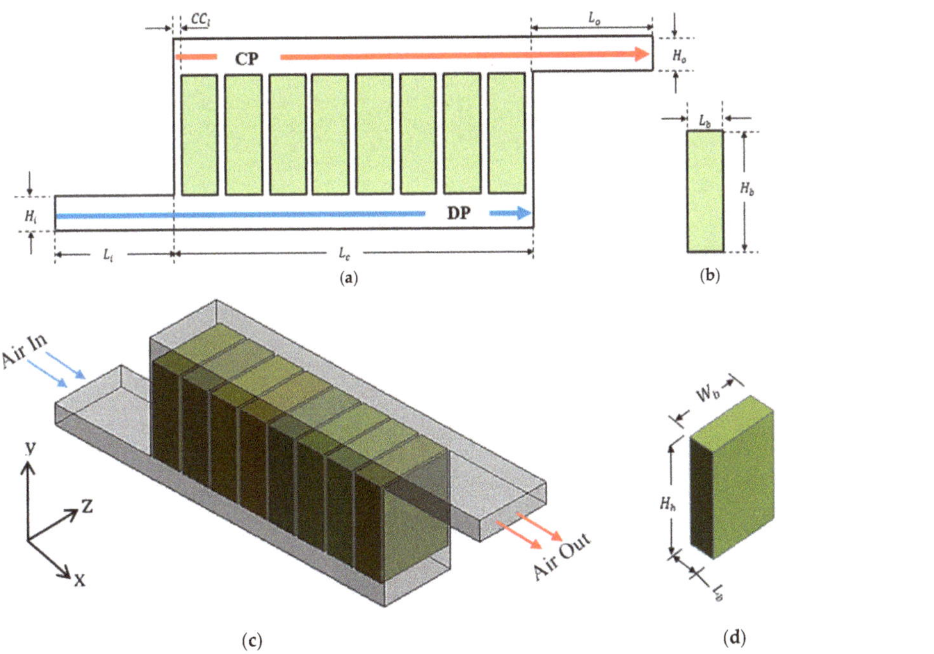

Figure 1. Z-type (**a**) 2D BTMS, (**b**) 2D battery, (**c**) 3D BTMS, and (**d**) 3D battery [34,40].

Table 1. Properties of the air and battery [34,37].

Property	Battery	Air	Unit
Specific heat (c_p)	1337	1005	J/kgK
Density (ρ)	1542.9	1.165	kg/m^3
Dynamic viscosity (μ)	-	1.86×10^{-5}	kg/ms
Thermal conductivity (k)	1.05 (x); 21.1 (y); 21.1 (z)	0.0267	W/mK

Table 2. Geometric and flow properties of the Z-type and Step-like BTMSs [34,37].

Description	Symbol	Value (mm)
Length of the cooling channel	CC_l	3
Length of the inlet/outlet section	L_i/L_o	100
Length of the cooling section	L_c	243
Length of the battery	L_b	27
Height of the inlet/outlet section	H_i/H_o	20
Height of the battery	H_b	90
Width of the inlet/outlet section	W_i/W_o	70
Width of the battery	W_b	70
Convergence plenum	CP	343
Divergence plenum	DP	343

2.1.2. Step-Like Plenum Geometry

The Step-like plenum geometry (Z-2) considered in this study was initially proposed by Oyewola et al. [37]. This geometry has further been studied by exploring some modifications to the step characteristics, such as increasing the number of steps (N_s) and varying the step height (H_s) [38,39]. Figure 2 shows a 2D schematic view of the Step-like plenum geometry and its characteristics, viz., the step number (N_s), height of the step (H_s), and length of the step (L_s). In the initial design by Oyewola et al. [37], it was reported that the step number is obtained by setting the step length to hover under each battery and CC along the length of the cooling section (L_c). Also, in order to ensure the heights of the steps are identical, the height must be an integer. The height of a step is calculated using Equation (3), while its corresponding length is calculated using Equation (4). The N_s values designed are 1, 3, 4, 7, 9, 15, and 19, denoted as Z-2-1, Z-2-3, Z-2-4, Z-2-7, Z-2-9, Z-2-15, and Z-2-19, respectively.

$$H_s = \frac{H_i}{N_s + 1} \tag{3}$$

$$L_s = \frac{L_c - CC_l}{N_s + 1} \tag{4}$$

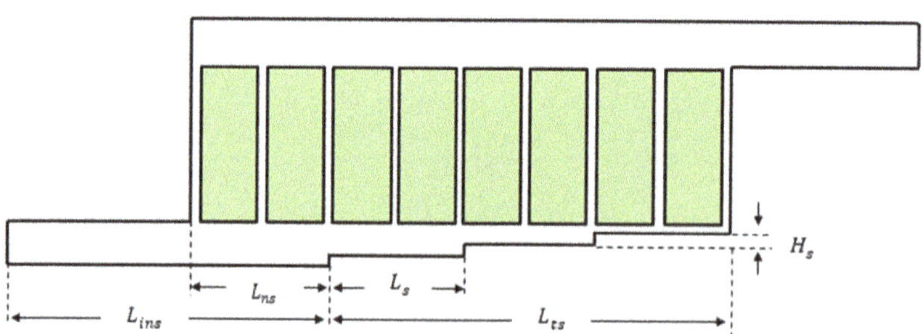

Figure 2. Two-dimensional schematic view of the Step-like plenum geometry.

2.1.3. One-Outlet Geometry

A one-outlet geometry Z-type design simply refers to a design with only one outlet section. In order to enhance the BTMSs' performances, some researchers have explored novel strategies to offset heat by introducing an extra outlet or creating vents [27,43]. For the one-outlet design, the location of the outlet section was varied, and four different locations, which have been investigated, were selected, whereby the cooling air in the outlet direction becomes perpendicular, parallel, and/or opposite to the cooling air's direction in the inlet section. These outlet directions have been investigated by researchers [17,31,44] with smooth and modified divergence plenums. The selected four outlet section directions in this study are shown in Figure 3, while their descriptions are presented in Table 3. In addition, each one-outlet geometry is installed with a Step-like design for each of the seven step numbers (1, 3, 4, 7, 9, 15, and 19).

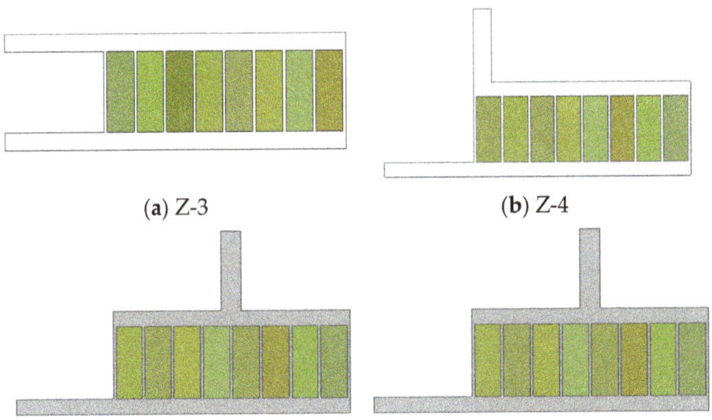

(**a**) Z-3 (**b**) Z-4

(**c**) Z-5 (**d**) Z-6

Figure 3. Schematics of one-outlet flow pattern designs.

Table 3. Descriptions of the one-outlet designs [28].

BTMS	Details
Z-3	Both the inlet and outlet sections are positioned perpendicular to CC on the same side.
Z-4	The outlet section is at the extreme edge of the plenum, perpendicular and close to the inlet section.
Z-5	The outlet section is at the midpoint of the plenum and perpendicular to the inlet section.
Z-6	The outlet section is at the extreme edge of the plenum, perpendicular to and farthest from the inlet section.

2.1.4. Two-Outlet Geometry

The two-outlet geometry is an approach with two outlet sections, which has attracted attention in the enhancement of the BTMS performance. From previous studies, the position of one outlet has been investigated on almost the entire perimeter of the geometry. However, researchers in recent times have been exploring adding one more outlet, to have two outlets, while interchanging their positions. This method has been explored, and the results have shown that it provides improvement in thermal and hydraulic performances [24,31]. Leveraging the potential of this method, this study investigated introducing two outlets to the Step-like design for each of the seven step numbers (1, 3, 4, 7, 9, 15, and 19). Table 4 presents the descriptions of the two-outlet designs, while Figure 4a–g shows the 2D geometries of the designs.

Table 4. Descriptions of the two-outlet geometries.

BTMS	Combination	Details
Z-7	Z-1/Z-3	This design basically has two outlets by adopting and combining the Z-1 and Z-3 outlet positions.
Z-8	Z-1/Z-4	Similar to the Z-7 design, this design adopts and combines the Z-1 and Z-4 outlet positions.
Z-9	Z-1/Z-5	Similarly, this design adopts and combines the Z-1 and Z-5 outlet positions.
Z-10	Z-1/Z-6	Similarly, this design adopts and combines the Z-1 and Z-6 outlet positions.
Z-11	Z-3/Z-4	This design adopts and combines the Z-3 and Z-4 outlet positions.
Z-12	Z-3/Z-5	Similarly, this design adopts and combines the Z-3 and Z-5 outlet positions.
Z-13	Z-3/Z-6	Similarly, this design adopts and combines the Z-3 and Z-6 outlet positions.

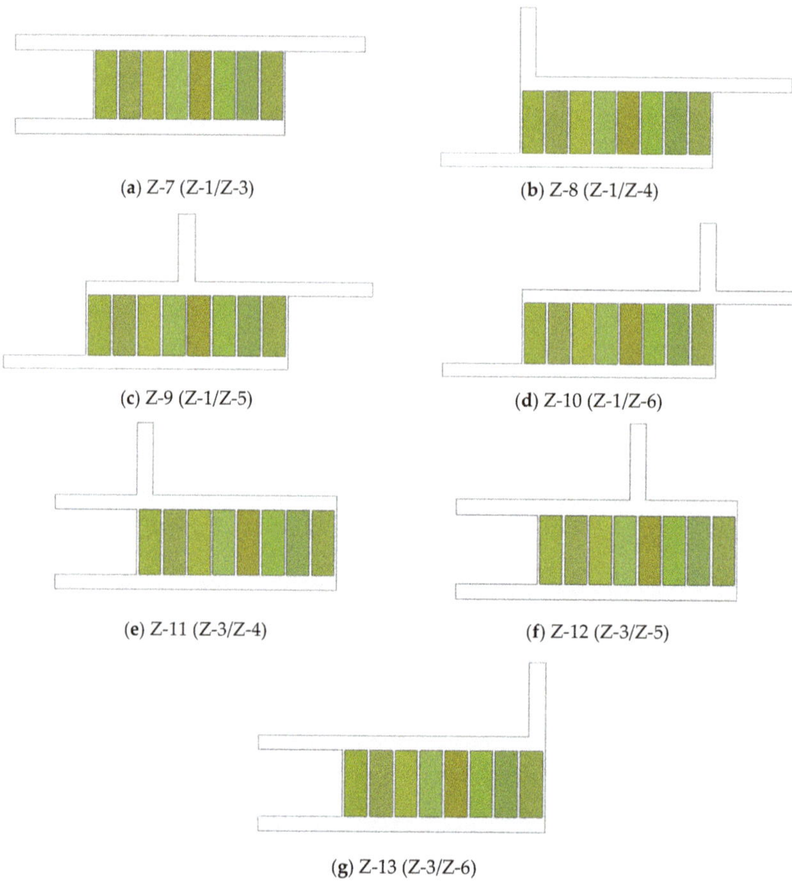

Figure 4. Schematics of two-outlet flow pattern geometric designs.

2.1.5. Combined-Strategy Geometries

The idea of combining different strategies to develop a new BTMS design is an interesting design concept with promising benefits, especially when seeking to enhance thermal performance without significant increases in the overall weights of the systems. In this category, each of the one-outlet and two-outlet geometries is combined with each of the seven Step-like plenum geometries. Figure 5a,b shows typical combined-strategy geometries with one outlet and two outlets, respectively, and four steps.

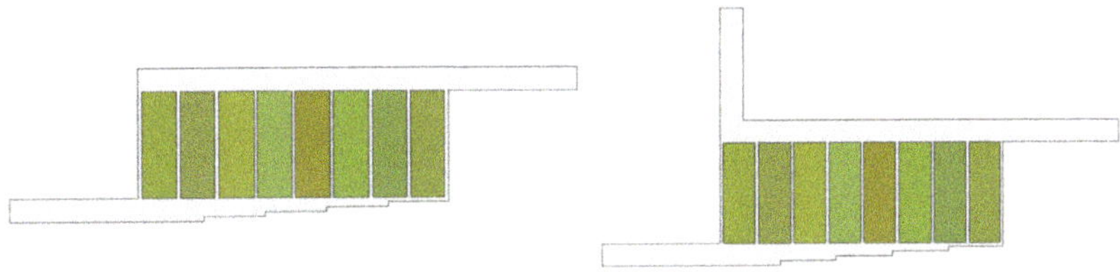

Figure 5. Selected combined-strategy geometries: (**a**) one outlet and (**b**) two outlets.

2.2. Numerical Simulations

In this study, the numerical simulations were based on the 2D geometries of the BTMSs. These simulations can be carried out for either steady-state or unsteady (transient)-state conditions, depending on the results targeted. When modeling Z-type BTMSs in a transient state, the thermal analysis will reach an equilibrium state (stable state) after a period of a continuous cooling time such that further increases in the cooling time will result in no change in the temperature of the batteries at a constant heat generation rate. This phenomenon was illustrated graphically during the transient cooling process of BTMSs in previous studies [33,45]. In cases where the numerical simulation results in the equilibrium state are of interest, a steady-state simulations is employed because a transient simulations takes a longer time to complete. An experimental study by Chen et al. [40] considered transient thermal analysis for both experimental and numerical simulation studies. Similar geometric and operational conditions utilized by Chen et al. [40] were successfully modeled in the steady state by Wang et al. [34] and Oyewola et al. [35]. Hence, all the solutions in this study were set in the steady state to minimize the computation time and cost. This study also employed governing equations, found in most commercial CFD tools, which have been adopted in modeling Z-type BTMSs [35,40], such as the continuity Equation (5) and the momentum Equation (6), for solving the flow problem, and the energy conservation Equation (7), for solving the thermal problem.

$$\nabla\cdot\vec{v} = 0 \tag{5}$$

$$\rho_a\frac{d\vec{v}}{dt} = -\nabla p + \mu_a\nabla^2\vec{v} \tag{6}$$

$$\rho_a c_{pa}\frac{\partial T_a}{\partial t} + \nabla\cdot\left(\rho_a c_{pa}\vec{v}\,\mathrm{T}_a\right) = \nabla\cdot(k_a\nabla T_a) \tag{7}$$

The Reynolds number, *Re*, is expressed using the equation:

$$Re = \frac{\rho_a V_i D_h}{\mu_a} \tag{8}$$

The hydraulic diameter is expressed using the equation:

$$D_h = \frac{2W_iH_i}{W_i + H_i} \tag{9}$$

The Reynolds number, *Re*, expressed using Equation (8), was estimated when the air's inlet velocity, V_i, was 3 m/s, which results in *Re* = 5825. Because the minimum estimated *Re* is greater than 3000, the flow is considered to be turbulent. Hence, the turbulence model (a standard $k-\varepsilon$ model) was selected, expressed as Equations (10) and (11), for the turbulent kinetic energy and kinetic energy dissipation, respectively, while employing an enhanced wall treatment [35,40].

Turbulent kinetic energy equation, k:

$$\frac{\partial}{\partial t}(\rho k) + \frac{\partial}{\partial x_i}(\rho k u_i) = \frac{\partial}{\partial x_j}\left[\left(\mu + \frac{\mu_t}{\sigma_k}\right)\frac{\partial k}{\partial x_j}\right] + G_k + G_b - \rho\varepsilon - Y_M + S_k \quad (10)$$

Turbulent energy dissipation equation, ε:

$$\frac{\partial}{\partial t}(\rho\varepsilon) + \frac{\partial}{\partial x_i}(\rho\varepsilon u_i) = \frac{\partial}{\partial x_j}\left[\left(\mu + \frac{\mu_t}{\sigma_\varepsilon}\right)\frac{\partial \varepsilon}{\partial x_j}\right] + C_{1\varepsilon}\frac{\varepsilon}{k}(G_k + C_{3\varepsilon}G_b) - C_{2\varepsilon}\rho\frac{\varepsilon^2}{k} + S_\varepsilon \quad (11)$$

The turbulent viscosity is given as follows:

$$\mu_t = \rho C_\mu \frac{k^2}{\varepsilon} \quad (12)$$

The values of the model constants used in Equations (10)–(12) are $C_{1\varepsilon} = 1.44$, $C_{2\varepsilon} = 1.92$, $C_\mu = 0.09$, $\sigma_k = 1.0$, and $\sigma_\epsilon = 1.3$.

The SIMPLE algorithm was selected for solving the governing equations. Additionally, convective and diffusive terms were discretized using central differencing and second-order terms, respectively. Convergence iteration residuals were set at 10^{-7} and 10^{-8} for flow terms and energy terms, respectively. The solution generally considered the following assumptions [33,35]:

1. The properties of the batteries and air are constant;
2. The air inlet flow is stable and incompressible;
3. The effect of the radiation is negligible;
4. The inlet air's velocity and temperature are constant.

The battery energy conservation Equation (13) [40,42] is

$$\rho_b c_{pb}\frac{\partial T}{\partial t} = \nabla\cdot(k_b \nabla T) + Q_b \quad (13)$$

In this study, the BTMSs' performances were generally evaluated by comparing the average temperature of each battery in the pack. Previous studies on Z-type BTMSs have measured the maximum temperature of the second battery (T_2), denoted as "T_{max}", while the minimum temperature was at the eighth battery (T_8), denoted as T_{min} [34,35,37,40]. The difference between T_{max} and T_{min}, also known as the maximum temperature difference of the batteries, ΔT_{max}, is expressed as in Equation (14).

The battery's maximum temperature difference is expressed as follows:

$$\Delta T_{max} = T_{max} - T_{min} \quad (14)$$

The pressure drops (ΔP) across the BTMSs were also estimated using Equation (15). It is the difference between the average pressures at the inlet and outlet(s).

$$\Delta P = P_i - P_o \quad (15)$$

The ΔP value of the BTMS is associated with the pumping power, which, consequently, may result in increasing the pumping cost.

2.3. Boundary Conditions

For the inlet section, the air velocity (3 m/s) and temperature (299.15 K) boundary conditions were initially assigned to be constant [34,37]. Additional solutions were carried out at inlet velocities of 3.5 m/s and 4 m/s, while the other settings remained the same. In general, the following assumptions were considered for the simulations [34,35,37]:

1. The heat generation rate assigned to the battery's surface was constant;
2. The zero (0)-gauge outlet pressure condition was assigned;
3. On the battery walls and inner walls of the systems, the no-slip condition was assigned;
4. An adiabatic wall was assigned to the entire surrounding wall of the systems;
5. The surrounding temperature and pressure were constant;
6. The contacting surfaces were set as coupled.

2.4. Meshing and Independence Study

All the geometries considered in this study were meshed on the ANSYS meshing platform. When setting up the meshing, five inflation layers were assigned at the boundary walls of the batteries, air domain, and system, while a height of 0.1 mm was set for the first layer (y^+) [33,34,37].

The mesh size can be improved by varying the element size and selecting several other options available in ANSYS to produce coarse or fine meshes, which can significantly influence the result obtained. However, by taking into consideration the available computation resources and time availability, a mesh independence study was carried out. This ensures that the solution is accurate at the minimum computational time possible. This targets the minimum possible number of elements from a meshed design, which results will have a negligible difference when results with a high number of elements are used [33,34,37]. Meshes were compared for seven element sizes ranging from 0.00065 to 0.00035 at 0.00005 intervals, and the corresponding numbers of elements for these element sizes are 121,383; 140,625; 166,368; 198,584; 244,027; 308,204; and 400,154, respectively. The independence study performances are further presented graphically in Figure 6a,b. As the number of elements increases from 308,204 to 400,154, the variation in the parameters was negligible, with differences of 0.02 K, 0.01 K, 0.03 K, and 0 Pa for T_{max}, T_{min}, ΔT_{max}, and ΔP, respectively. Therefore, the mesh setup that produced a mesh size of 400,154 was used for simulating other models.

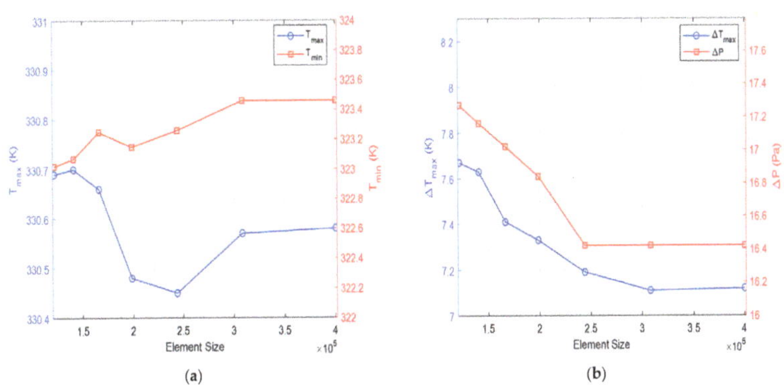

Figure 6. Mesh independence tests: (**a**) T_{max} and T_{min} and (**b**) ΔT_{max} and ΔP.

2.5. CFD Method Validation

The results obtained from the simulations were compared with the experimental results, which has also been performed in previous studies for the original Z-type BTMS [28,33,34,37], by comparing the simulation results with the experimental results of Chen et al. [40]. The primary parameters used for the comparison are the average battery temperatures measured for the air's inlet velocities of 3, 3.5, and 4 m/s. In this study, the average battery temperatures were measured, which also show very good agreement with the experimental results of Chen et al. [40]. Figure 7a,b presents the comparison for the maximum (T_{max}) and minimum (T_{min}) values, which were also measured for the second and eighth batteries, respectively. Figure 8a–d illustrates the 2D physical geometry, selected mesh, temperature contour, and air-velocity contour of the reference Z-type BTMS. From the comparison, maximum differences of 1.6 K (0.5%) and 0.65 K (0.2%) were obtained for T_{max} and T_{min}, respectively. Zhang et al. [46], reported that differences of up to 2% can be considered as negligible. Interestingly, the estimated values obtained in this study are below 2%. Therefore, the CFD method is considered as acceptable.

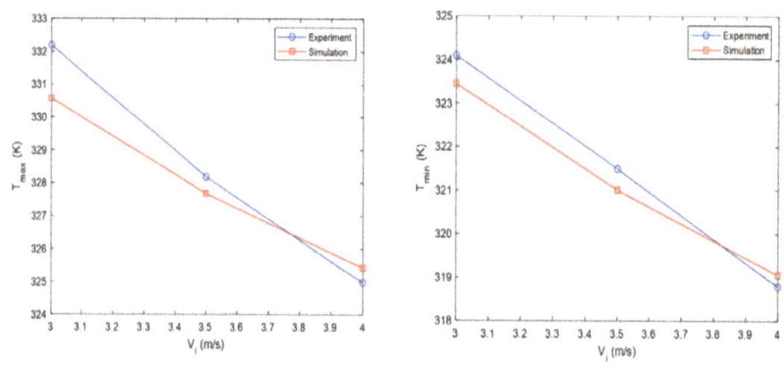

(a) (b)

Figure 7. Comparison between experiments [40] and the CFD method (the current study) for (a) T_{max} and (b) T_{min}.

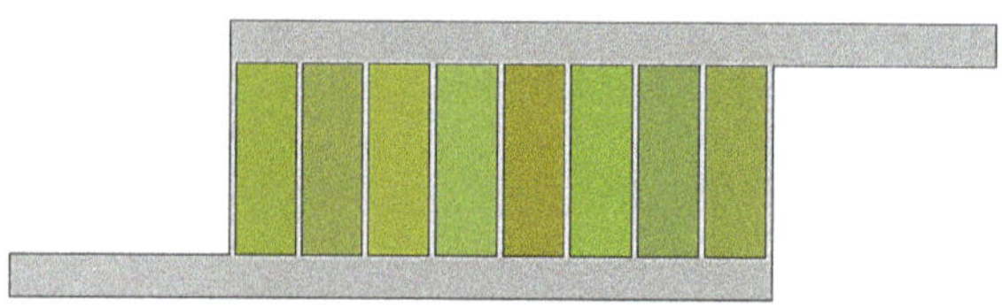

(a)

(b)

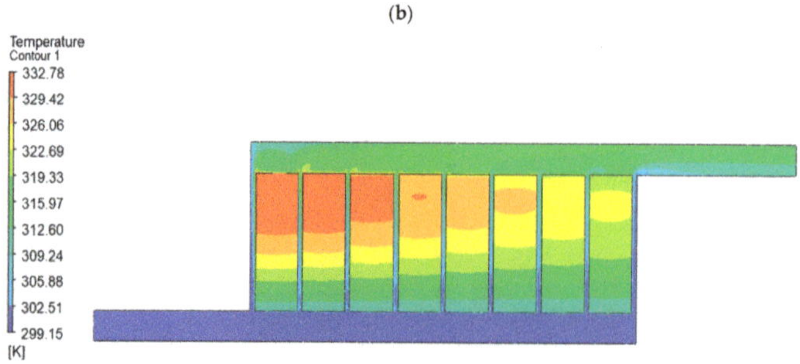

(c)

Figure 8. *Cont.*

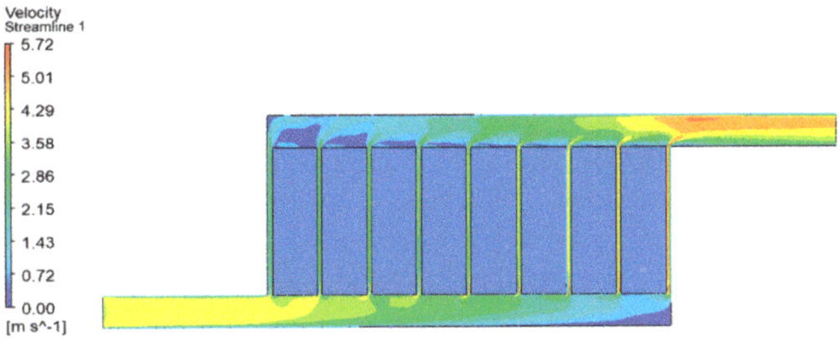

(d)

Figure 8. The reference Z-type BTMS's (**a**) 2D geometry, (**b**) selected mesh, (**c**) temperature contour, and (**d**) air-velocity contour.

3. Results and Discussion

In the development of BTMSs where air is the cooling fluid, it is important to minimize ΔP while maintaining the battery temperature or possibly minimizing the temperature. ΔP minimization is necessary because it corresponds to a reduction in the pumping cost. When focusing on minimizing the battery temperature, it is also important to ensure that it does not go below the recommended temperature. Hence, ΔT_{max} among the batteries is evaluated, which compares the thermal uniformity or homogeneity of the batteries in a BTMS. Based on these criteria, the designs will be compared based on T_{max}, ΔT_{max} , and ΔP. In general, a total of 95 geometries were developed, and the simulation results for each geometry were obtained at the air's inlet velocity and a temperature of 3 m/s and 299.15 K, respectively.

3.1. *Evaluation of T_{max} Performances*

One of the parameters considered for the thermal performances of BTMSs is T_{max}. In order to avoid the overheating of batteries and thermal runaway, it is imperative to design a BTMS such that T_{max} is minimized and maintained below the maximum recommended working temperature. In this section, the T_{max} values of the different designs are compared.

3.1.1. T_{max} Performances of Step-like Plenum, One-Outlet, and Two-Outlet Geometries

In this subsection, the T_{max} values for the Step-like plenum, one-outlet, and two-outlet designs are compared. Afterward, the best-performing designs (with the minimum T_{max} values) from each group are also compared. Figure 9a–c shows the variations in T_{max} for the Step-like plenum, one-outlet, and two-outlet designs, respectively, while Figure 9d shows the variation in T_{max} among the best-performing designs in the three aforementioned groups. Furthermore, Figure 10a–d shows the temperature contours of the Z-1, best Step-like plenum, best one-outlet, and best two-outlet designs, respectively. The contours show the temperature variation of the cooling air and the batteries in the systems. For instance, in Figure 10a, the inlet section is dominated by cold air, and in its legend, the "blue" color corresponds to the lower-temperature section. However, the most important temperature distributions are those observed among the batteries. Based on the legend's color, is it obvious that the highest temperature in the system was recorded for the batteries, and this was observed for only four batteries. This performance implies that there will be a significant difference in temperature between the first battery and the eighth battery, a phenomenon that indicates poor thermal homogeneity and excess heat accumulation among the four batteries. In order to better understand why Figure 10a shows this poor performance, the distribution of the temperature on the BTMS's contour in Figure 10b is described. In this case, the temperature spreads across the eight batteries, indicating very good thermal homogeneity and the eventual reduction in the maximum temperature, as shown in the two legends. Similar performance distribution trends are used for other designs, and by comparing the contours in Figure 10 with the results in Figure 9d, it can be seen that the contours provide good insight into the performance of this system. First, in the Step-like plenum category in Figure 9a, where seven different numbers of steps (1, 3, 4, 7, 9, 15, and 19) were investigated, the seven-step design denoted as "Z-2-7" produced the best $T_{max} = 327.13\ K$, with a reduction by 3.46 K compared with that of the original Z-type design, denoted as "Z-1", in this study. Designs Z-2-3 (three steps) and Z-2-4 (four steps) also produced good performances, both with reductions in T_{max} by 3.30 K. The poorest of the Step-like plenum designs is Z-2-19 (nineteen steps), which happens to be the only Step-like plenum design that increased T_{max}, with an increase of 0.29 K compared to that of the original Z-type (Z-1) design. For a step design, any modification to another part of the structure, aside from the step, will affect the performance of the BTMS for the same number of steps. Previous studies on the step design have revealed that increasing the number of steps serially does not always result in corresponding increases or decreases in the temperature or pressure drop. However, higher numbers of steps, e.g., 15 and 19, have shown poor performances, which has been reported in previous studies [37,39]. An explanation for this is the fact that the concept of the step design for one and two steps produced poor thermal homogeneity, while for fifteen and nineteen steps, the plenum's area decreases; hence, the air becomes insufficient to cool down the batteries at that end of the plenum. Figure 9b presents the comparison between the one-outlet designs. From the figure, it can be seen that design Z-5 performed better than the other designs, with a reduction in T_{max} by 1.9 K compared with that of the Z-1 design.

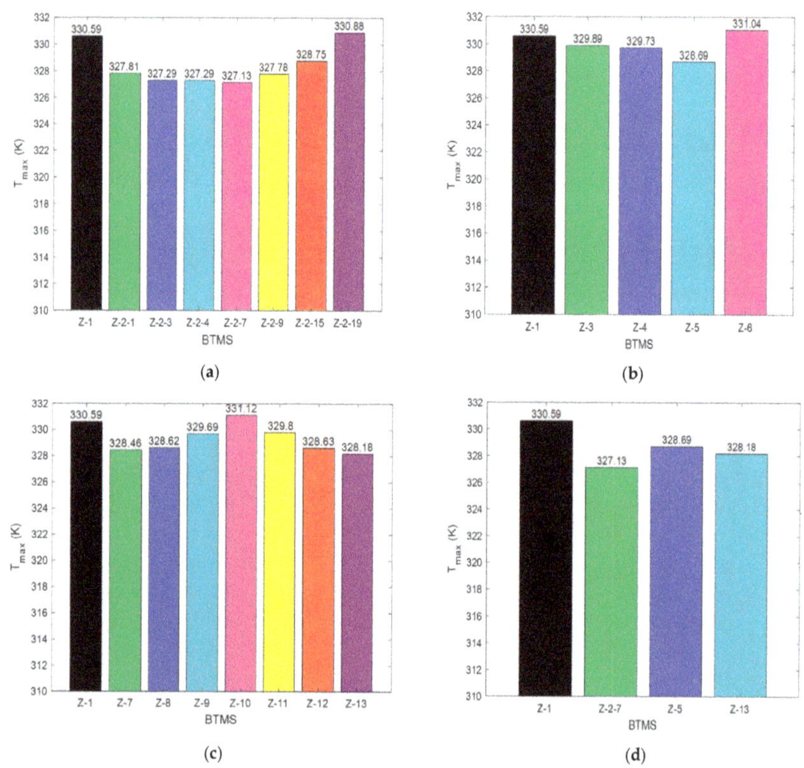

Figure 9. T_{max} performances of BTMSs: (**a**) Step-like plenum, (**b**) one outlet, (**c**) two outlets, and (**d**) the best in groups (**a**–**c**).

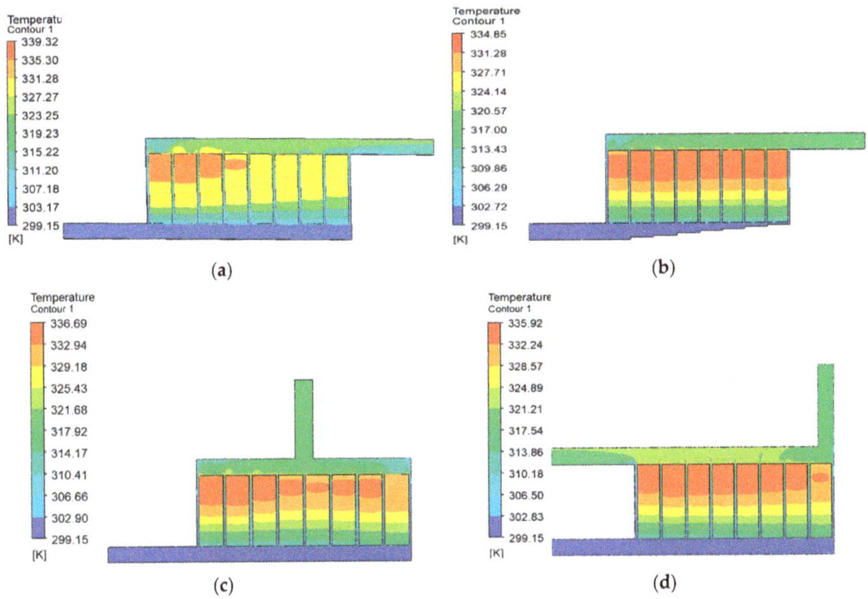

Figure 10. Temperature contours of (**a**) Z-1, (**b**) Z-2-7, (**c**) Z-5, and (**d**) Z-13 BTMSs.

Furthermore, compared to the Z-1 design, the Z-3 and Z-4 designs also produced better performances, with reductions in T_{max} by 0.7 K and 0.86 K, respectively, while design Z-6 produced a poor performance, with an increase in T_{max} by 0.45 K. Figure 9c presents the comparison of the T_{max} performances for the two-outlet designs. Out of the seven designs considered in this category, only Z-10 (Z-1/Z-6) failed to produce a better performance than that of the Z-1 design, increasing T_{max} by 0.53 K. The other six designs produced reductions in T_{max}, with the Z-13 (Z-3/Z-6) design having the best performance, with a reduction in T_{max} by 2.41 K compared to that of the Z-1 design. The best-performing designs from the three categories were compared and are presented graphically in Figure 9d. From the figure, it can be seen that Z-2-7 produced the best T_{max} performance. Interestingly, in the Z-13 design, a combination of two different outlets (Z-3/Z-6) performed better than the Z-5 design, which happens to be the best-performing design among the one-outlet designs, with a reduction in T_{max} by 0.51 K.

3.1.2. T_{max} Performances of the Combined-Strategy Geometries

The combined-strategy geometries explore the combination of Step-like plenum designs for the considered numbers of steps with the one-outlet designs and two-outlet designs. This implies that each of the four one-outlet designs was investigated when 1, 3, 4, 7, 9, 15, and 19 steps were installed, summing up to twenty-eight designs. Similarly, each of the seven two-outlet designs was investigated for the seven different numbers of steps, summing up to forty-nine designs. These designs were compared, and the best five designs in the two combination groups were selected and are presented in Table 5. Figure 11a,b shows the temperature distribution of the batteries for the combined-strategy geometries with one outlet and two outlets, respectively. The comparison of T_{max} values for the best five performing designs for one outlet and two outlets is presented in Figure 12a,b, respectively. Lastly, the temperature contours of the optimal designs in the one-outlet and two-outlet groups are presented in Figure 13a,b. In Figure 11a,b, the reference Z-type design, denoted as Z-1, is compared with the designs in the two combination groups. The temperature distribution reveals the temperature uniformity of the best-performing design when compared with those of the reference Z-1 designs.

Table 5. The best five designs of the combined-strategy geometries in terms of T_{max}.

S/N	Step-like Plenum and One-Outlet Combination			Step-like Plenum and Two-Outlet Combination		
	Step	Outlet	T_{max} (K)	Step	Outlet	T_{max} (K)
1.	Z-2-7	Z-6	327.07	Z-2-7	Z-1/Z-6	327.14
2.	Z-2-3	Z-6	327.14	Z-2-3	Z-1/Z-6	327.26
3.	Z-2-4	Z-6	327.63	Z-2-4	Z-1/Z-6	327.59
4.	Z-2-9	Z-6	328.07	Z-2-9	Z-1/Z-6	327.88
5.	Z-2-1	Z-5	328.69	Z-2-7	Z-1/Z-3	328.21

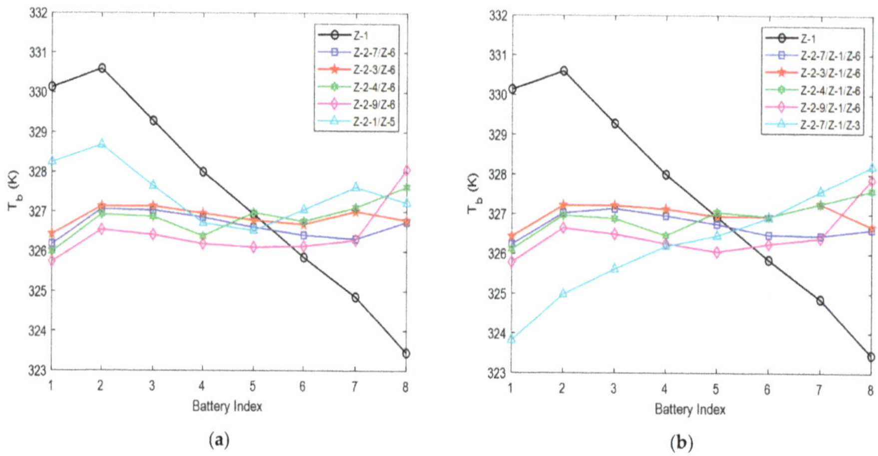

Figure 11. T_b values of combined-strategy designs with (**a**) one outlet and (**b**) two outlets.

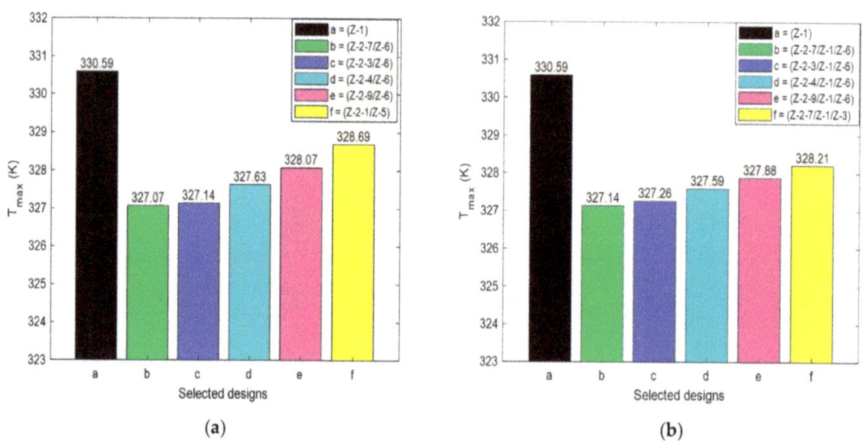

Figure 12. T_{max} values of the batteries for the best five designs with (**a**) one outlet and (**b**) two outlets.

Figure 12a,b further presents the comparison between the designs in terms of their T_{max} values. Comparing the one-outlet designs, shown in Figure 12a, the best-performing design is Z-2-7/Z-6, with a minimum T_{max} value of 327.07 K. When compared with the Z-1 designs, it produced a reduction in T_{max} by 3.52 K. Considering the two-outlet designs, presented in Figure 12b, Z-2-7/Z-1/Z-6 produced a minimum T_{max} value of 327.14 K, and when compared with the reference Z-1 designs, it produced a reduction in T_{max} by 3.45 K. These performances show that both one outlet and two outlets can be combined with a Step-like plenum design to achieve a good reduction in T_{max}. Another intriguing performance shows that Z-2-7/Z-6 also performed better than the initial best Step-like plenum design, which was earlier reported to have performed better than the reference Z-type design without steps. The Z-2-7/Z-6 design reduced the T_{max} of the best reference Step-like plenum design by 0.06 K.

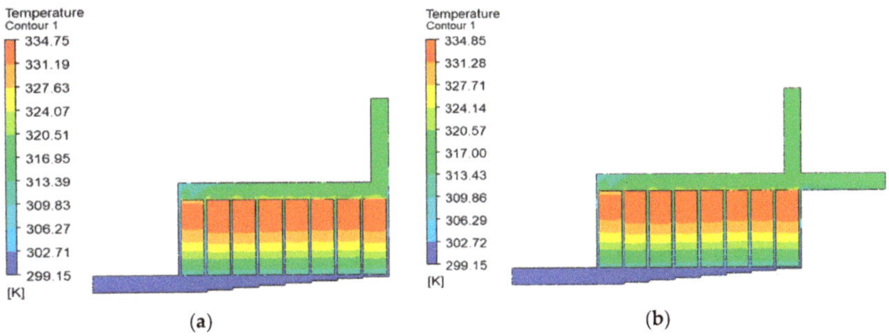

Figure 13. Temperature contours of T_{max} for the best-performing designs with (**a**) one outlet and (**b**) two outlets.

3.2. Evaluation of ΔT_{max} Performances

ΔT_{max} is also an important parameter used in evaluating the performance of a BTMS with more than one battery in a pack. This parameter shows the ability of a design to uniformly distribute heat generation in the battery pack, thereby avoiding the overheating of one battery, while the other batteries are at a significantly lower temperature. Such a phenomenon is often regarded as thermal homogeneity. Previous studies in the literature have shown that designs with low ΔT_{max} values produce thermal performances with low T_{max} values. However, one design can produce a lower ΔT_{max} value compared to that of another design under the same operating conditions while still having a higher T_{max} value, although the difference between the ΔT_{max} values may not be significant [28,34,35,44,47]. A better ΔT_{max} value is, however, very important in designs where the cooling air enters the system at extremely low temperatures because any small irregularity of a battery can affect the chemical reaction through a sharp decline in its operating temperature.

3.2.1. ΔT_{max} Performances of Step-like Plenum, One-Outlet, and Two-Outlet Geometries

This subsection focuses on the comparison of the ΔT_{max} values for the Step-like plenum, one-outlet, and two-outlet designs. The optimal designs (with the minimum ΔT_{max} values) in all the design groups are then considered for further comparison. Variations in ΔT_{max} for the Step-like plenum, one outlet, and two outlets, are shown in Figure 14a–c, respectively, while Figure 14d presents the variation of ΔT_{max} between the best-performing designs in the earlier-mentioned groups. In addition, the temperature contours of the Z-1, best Step-like plenum, best one-outlet, and best two-outlet designs are shown in Figure 15a–d, respectively.

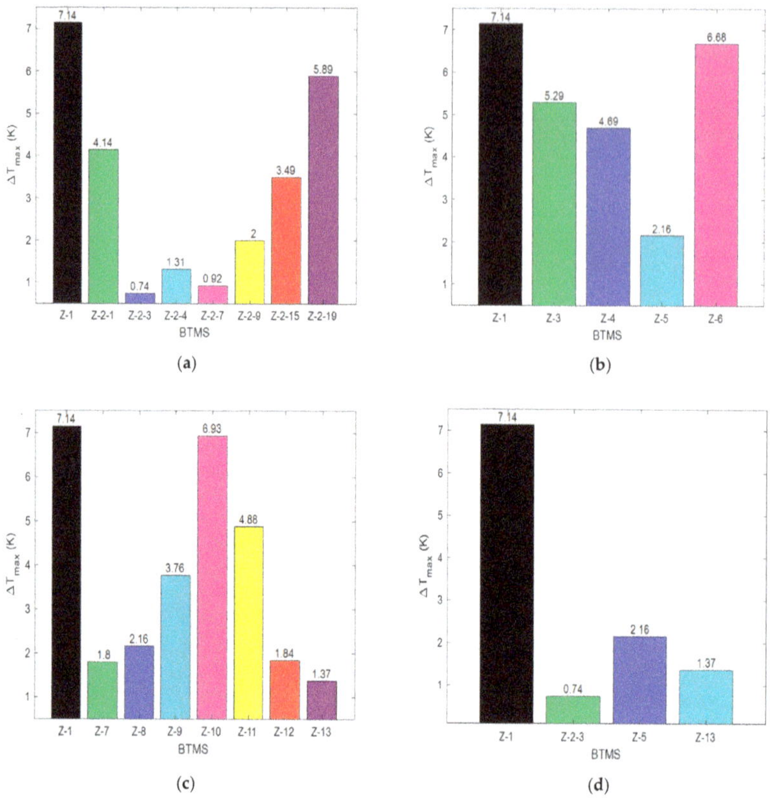

Figure 14. ΔT_{max} performances of BTMSs: (**a**) Step-like plenum, (**b**) one outlet, (**c**) two outlets, and (**d**) the best in groups (**a**–**c**).

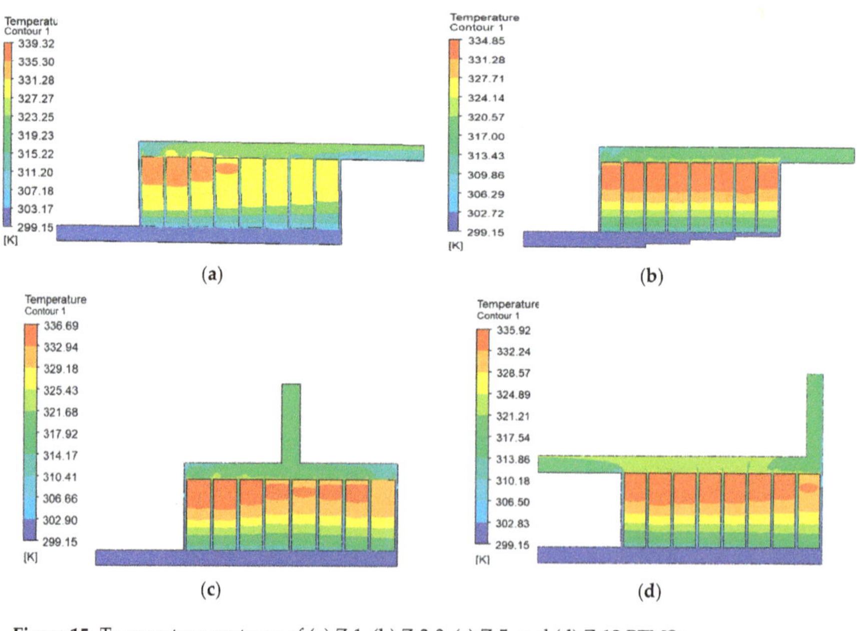

Figure 15. Temperature contours of (**a**) Z-1, (**b**) Z-2-3, (**c**) Z-5, and (**d**) Z-13 BTMSs.

For the Step-like plenum designs in Figure 14a, the three-step design, denoted as "Z-2-3", produced the best ΔT_{max} performance, with a recorded value of 0.74 K. When compared with the ΔT_{max} value of the Z-1 design, the ΔT_{max} value of Z-2-3 reduced by 6.4 K. In general, as shown in Figure 14a, every Step-like plenum design produced a good ΔT_{max} performance, with Z-2-4 (four steps) and Z-2-7 (seven steps) producing reductions by 5.84 K and 6.22 K, respectively, compared with that of the Z-1 design. Figure 14b shows the performances of the one-outlet designs. Design Z-5 has the minimum ΔT_{max} value, recorded at 2.16 K, and compared with the ΔT_{max} value of the reference Z-1 design, produced a reduction by 4.98 K. The remaining one-outlet flow designs also produced ΔT_{max} reductions by 2.45 K, 1.85 K, and 0.46 K for Z-3, Z-4, and Z-6, respectively. Figure 14c shows the performance comparison of the two-outlet designs. The results revealed that design Z-10 produced the poorest performance, while design Z-13 produced the best, with ΔT_{max} values of 6.93 K and 1.37 K, respectively, and reductions by 0.21 K and 5.77 K, respectively, compared with that of the reference Z-1 design. The best-performing designs selected from Figure 14a–c were also compared and are presented in Figure 14d. The results revealed that design Z-2-3 has the best performance, with a ΔT_{max} value of 0.74 K, and when compared with the performances of the Z-5 and Z-13 designs, produced reductions by 1.42 K and 0.63 K, respectively.

3.2.2. ΔT_{max} Performances of the Combined-Strategy Geometries

The ΔT_{max} performances in this section also consider the same geometries for comparison as those in the case of T_{max}. The designs were compared, and the best five in the two groups of combinations are presented in Table 6. Figure 16a,b shows the comparison of the best-performing designs in terms of ΔT_{max} for the one-outlet and two-outlet designs, respectively. Additionally, the temperature contours of the best-performing designs in the one-outlet and two-outlet groups are presented in Figure 17a,b. Figure 16a,b presents the comparison of the reference Z-1 designs with the designs in the two combination groups.

Table 6. The best five designs of combined-strategy geometries in terms of ΔT_{max}.

S/N	Step-like Plenum and One-Outlet Combination			Step-like Plenum and Two-Outlet Combination		
	Step	Outlet	ΔT_{max} (K)	Step	Outlet	ΔT_{max} (K)
1.	Z-2-3	Z-6	0.70	Z-2-3	Z-1/Z-6	0.82
2.	Z-2-7	Z-6	0.88	Z-2-7	Z-1/Z-6	0.90
3.	Z-2-4	Z-6	1.63	Z-2-4	Z-1/Z-6	1.47
4.	Z-2-1	Z-5	2.16	Z-2-1	Z-1/Z-5	1.94
5.	Z-2-9	Z-6	2.32	Z-2-9	Z-1/Z-6	2.09

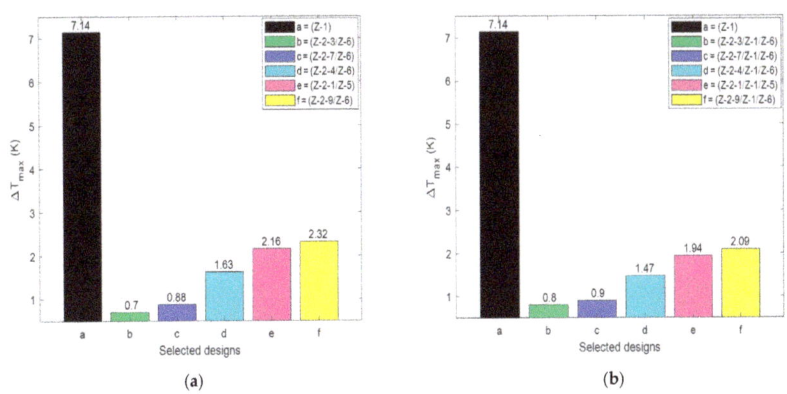

Figure 16. ΔT_{max} values of the batteries for the best five designs with (**a**) one outlet and (**b**) two outlets.

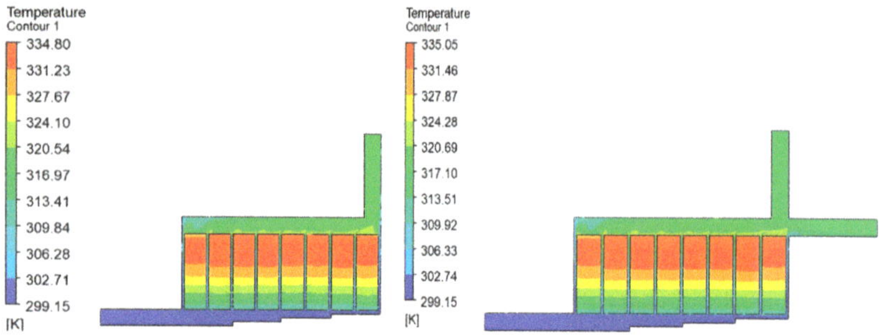

(**a**) (**b**)

Figure 17. Temperature contours of ΔT_{max} for the best-performing designs with (**a**) one outlet and (**b**) two outlets.

Comparing the one-outlet designs presented in Figure 16a, design Z-2-3/Z-6 produced the best performance, with a ΔT_{max} value of 0.7 K, while the poorest was produced by design Z-2-9/Z-6, with a ΔT_{max} value of 2.32 K. Both the best- and poorest-performing designs performed significantly better than the reference Z-1 design, with reductions by 6.44 K and 4.82 K, respectively. In Figure 16b, the performances of the best five two-outlet designs show that these five designs also performed significantly better than the reference Z-1 design. Furthermore, the Z-2-3/Z-1/Z-6 design produced the best performance, while design Z-2-9/Z-1/Z-6 produced the poorest. Additionally, it was observed that the ΔT_{max} values were reduced by 6.34 K and 5.05 K, when compared with those of Z-2-3/Z-1/Z-6 and Z-2-9/Z-1/Z-6, respectively.

3.3. Evaluation of ΔP Performances

In air-cooled BTMSs, especially where external devices are used to pump the cooling air, it is necessary to take into consideration the pumping cost, which is related to the pressure drop. Previous studies have shown that the structural redesign of BTMSs and increases in the cooling air's velocity can minimize both T_{max} and ΔT_{max}. However, this comes with the drawback of an increase in ΔP. With continuous research in the redesign of BTMSs, both T_{max} and ΔT_{max} are being enhanced, with minimal ΔP increases and without the need to increase the air's velocity.

3.3.1. ΔP Performances of the Step-like Plenum, One-Outlet, and Two-Outlet Geometries

This section considered the ΔP performances of the designs with Step-like plenums, one outlet, and two outlets for comparison. The best-performing designs (with the minimum ΔP values) in each group were then compared. Figure 18a–c presents the variations in ΔP for the Step-like plenum, one-outlet, and two-outlet designs, respectively, while Figure 18d shows the variation in ΔP between the best-performing designs in each of the three groups. Lastly, Figure 19a–d shows the pressure contours of the Z-1, best Step-like plenum, best one-outlet, and best two-outlet designs, respectively.

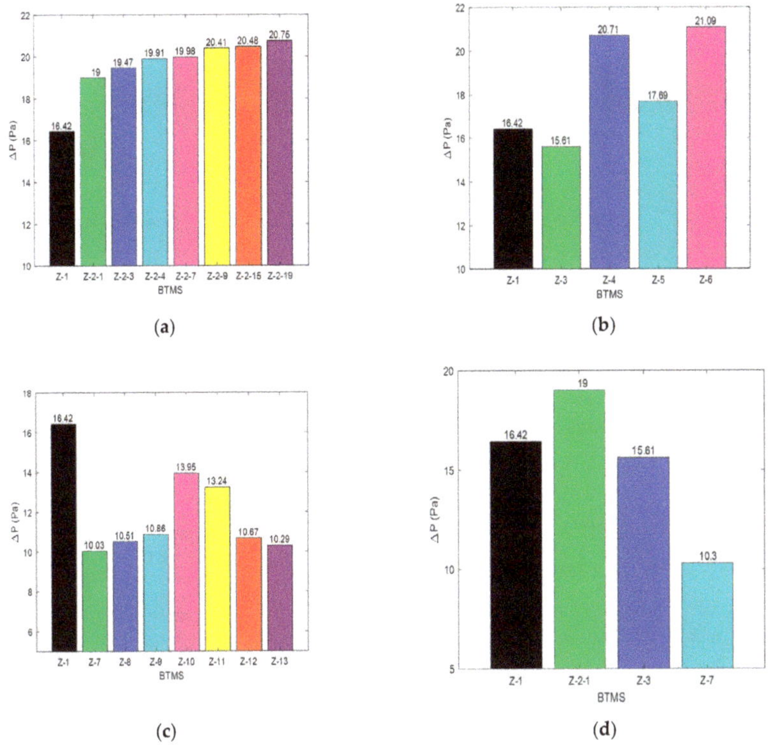

Figure 18. ΔP performances of BTMSs: (**a**) Step-like plenum, (**b**) one outlet, (**c**) two outlets, and (**d**) the best in groups (**a**–**c**).

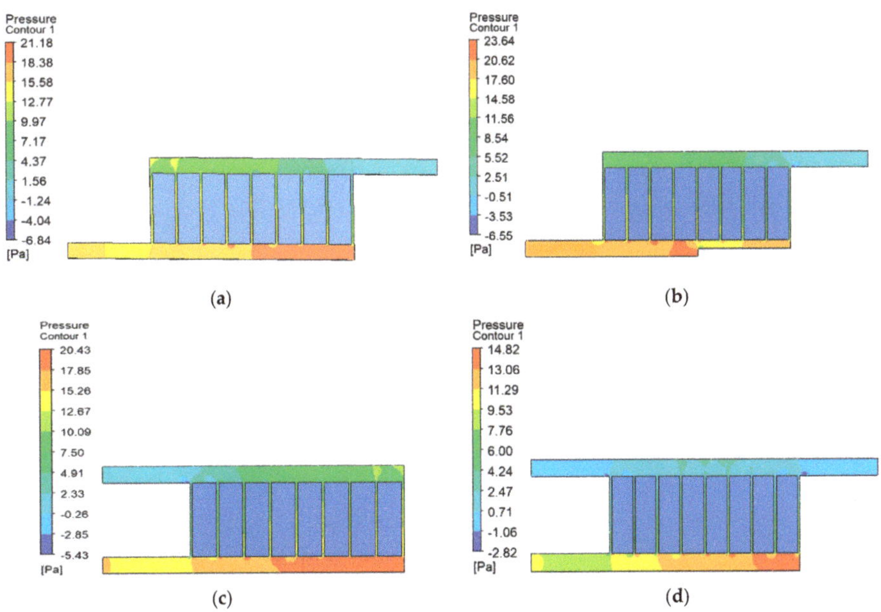

Figure 19. Temperature contours of (**a**) Z-1, (**b**) Z-2-1, (**c**) Z-3, and (**d**) Z-7 BTMSs.

The Step-like plenum designs are first considered and presented in Figure 18a. From the figure, it can be seen that the Z-1 design performed better than all the other designs, while the Z-2-1 design produced the minimum ΔP value among the Step-like plenum designs. The ΔP performances of the Step-like plenum designs also show increases in ΔP with increases in the number of steps. Z-2-1, being the best-performing design, produced an increase in ΔP by 2.58 Pa, when compared with that of the Z-1 design. The poorest performance was observed for design Z-2-19, which produced an increase in ΔP by 4.33 Pa. Next, are the performances of the one-outlet designs presented in Figure 18b. The results reveal that design Z-3 produced the best performance, with a reduction in ΔP by 0.81 Pa when compared with that of the reference Z-1 design. However, the remaining one-outlet designs did not perform better than the Z-1 design, recording increases in ΔP by 4.29 Pa, 1.27 Pa, and 4.67 Pa for Z-4, Z-5, and Z-6, respectively. The performances of the two-outlet designs were also considered and are presented in Figure 18c. First, the performances show that all the two-outlet designs produced lower ΔP values compared with that of the reference Z-1 design. Second, Z-7 produced the minimum value, and compared with the Z-1 design, yielded a reduction in ΔP by 6.39 Pa. Figure 18d further presents the performances of the best-performing designs in each of the three groups, and design Z-7 clearly produced the best performance.

3.3.2. ΔP Performances of the Combined-Strategy Geometries

This section explores the performances of the combined-strategy designs with respect to ΔP. The designs were compared, and the best five in the two groups of combinations are presented in Table 7. Figure 20a,b shows the comparison of the best-performing designs in terms of ΔP for one outlet and two outlets, respectively. Additionally, the pressure contours of the best-performing designs with one outlet and two outlets are presented in Figure 21a,b Figure 20a presents the comparison among the one-outlet designs. From the figure, the Z-1 design performed better than all the best five one-outlet designs. When focusing on the one-outlet combinations only, design Z-2-1/Z-3 produced the best performance, with a ΔP value of 17.13 Pa, while the poorest-performing design, Z-2-4/Z-3, produced a ΔP value of 18.71 Pa. Figure 20b further presents the performances of the two-outlet designs. From the figure, all the five best-performing designs performed better than the Z-1 design. Among the two-outlet designs, Z-2-1/Z-1/Z-4 and Z-2-1/Z-1/Z-5 produced the best and poorest performances, respectively, reducing ΔP by 5.91 Pa and 3.66 Pa, respectively, when compared with the ΔP value of the reference Z-1 design.

Table 7. The best five designs of the combined-strategy geometries in terms of ΔP.

S/N	Step-like Plenum and One-Outlet Combination			Step-like Plenum and Two-Outlet Combination		
	Step	Outlet	ΔP (Pa)	Step	Outlet	ΔP (Pa)
1.	Z-2-1	Z-3	17.13	Z-2-1	Z-1/Z-4	10.51
2.	Z-2-1	Z-5	17.69	Z-2-1	Z-1/Z-3	11.67
3.	Z-2-3	Z-3	18.27	Z-2-1	Z-3/Z-6	11.97
4.	Z-2-7	Z-3	18.68	Z-2-1	Z-3/Z-5	12.03
5.	Z-2-4	Z-3	18.71	Z-2-1	Z-1/Z-5	12.76

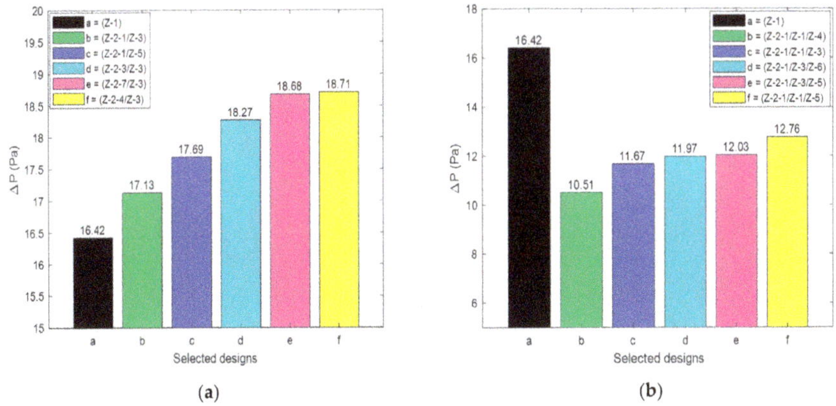

Figure 20. ΔP values of the batteries for the best five designs with (**a**) one outlet and (**b**) two outlets.

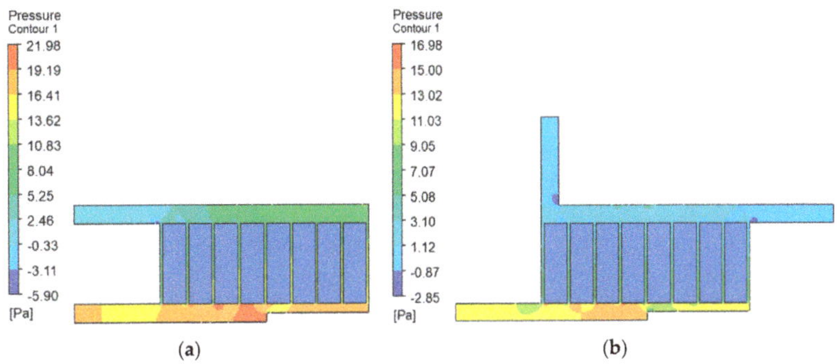

Figure 21. Pressure contours of ΔP for the best-performing designs with (**a**) one outlet and (**b**) two outlets.

4. Conclusions

The redesign of the Z-type BTMS has brought about a drop in T_{max}, though with a significant increase in ΔP, hence requiring studies to focus on means of decreasing the pressure drop by exploring new designs. In this study, the Step-like plenum design was further investigated by observing the effects of two outlets on the performances of the systems with the aim of reducing or maintaining T_{max} and minimizing ΔP. The CFD approach was used to compute the performances of the designs. The designs explore combinations of Step-like plenum designs having different numbers of steps (N_s) = 1, 3, 4, 7, 9, 15, and 19 with one and two air outlets. From the results obtained, the following conclusions were drawn:

1. The initial Step-like plenum design with seven steps (Z-2-7) produced the best T_{max} (327.13 K) among the Step-like plenum categories of the reference Z-type BTMSs, with a reduction by 3.46 K compared to that of the original Z-type design without any steps installed. For the one-outlet designs, the best-performing design is Z-2-7/Z-6, with a minimum T_{max} value of 327.07 K, and compared with the original Z-type designs, it produced a reduction in T_{max} by 3.52 K. For the two-outlet designs, design Z-2-7/Z-1/Z-6 produced a minimum T_{max} value of 327.14 K, and when compared to the original Z-1 designs, it produced a reduction in T_{max} by 3.45 K. The Z-2-7/Z-6 design reduced the T_{max} value of the best Step-like plenum design by 0.06 K;
2. By comparing the Step-like plenum, one-outlet, and two-outlet designs, the Z-2-3 design has the best performance, with a ΔT_{max} value of 0.74 K, and compared with

the Z-5 and Z-13 designs, it produced reductions in ΔT_{max} by 1.42 K and 0.63 K, respectively. Considering the best five two-outlet designs, with respect to the ΔT_{max} performance, significant performance improvements were observed when compared with that of the reference Z-type design. The Z-2-3/Z-1/Z-6 design produced the best performance, while design Z-2-9/Z-1/Z-6 produced the poorest. It was observed that the ΔT_{max} value reduced by 6.34 K and 5.05 K when compared with those of Z-2-3/Z-1/Z-6 and Z-2-9/Z-1/Z-6, respectively;

3. For the designs in the three groups, Z-7 produced the minimum value, and compared with the Z-1 design, it yielded a reduction in ΔP by 6.39 Pa. In terms of ΔP, the Step-like plenum and one-outlet combination designs produced good performances. The Z-2-1/Z-3 design produced the best performance, with a ΔP value of 17.13 Pa, while the poorest-performing design, Z-2-4/Z-3, produced a ΔP value of 18.71 Pa. Among the best five two-outlet designs with and without a Step-like plenum structure, Z-2-1/Z-1/Z-4 and Z-2-1/Z-1/Z-5 produced the best and poorest performances, reducing ΔP by 5.91 Pa and 3.66 Pa, respectively.

Author Contributions: Conceptualization, O.M.O.; Methodology, O.M.O., E.T.I., M.J.L., M.C.H. and M.L.D.; Software, O.M.O., E.T.I., M.J.L., M.C.H. and M.L.D.; Validation, O.M.O., E.T.I., M.J.L., M.C.H. and M.L.D.; Formal analysis, O.M.O., E.T.I., M.J.L., M.C.H. and M.L.D.; Investigation, O.M.O., E.T.I., M.J.L., M.C.H. and M.L.D.; Resources, O.M.O.; Data curation, O.M.O.; Writing—original draft, O.M.O., E.T.I., M.J.L., M.C.H. and M.L.D.; Writing—review & editing, O.M.O., E.T.I., M.J.L., M.C.H. and M.L.D.; Visualization, O.M.O.; Supervision, O.M.O.; Project administration, O.M.O. All authors have read and agreed to the published version of the manuscript.

Funding: This research received no external funding.

Data Availability Statement: The original contributions presented in the study are included in the article, further inquiries can be directed to the corresponding author.

Conflicts of Interest: The authors declare no conflict of interest.

Nomenclature

Acronym		**Unit**
SOC	State of charge	-
TMS	Thermal management system	-
BTMS	Battery thermal management system	-
CFD	Computational fluid dynamics	-
HTMS	Hybrid thermal management system	-
PCM	Phase-change material	-
EV	Electric vehicle	-
Symbol(s)		
L_s	Step length	mm
H_s	Step height	mm
L_{ns}	Non-stepped length	mm
L_{ts}	Stepped length	mm
L_{ins}	Initial non-stepped length	mm
P_i and P_o	Average pressures of cooling air at the inlet and outlet(s)	Pa
T_{max}	The maximum temperature of the batteries	K
T_{min}	The minimum temperature of the batteries	K
ΔT_{max}	The maximum temperature difference of the batteries	K
ΔP	Pressure drop	Pa

A_i	Inlet sections' cross-sectional area	m^2
$\dot{V}$	Volume flow rate of the air	m^3/s
u_i	Velocity vector's ith component	-
x, y, and z	x-, y-, and z-coordinates, respectively	-
Re	Reynolds number	-
T_b	Temperature of the battery	K
I	Discharge current	A
R_b	Battery's equivalent resistance	Ω
Q_b	Battery's heat generation rate	W/m^3
V_b	Volume of the battery	m^3
k_a	Thermal conductivity (air)	W/m.K
k_b	Thermal conductivity (battery)	W/m.K
c_a	Specific heat capacity (air)	J/kg.K
c_b	Specific heat capacity (battery)	J/kg.K
D_h	Inlet's hydraulic diameter	mm
G_k and G_b	Average velocity gradients and buoyancy effects, respectively, for turbulence kinetic energy generation	-
Y_M	Contribution of the fluctuating dilatation in compressible turbulence to the overall dissipation rate	-
$C_{1\varepsilon}$, $C_{2\varepsilon}$, $C_{3\varepsilon}$	Model constants	-
S_k and S_ϵ	Source terms	-
Greek letter(s)		
ρ_a	Density (air)	kg/m^3
ρ_b	Density (battery)	kg/m^3
μ_a	Dynamic viscosity (air)	kg/m.s
μ_t	Turbulent dynamic viscosity	kg/m.s
σ_k and σ_ϵ	k- and ε-turbulences' Prandtl numbers	-

References

1. Mo, C.; Xie, J.; Zhang, G.; Zou, Z.; Yang, X. All-climate battery thermal management system integrating units-assembled phase change material module with forced air convection. *Energy* **2024**, *294*, 130642. [CrossRef]
2. Na, X.; Kang, H.; Wang, T.; Wang, Y. Reverse layered air flow for Li-ion battery thermal management. *Appl. Therm. Eng.* **2018**, *143*, 257–262. [CrossRef]
3. Zhong, G.; Zhang, G.; Yang, X.; Li, X.; Wang, Z.; Yang, C.; Yang, C.; Gao, G. Researches of composite phase change material cooling/resistance wire preheating coupling system of a designed 18650-type battery module. *Appl. Therm. Eng.* **2017**, *127*, 176–183. [CrossRef]
4. Dan, D.; Yao, C.; Zhang, Y.; Zhang, H.; Zeng, Z.; Xu, X. Dynamic thermal behavior of micro heat pipe array-air cooling battery thermal management system based on thermal network model. *Appl. Therm. Eng.* **2019**, *162*, 114183. [CrossRef]
5. Rao, Z.; Wang, S. A review of power battery thermal energy management. *Renew. Sustain. Energy Rev.* **2011**, *15*, 4554–4571. [CrossRef]
6. Yong, S.C.; Mo, K.D. Prediction of thermal behaviors of an air-cooled lithium-ion battery system for hybrid electric vehicles. *J. Power Sources* **2014**, *270*, 273–280. [CrossRef]
7. Mohammadian, S.K.; Zhang, Y. Thermal management optimization of an air-cooled Li-ion battery module using pin fin heat sinks for hybrid electric vehicles. *J. Power Sources* **2015**, *273*, 431–439. [CrossRef]
8. Yang, X.; Tan, S.; Liu, J. Thermal management of li-ion battery with liquid metal. *Energy Convers. Manag.* **2016**, *117*, 577–585. [CrossRef]
9. Wang, C.; Zhang, G.; Meng, L.; Li, X.; Situ, W.; Lv, Y.; Rao, M. Liquid cooling based on thermal silica plate for battery thermal management system. *Int. J. Energy Res.* **2017**, *41*, 2468–2479. [CrossRef]
10. Liang, J.; Gan, Y.; Li, Y. Investigation on the thermal performance of a battery thermal management system using heat pipe under different ambient temperatures. *Energy Convers. Manag.* **2018**, *155*, 1–9. [CrossRef]
11. Liang, Z.; Wang, R.; Malt, A.H.; Souri, M.; Esfahani, M.; Jabbari, M. Systematic evaluation of a flat-heat-pipe-based thermal management: Cell-to-cell variations and battery ageing. *Appl. Therm. Eng.* **2021**, *192*, 116934. [CrossRef]

12. Rao, Z.; Wang, Q.; Huang, C. Investigation of the thermal performance of phase change material/mini-channel coupled battery thermal management system. *Appl. Energy* **2016**, *164*, 659–669. [CrossRef]
13. Samimi, F.; Babapoor, A.; Azizi, M.; Karimi, G. Thermal management analysis of a li-ion battery cell using phase change material loaded with carbon fibers. *Energy* **2016**, *96*, 355–371. [CrossRef]
14. Zhao, Y.; Li, Q.; Zou, B.; Zhang, T.; Jin, L.; Qiao, G.; Nie, B.; Huang, Y.; Ding, Y. Performance of a liquid cooling-based battery thermal management system with a composite phase change material. *Int. J. Energy Res.* **2020**, *44*, 4727–4742. [CrossRef]
15. Mashayekhi, M.; Houshfar, E.; Ashjaee, M. Development of hybrid cooling method with PCM and Al2O3 nanofluid in aluminium minichannels using heat source model of Li-ion batteries. *Appl. Therm. Eng.* **2020**, *178*, 115543. [CrossRef]
16. El Idi, M.M.; Karkri, M.; Tankari, M.A.; Vincent, S. Hybrid cooling based battery thermal management using composite phase change materials and forced convection. *J. Energy Storage* **2021**, *41*, 102946. [CrossRef]
17. Li, Y.; Kong, B.; Qiu, C.; Li, Y.; Jiang, Y. Numerical study on air-cooled battery thermal management system considering the sheer altitude effect. *Appl. Therm. Eng.* **2025**, *258*, 124707. [CrossRef]
18. Fan, Y.; Zuo, X.; Zhan, D.; Zhao, J.; Zhang, G.; Wang, H.; Wang, K.; Yang, S.; Tan, X. A novel control strategy for active battery thermal management systems based on dynamic programming and a genetic algorithm. *Appl. Therm. Eng.* **2023**, *233*, 121113. [CrossRef]
19. Qin, P.; Liao, M.; Zhang, D.; Liu, Y.; Sun, J.; Wang, Q. Experimental and numerical study on a novel hybrid battery thermal management system integrated forced-air convection and phase change material. *Energy Convers. Manag.* **2019**, *195*, 1371–1381. [CrossRef]
20. Behi, H.; Karimi, D.; Behi, M.; Ghanbarpour, M.; Jaguemont, J.; Sokkeh, M.A.; Gandoman, F.H.; Berecibar, M.; Mierlo, J.V. A new concept of thermal management system in Li ion battery using air cooling and heat pipe for electric vehicles. *Appl. Therm. Eng.* **2020**, *174*, 115280. [CrossRef]
21. Zhang, N.; Zhang, Z.; Li, J.; Cao, X. Performance analysis and prediction of hybrid battery thermal management system integrating PCM with air cooling based on machine learning algorithm. *Appl. Therm. Eng.* **2024**, *257*, 124474. [CrossRef]
22. Zhou, L.; Li, S.; Jain, A.; Chen, G.; Guo, D.; Kang, J.; Zhao, Y. Lithium Battery Thermal Management Based on Liightweight Stepped-Channel Liquid Cooling. *ASME J. Electrochem. Energy Convers. Storage* **2024**, *21*, 031012-1. [CrossRef]
23. Song, K.; He, R.; Gao, C.; Pea, H.J.M.; He, A.; Zhang, Q.; Zhang, K.; An, Z. Performance of a combined battery thermal management system with dual-layer phase change materials and air cooling technologies. *Appl. Therm. Eng.* **2024**, *254*, 123865. [CrossRef]
24. Zhu, X.; Liao, X.; Kang, S.; Sun, L.; Zhao, Y. Optimization design of lithium battery management system based on Z-F composite air cooling structure. *J. Energy Storage* **2024**, *102 Pt A*, 114068. [CrossRef]
25. Lamba, R.; Montero, F.J.; Rehman, T.; Singh, S.; Manikandan, S. PCM-based hybrid thermal management system for photovoltaic modules: A comparative analysis. *Environ. Sci. Pollut. Res.* **2024**, *31*, 46397–46416. [CrossRef] [PubMed]
26. Moaveni, A.; Siavashi, M.; Mousavi, S. Passive and hybrid battery thermal management system by cooling flow control, employing nano-PCM, fins, and metal foam. *Energy* **2024**, *288*, 129809. [CrossRef]
27. Argade, S.; De, A. Optimization study of a Z-type airflow cooling system of a lithium-ion battery pack. *Phys. Fluids* **2024**, *36*, 067119. [CrossRef]
28. Oyewola, O.M.; Idowu, E.T. Wave and straight plenum effects on thermal management system performance. *Int. J. Thermofluids* **2024**, *22*, 100678. [CrossRef]
29. Suo, Y.; Tang, C.; Han, E.; Chen, Z. Comparative assessment and optimization among several plenum shapes and positions for the forced air-cooled battery thermal management system. *J. Energy Storage* **2024**, *99 Pt B*, 113319. [CrossRef]
30. Chen, K.; Zhang, Z.; Wu, B.; Song, M.; Wu, X. An air-cooled system with a control strategy for efficient battery thermal management. *Appl. Therm. Eng.* **2024**, *236 Pt B*, 121578. [CrossRef]
31. Alzwayi, A.; Paul, M.C. Effective battery pack cooling by controlling flow patterns with vertical and spiral fins. *Therm. Sci. Eng. Prog.* **2024**, *55*, 102907. [CrossRef]
32. Qin, P.; Liao, M.; Mei, W.; Sun, J.; Wang, Q. The experimental and numerical investigation on a hybrid battery thermal management system based on forced-air convection and internal finned structure. *Appl. Therm. Eng.* **2021**, *195*, 117212. [CrossRef]
33. Wang, N.; Li, C.; Huang, M.; Qi, D. Effect analysis on performance enhancement of a novel air cooling battery thermal management system with spoilers. *Appl. Therm. Eng.* **2021**, *192*, 116932. [CrossRef]
34. Wang, M.; Hung, C.; Xi, H. Numerical Study on Performance Enhancement of the Air Cooled Battery Thermal Management System by Adding Parallel Plates. *Energies* **2021**, *14*, 3096. [CrossRef]
35. Oyewola, O.M.; Idowu, E.T.; Drabo, M.L. Influence of straight and inclined baffles on enhancement of battery thermal management system performance. *Heliyon* **2024**, *10*, e38585. [CrossRef] [PubMed]
36. Alzwayi, A.; Paul, M.C. Heat transfer enhancement of a lithium-ion battery cell using vertical and spiral cooling fins. *Therm. Sci. Eng. Prog.* **2024**, *47*, 102304. [CrossRef]

37. Oyewola, O.M.; Awonusi, A.A.; Ismail, O.S. Design optimization of Air-Cooled Li-ion battery thermal management system with Step-like divergence plenum for electric vehicles. *Alex. Eng. J.* **2023**, *71*, 631–644. [CrossRef]
38. Oyewola, O.M.; Awonusi, A.A.; Ismail, O.S. Performance Optimization of Step-like Divergence Plenum Air-Cooled Li-Ion Battery Thermal Management System Using Variable-Step-Height Configuration. *Emerg. Sci. J.* **2024**, *8*, 795–814. [CrossRef]
39. Oyewola, O.M.; Idowu, E.T. Numerical and artificial neural network inspired study on Step-like-plenum battery thermal management system. *Int. J. Thermofluids* **2024**, *24*, 100897. [CrossRef]
40. Chen, K.; Wu, W.; Yuan, F.; Chen, L.; Wang, S. Cooling efficiency improvement of air-cooled battery thermal management system through designing the flow pattern. *Energies* **2019**, *167*, 781–790. [CrossRef]
41. Chen, K.; Hou, J.; Wu, X.; Chen, Y.; Song, M.; Wang, S. Design of flow pattern in air-cooled battery thermal management system. *Int. J. Energy Res.* **2021**, *45*, 9541–9554. [CrossRef]
42. Wang, C.; Xi, H.; Wang, M. Investigation of forced air-cooling strategy of battery thermal management system considering the inconsistency of battery cells. *Appl. Therm. Eng.* **2022**, *214*, 118841. [CrossRef]
43. Hong, S.; Zhang, X.; Chen, K.; Wang, S. Design of flow configuration for parallel air-cooled battery thermal management system with secondary vent. *Int. J. Heat Mass Transf.* **2018**, *116*, 1204–1212. [CrossRef]
44. Oyewola, O.M.; Idowu, E.T. Effects of Step-like plenum, flow pattern and inlet flow regime on thermal management system. *Appl. Therm. Eng.* **2024**, *243*, 122637. [CrossRef]
45. Yan, H.; Ma, X.; Chen, Y.; Tao, Q.; Song, M. Performance analysis of a wet pad assisted air cooled battery thermal management system with varying number of battery cells. *Appl. Therm. Eng.* **2024**, *259*, 124747. [CrossRef]
46. Zhang, F.; Lin, A.; Wang, P.; Liu, P. Optimization design of a parallel air-cooled battery thermal management system with spoilers. *Appl. Therm. Eng.* **2021**, *182*, 116062. [CrossRef]
47. Wang, M.; Teng, S.; Xi, H.; Li, Y. Cooling performance optimization of air-cooled battery thermal management system. *Appl. Therm. Eng.* **2021**, *195*, 117242. [CrossRef]

batteries

MDPI

Article

Enhancing Battery Pack Cooling Efficiency Through Graphite-Integrated Hybrid-Battery Thermal Management Systems

Amin Rahmani, Mahdieh Dibaj and Mohammad Akrami *

Department of Engineering, University of Exeter, Exeter EX4 4QF, UK; ar949@exeter.ac.uk (A.R.); m.dibaj3@exeter.ac.uk (M.D.)
* Correspondence: m.akrami@exeter.ac.uk

Abstract: This study investigates a hybrid-battery thermal management system (BTMS) integrating air-cooling, a cold plate, and porous materials to optimize heat dissipation in a 20-cell battery pack during charging and discharging cycles of up to 5C. A computational fluid dynamics (CFD) model based on the equivalent circuit model (ECM) is developed to simulate battery pack behavior under various cooling configurations, including different porous media and vortex generators placed between cells. The impact of battery pack configurations on heat generation is analyzed, and five different porous materials are tested for their cooling performance. The results reveal that, among the examined materials, graphite is the most effective in maintaining the battery temperature within an acceptable range, particularly during high C-rate charging. Graphite integration significantly reduces the thermal stabilization time from over an hour to approximately 600 s. Additionally, our parametric experiment evaluates the influence of ambient temperature, airflow velocity, and cold-plate temperature on the system's cooling efficiency. The findings demonstrate that maintaining the cold-plate temperature between 300 K and 305 K minimizes the temperature gradient, ensuring uniform thermal distribution. This research highlights the potential of hybrid BTMS designs incorporating porous media and cold plates to enhance battery performance, safety, and lifespan under various operational conditions.

Keywords: hybrid-battery thermal management system; cold plate; air-cooling; graphite; computation fluid dynamics (CFD); equivalent circuit model (ECM)

1. Introduction

The rise in electric vehicles (EVs) has significantly transformed the automotive industry, playing a crucial role in reducing carbon emissions and promoting sustainability [1]. However, thermal runaway in lithium-ion batteries remains a major safety concern due to factors like rapid charging, extreme temperatures, and manufacturing defects [2,3]. This phenomenon, which can lead to overheating, fires, and explosions, highlights the importance of effective thermal management systems [4]. Advanced cooling technologies, thermal insulation, and monitoring systems are essential to ensuring the safety and reliability of EVs, ultimately supporting their widespread adoption.

To enhance EV efficiency and safety, battery packs require proper cooling through active and passive methods [5]. Active cooling systems, such as liquid cooling and air conditioning, use external technologies to dissipate heat, while passive systems rely on natural mechanisms like heat sinks and phase change materials [6–8]. Among these, air-cooling is widely used due to its affordability, reliability, and ease of implementation [9].

By using fans or blowers to regulate battery temperature, air-cooling systems improve heat dissipation efficiently, making them a preferred choice in the automotive industry [10,11].

Incorporating air-cooling techniques into commercial hybrid electric vehicles indicates a noteworthy advancement in battery thermal management systems. Both vehicles, recognized as pioneering innovations in the industry, employed air-cooling for their battery packs [12]. They accomplished this by utilizing cabinet-heated air to facilitate the cooling of the modules, eventually releasing the heat they had absorbed into the atmosphere. This approach has been accepted as the standard for future electric car models. The widespread adoption of air conditioning in commercial electric vehicles serves as proof of its effectiveness as a reliable and cost-effective solution, ensuring optimal battery temperatures and facilitating a wider acceptance of electric mobility in the interest of sustainable transportation alternatives [13]. Air-cooling battery thermal management systems (BTMSs) are divided into three different aspects, including the investigation of design parameters, flow parameters, and operational parameters [14]. Channel size and inlet/outlet position [15], cell arrangement [16], and cell spacing [17] are paramount design factors in air-cooling battery pack design.

Recent advancements in air-cooling battery thermal management systems (BTMS) have led to innovative designs aimed at improving cooling efficiency. Early models, such as the Z-type [18] and U-type [19] configurations, provided the foundation for further developments. Wang et al. [20] explored nine BTMS designs with different inlet and outlet placements, demonstrating that optimizing airflow direction could lower the maximum battery temperature by up to 4.9 K. Building on these models, Zhang et al. [21] introduced the parallel T-type BTMS, which significantly reduced Tmax, ΔTmax, and power consumption by 3.5%, 67.7%, and 19.0%, respectively, compared to Z-type and U-type configurations.

Further improvements include the use of baffles in cooling channels, as investigated by Furen et al. [22], who reported reductions of 12.7 K in Tmax and 4.49 K in ΔTmax with a 1.1 mm baffle. Additionally, Kang et al. [23] studied different cell arrangements in battery modules, highlighting that, even with the same number of cells, the temperature distribution can vary significantly based on module design. These findings underscore the importance of optimizing airflow pathways and structural configurations in BTMS to enhance cooling performance.

Recent advancements in forced-air BTMS have focused on enhancing cooling performance through innovative airflow modifications. Shahid et al. [24] demonstrated that introducing a secondary inlet to alter airflow direction improved maximum cell temperature by 4% and temperature differential by 39%. Oyewola et al. [25] optimized a Z-type BTMS using a step-like divergence plenum, reducing maximum temperature by 3.94 K and temperature difference by 5.93 K. Jiang et al. [26] explored a reciprocating mist cooling system for 18650 lithium-ion battery packs, identifying ideal operating parameters. Verma and Saraswati [27] analyzed different inlet flow characteristics in air-cooling BTMS, reducing the maximum temperature difference from 22.14 K to 0.21 K.

To overcome the limitations of air-cooling systems, hybrid BTMS integrating liquid cooling [28], passive cooling [29], PCM, and porous mini-channels have been proposed. Kavasogullari et al. [30] found that a 9 mm PCM layer effectively maintains ideal battery temperatures. Jiang et al. [31] used porous mini-channels, reducing maximum temperature by 5 K at varying inlet velocities. Sharma et al. [32] tested PCM–heat-pipe hybrid designs for high ambient temperatures, keeping battery temperatures within safe limits. Majid et al. [33] introduced a thermoelectric–air-cooling hybrid BTMS that lowered temperature by 7 K, while Moaveni et al. [34] showed that metal foam and fins reduced peak temperatures by 10.36 K and 9 K, respectively. Other hybrid BTMS solutions include magnetohydrodynamics [35], thermoelectric cooling [36], mini-channel heat sinks [37], evaporative cooling [38], and fin-based systems [39]. Bargal et al. [40] reviewed liquid-cooling techniques for PEMFC stacks, emphasizing nanofluids and coolant channel optimization for improved thermal management.

Heat generation modeling is essential in battery simulations, with various methods developed to balance accuracy and complexity. The conjugate heat transport (CHT) model provides a basic approach by assuming a constant heat source, while the NTGK model [41,42] estimates heat generation using voltage–current response curves but struggles with sudden load variations.

The equivalent circuit model (ECM) offers a more accurate and adaptable solution by simulating electrical behavior, making it suitable for both single cells and battery packs. While it effectively captures battery property evolution, its complexity limits its practicality in engineering applications.

Research on the role of busbars in battery packs is limited, but studies highlight their impact on thermal behavior. Yetik and Karakok [43] emphasized busbars' significance, noting temperature challenges in series-connected prismatic batteries. Further investigations [44,45] analyzed material and design effects, with Chen et al. [46] demonstrated that optimized busbars can reduce maximum battery temperature by 10.06 K and temperature differential by 9.75 K at a 3C discharge rate.

This study simulates a hybrid battery pack with 20 cells, addressing the gap in the literature on the impact of different configurations (1P20S to 20P1S) on heat generation. A virtual connection is implemented to account for busbar effects. Given that excessive heat generation can lead to thermal runaway, effective cooling strategies are essential for maintaining battery safety. To enhance air-cooling efficiency, this study introduces modified inlet and outlet channels, cold plates, and porous media between cells. Additionally, aluminum vortex generators are used to optimize airflow. Simulations are conducted under various charge–discharge rates and ambient conditions to assess the cooling system's performance. The results demonstrate that the proposed system effectively regulates temperature and reduces thermal risks, making it suitable for different environmental conditions. The simulations were performed using ANSYS Fluent 2023 R1.

2. Problem Description

Figure 1 shows the battery pack geometrical model of hybrid BTMS using 18650 cylindrical Li-ion batteries. In this configuration, air enters the battery pack from the top and exits through the side walls. The optimal gap between batteries is 2 mm, which is considered in this study [47]. Porous materials create a porous medium in the spaces between cells. Twenty 18650 cylindrical cells are in the battery pack, and the battery properties are presented in Table 1. To circulate air in the battery pack in a more efficient way, some aluminum vortex generators are inserted inside the battery pack. A 10 mm distance is considered between batteries and cold plate.

Table 1. Properties of the battery used in this study [8,48].

Properties	Unit	Value
Cell diameter	mm	18
Cell length	mm	65
Nominal capacity	mAh	3500
Nominal voltage	V	3.56
ρ	kg/m^3	2092
c_p	j/kg·K	678
k	W/m·K	18.2
Operating temperature	K	Charge: 273–318 K Discharge: 253–333 K

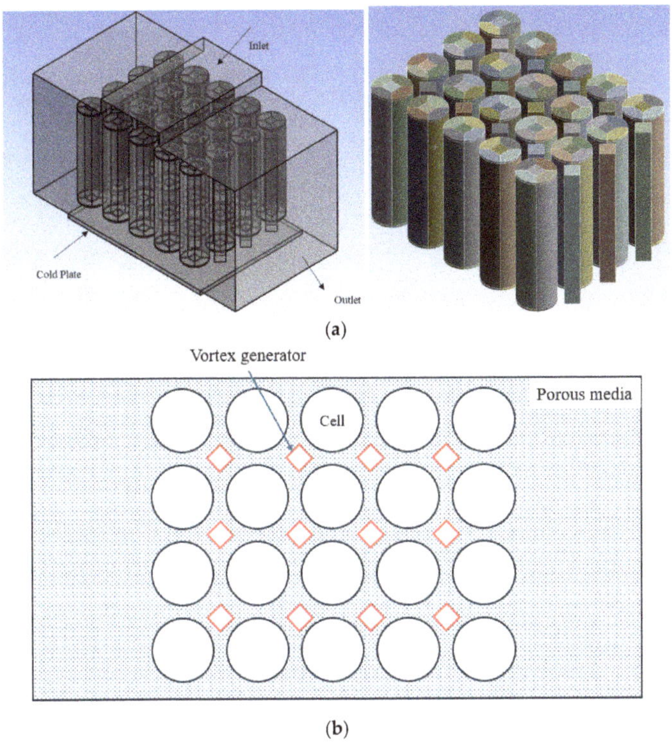

Figure 1. (**a**) Three-dimensional geometry of the Li-ion battery pack with a cold plate and airflow pathways. (**b**) Top-view schematic of the battery pack with porous media and vortex generators.

Four zones are described in the geometry: batteries, porous domain, positive and negative tabs, and cold plates. The lithium-ion battery's anode is made of graphite, while the cathode is made of $LiFePO_4$. The battery material is isotropic, so the thermal conductivity and specific heat transfer are considered constant. Lithium-ion batteries are made of several materials. The thermophysical properties of a Li-ion battery can be calculated using Equation (1). For instance, for thermal conductivity, the following applies:

$$k = \frac{\sum_{i=1}^{n} k_i \delta_i}{\sum_{i=1}^{n} \delta_i}, \tag{1}$$

where δ_i is the thickness of the ith layer, and k is thermal conductivity.

3. Governing Equations

3.1. Porous Domain

This study employs the computational fluid dynamics (CFD) technique to determine the distribution of temperature and investigate the impact of various components that influence it. The objective is to assess the impact of various porous materials on the air-cooling battery thermal management system. The fluid flow in the porous domain is characterized using mass, momentum, and energy conservation equations. The conversion of the mass equation is shown as follows [20]:

$$\frac{\partial \rho}{\partial t} + \nabla.\left(\rho \vec{v}\right) = 0 \tag{2}$$

The momentum equation is expressed as follows:

$$\frac{\partial}{\partial t}(\rho\vec{v}) + \nabla\cdot(\rho\vec{v}\vec{v}) = -\nabla p + \nabla\cdot(\tau) + \rho\vec{g} + \vec{S} \quad (3)$$

$\vec{S}$, which is defined as the external force, can also contain source terms such as porous media. $\rho\vec{g}$ is the gravitational body force, and τ is the stress tensor. $\vec{S}$ and τ are defined as follows:

$$S_i = -\left(\sum_{j=1}^{3} D_{ij}\mu v_j + \sum_{j=1}^{3} C_{ij}\frac{1}{2}\rho|v|v_j\right) \quad (4)$$

$$\tau = \mu\left[\left(\nabla\vec{v} + \nabla\vec{v}^{r}\right) - \frac{2}{3}\nabla\cdot\vec{v}r\right] \quad (5)$$

D_{ij} and C_{ij} are the viscous resistance term and internal resistance term, which can be defined as follows:

$$D = \frac{\varepsilon^2}{K} \quad (6)$$

$$C = \frac{2C_F\varepsilon^3}{\sqrt{K}} \quad (7)$$

where ε is porosity and K is defined as a function of porosity [20].

$$K = 0.00073(1-\varepsilon)^{-0.224}\left(\frac{1.18}{1-e^{-\frac{1-\varepsilon}{0.04}}}\sqrt{\frac{1-\varepsilon}{3\pi}}\right)^{-1.11} d_m^2 \quad (8)$$

$d_m = 0.0254/PPI$, where PPI is the number of pores per inch, and d_m is the pore diameter. Equation (8) primarily considers pore diameter as a key parameter in characterizing the structural effects on thermal transport. Although permeability is not explicitly included in the equation, it is inherently linked to porosity ε and pore size. Since permeability is a function of these parameters, its influence is indirectly captured within the formulation. For highly porous materials, empirical correlations such as this provide a reasonable estimation of thermal transport characteristics without requiring permeability as a direct input.

The energy equation is defined as follows:

$$\frac{\partial}{\partial t}(\rho E) + \nabla(\nu(\rho E + p)) = \nabla\cdot\left(k_e\nabla T - \sum h_i\vec{J_i} + (\tau_e\cdot\nu)\right) + S_h \quad (9)$$

E is defined as $E = h - \frac{p}{\rho} + \frac{v^2}{2}$, where h is the enthalpy of the fluid. S_h is the heat generation in the porous domain which can be neglected. k_e is the effective thermal conductivity, and $\left(k_e\nabla T - \sum h_i\vec{J_i} + (\tau_e\cdot\nu)\right)$ represents energy transfer from thermal conduction. Other variables are defined in the nomenclature section.

3.2. Battery

Modeling a lithium-ion (Li-ion) battery is challenging due to its complex multi-domain, multi-physics nature, with significant variations in length scales. In thermal analysis, the objective is to ascertain the temperature distribution at the overall battery length scale. The multi-scale multi-domain (MSMD) approach is employed to address these challenges by managing different physics in separate solution domains. Heat is generated within the battery during the charge/discharge process. If a thermal management system exists, most of the heat will be transmitted from the interior part to the surface, where it will be dispersed by ambient air or another coolant. The battery contains the remaining energy.

As batteries are considered solid, there is no need to solve momentum equations for batteries. So, only thermal and electrical fields will be solved in the CFD domain using the following equations [49]:

$$\frac{\partial \rho C_p T}{\partial t} - \nabla \cdot (k \nabla T) = \sigma_+ |\nabla \varphi_+|^2 + \sigma_- |\nabla \varphi_-|^2 + \dot{q}_{ECh} + \dot{q}_{\text{short}} + \dot{q}_{\text{abuse}} \tag{10}$$

$$\nabla \cdot (\sigma_+ \nabla \varphi_+) = -(j_{\text{ECh}} - j_{\text{short}}) \tag{11}$$

$$\nabla \cdot (\sigma_- \nabla \varphi_-) = j_{\text{ECh}} - j_{\text{short}} \tag{12}$$

In Equation (10), $\dot{q}_{\text{short}}$ and $\dot{q}_{\text{abuse}}$ can be neglected as the battery pack operates under controlled conditions without internal short circuits or external abuse scenarios, such as overcharging or mechanical damage, which would otherwise contribute to these heat generation sources. But $\dot{q}_{\text{ECh}}$ is the main source of heat generation in the battery where numerous electrochemical models, spanning from simplistic empirically derived to those grounded in fundamental physics, are accessible in the open literature. In this study, the equivalent circuit model is used as the electrochemical model.

In the case that thermal abuse condition is considered, $\dot{q}_{\text{abuse}}$ should be considered as follows:

$$\dot{q}_{\text{abuse}} = \text{HW} \cdot \text{A} \cdot exp\left(-\frac{\text{E}}{\text{RT}}\right) \alpha^{\text{m}} (1 - \alpha)^{\text{n}} \tag{13}$$

where α, A, E, R, T, m, and n are the degree of conversion, pre-exponential factor of the reaction, activation energy of the reaction, universal gas constant, temperature, and reaction order parameters. The values for these parameters are presented in Table 2.

Table 2. Values for heat generation in thermal abuse conditions.

Parameter	Unit	Value
α	-	1×10^{-6}
A	$10^{10} \times 1/\text{s}$	3.97×10^{-13}
HW	J/m^3	5.85×10^8
E	J/mol	−15,253
R	J/(K·mol)	8.314
m	-	0.9404
n	-	0.425

3.3. *Equivalent Circuit Model (ECM)*

The equivalent circuit model (ECM) emulates the electrical behavior of a battery through the representation of an electrical circuit. This model incorporates a circuit configuration comprising an open-circuit voltage (OCV), an internal resistance (R_s), and some resistor–capacitor pairs. A battery can be precisely modeled using an infinite number of resistor–capacitor pairs, but a battery can also be highly accurately modeled with the assumption of two R-C pairs. This assumption leads to a model called the 'Thévenin' model. Figure 2 illustrates the specific electrical circuit employed in the ECM.

$$V = V_{OCV}(SOC) + V_1 + V_2 - R_S(SOC)I(t) \tag{14}$$

$$\frac{\text{dV}_1}{\text{dt}} = -\frac{1}{\text{R}_1(\text{SOC})\text{C}_1(\text{SOC})}\text{V}_1 - \frac{1}{\text{C}_1(\text{SOC})}\text{I(t)} \tag{15}$$

$$\frac{\text{dV}_2}{\text{dt}} = -\frac{1}{\text{R}_2(\text{SOC})\text{C}_2(\text{SOC})}\text{V}_2 - \frac{1}{\text{C}_2(\text{SOC})}\text{I(t)} \tag{16}$$

$$\frac{\text{d(SOC)}}{\text{dt}} = \text{I(t)}/3600\text{Q}_{\text{ref}} \tag{17}$$

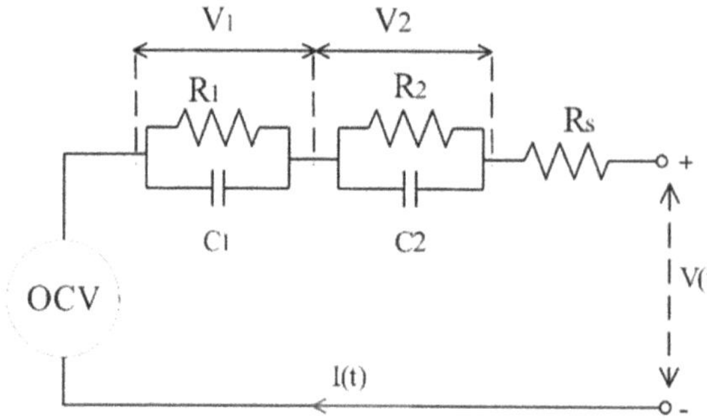

Figure 2. Schematic of double R-C pair equivalent circuit model (ECM).

Running some static and dynamic experiments, the values of OCV, R_s, R_1, R_2, C_1, and C_2 can be found as a function of state of charge (SOC) and battery's temperature.

Electrochemical heat generation ($\dot{q}_{ECh}$) in Equation (10) can be computed as follows [50]:

$$j_{ECh} = I\frac{Q_{Nominal}}{Q_{ref}\forall} \tag{18}$$

$$\dot{q}_{ECh} = j_{ECh}\left[V_{OCV} - V - T\frac{dU}{dT}\right] \tag{19}$$

It is notable that in the battery's active zone, Equations (10)–(12) should be solved, whereas, in passive zones, only the potential equation is solved, and the second equation is excluded by setting the electrical conductivity to zero.

To simulate a battery pack, it is necessary to consider the effect of busbars. Busbars are usually very thin and difficult to mesh. So, to solve this problem, a virtual connection between batteries is considered, and Joule heating is neglected in these areas.

3.4. Assumptions

Some assumptions are considered in these simulations which are presented as follows:

1. All batteries are identical and have the same initial state of charge at the start of the simulations.
2. The porous media between batteries is assumed to be homogeneous.
3. The boundary condition in outlets is 'pressure outlet' and in inlets is 'velocity inlet'.
4. Natural convection is considered on the battery box, and the heat transfer coefficient is 5 W/m·K.

4. Grid Study

The time-step independence of the solution is essential to validate the accuracy of numerical studies, especially when a transient solution is carried out. Figure 3 displays a graph illustrating the effect of time step on cell heat generation, specifically for the started time steps. Based on the data, during the first 50 s, the results for heat generation exhibit a maximum 3% variations when using time steps of 1, 5, and 10. However, after 60 s, these differences become negligible. Therefore, to optimize computational efficiency while maintaining accuracy, a time step of 10 is chosen for studying heat generation beyond 60 s, while, in the first 60 s, the time step is considered as 1.

Prior to commencing problem-solving, it is essential to verify that the solution outcomes are unaffected by the mesh size. Enhancing the quantity of meshes enhances solution accuracy but also escalates computational expenses. Hence, the assessment of independence from the system needs to be conducted. Ansys Meshing 2023 software is utilized in this section to create a mesh for the specified geometry. The battery compartment is partitioned into five sections to enhance the quality of the meshes. The inflation method has been utilized to accurately represent the boundary layer along the walls around the solid area. Five distinct computer grids were tested in this part, which demonstrated that, once the number of meshes exceeded 2,753,472 elements, the variations in battery temperature during BTMS charging mode were negligible. Throughout this paper, 2,753,472 represents the number of meshes for BTMS gradient. Table 3 shows the number of mesh structures which are used in the grid study. The factors that affect the number of meshes in this study are fluid domain mesh size, inflation mesh size rate near battery walls, and mesh sizes on solid bodies including batteries. Figure 4 displays the grid utilized.

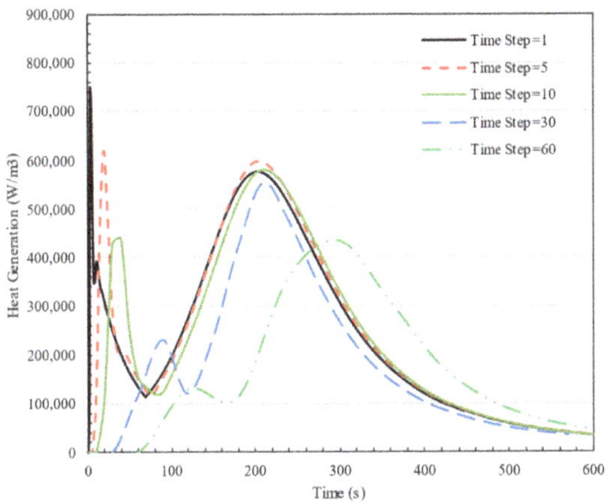

Figure 3. Time step study on heat generation of a cell in charge rate 3C for different time steps from 1 s to 60 s.

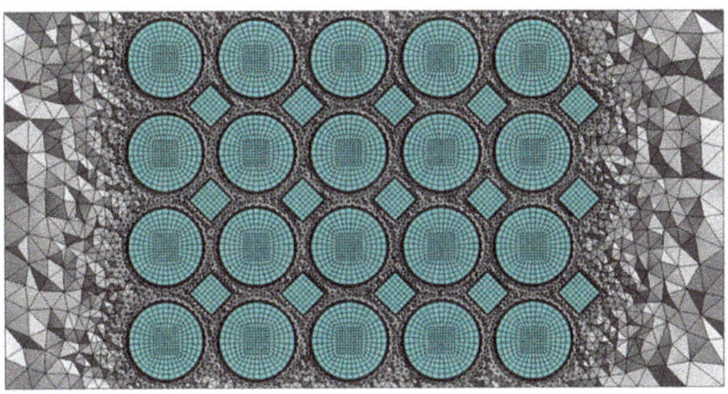

Figure 4. Mesh used in this study.

Table 3. Grid study.

Number of Total Elements	Maximum Peak Temperature	Error
563,156	313.67	-
991,325	315.34	1.67 K
1,511,644	316.75	1.41 K
2,753,472	316.98	0.23 K
14,165,808	317.06	0.08 K

5. Validation

The performance of the equivalent circuit model for a cell is verified against the results of Li et al. [50] at different discharge rates (1C and 2C). Figure 5 shows that the simulation results agree well with the experimental data for an 18650 cylindrical battery with a nominal capacity of 2.5 Ah. The ambient temperature in this study is 298.15 K. To evaluate the accuracy of the numerical model, error metrics were computed between the simulation results and experimental data at different discharge rates. The Mean Absolute Error (MAE) was found to be 1.92 for C1 and 0.28 for C2. Additionally, the Mean Absolute Percentage Error (MAPE) remained below 1% in both cases, confirming the reliability of the numerical predictions.

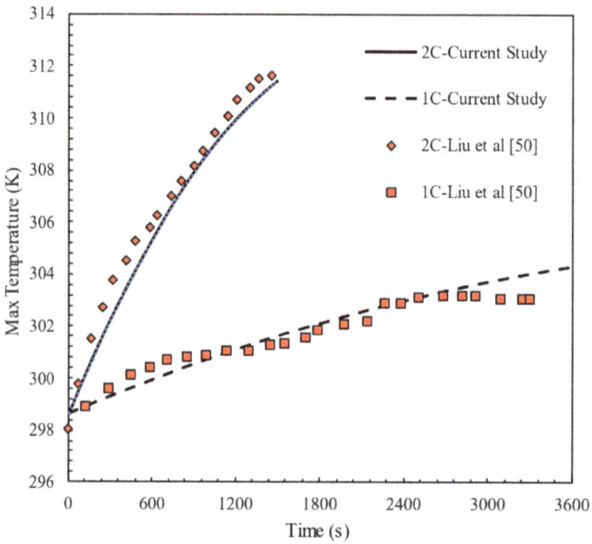

Figure 5. Model validation at single cell with experimental results of Li et al. [50].

The heat generation in a lithium-ion battery during operation can be mathematically described as $= I^2R_{int} + I \cdot \Delta V_{polarization}$, where Q is the total heat generated, I is the current, R_{int} is the internal resistance, and $\Delta V_{polarization}$ represents the polarization losses. The current I is proportional to the C-rate, defined as $= C \cdot C_{rate}$, where C is the battery's capacity. The first term, ANSYSI^2R_{int}, accounts for Joule heating due to internal resistance, while the second term, $I \cdot \Delta V_{polarization}$, reflects the heat generated from polarization effects, which become more pronounced at higher C-rates. In many cases, especially at moderate C-rates, the term $I \cdot \Delta V_{polarization}$ can often be small compared to the Joule heating term I^2R_{int} and may be neglected in thermal calculations.

Figure 6 illustrates the effect of charging and discharging rates on battery heat generation in a steady-state condition. The graph presents heat generation as a function of C-rate (ranging from 1C to 5C) for both charging and discharging phases, showing a quadratic relationship for each case. Furthermore, the graph shows the amount of heat generation at t = 1000 s at different C-rates. By fitting a quadratic equation to this graph, it is obvious that the amount of heat generation in both charging and discharging stages is directly proportional to the square of the C-rate. Equation (20). shows the relationship between heat generation and C-rate.

$$Q_{c-\text{rate}} = Q_{C1}C^2 \tag{20}$$

where Q_{C1} is heat generated in a cell at 1C when charging and discharging the battery pack. Also, Figure 6 shows that heat generation during charging is 7% higher when discharging the battery pack on average. The yellow and blue dotted lines represent the fitted curves corresponding to the charge and discharge data, respectively, illustrating the relationship between heat generation and C-rate.

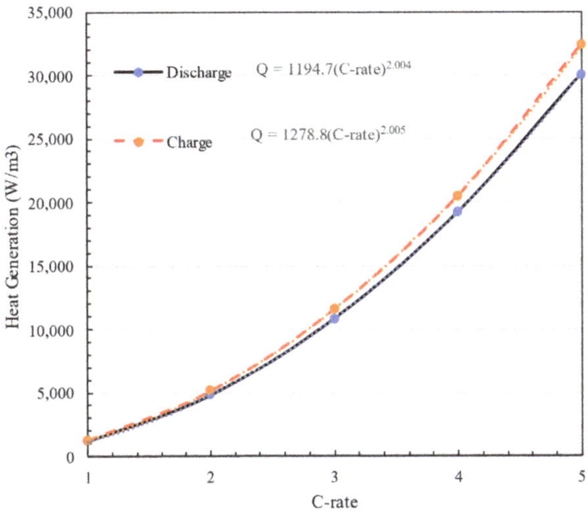

Figure 6. Heat generation amount at different C-rate = 1C–5C in charge and discharge modes.

The figure shows a clear relationship between heat generation and C-rate for both charge and discharge phases, and it quantitatively represents the heat generated at various C-rates using fitted power laws.

6. Results

6.1. Heat Generation in Battery Pack

In this section, the heat generation of a cell in a battery pack set at different C-rates and different configurations of 20 batteries in parallel and series in two charging and discharging modes has been investigated. It is expected that the rate of heat generation in the battery will increase with the increase in charge rate because the current passing through the battery will increase with the increase in charge rate. Figure 7 shows how heat generation changes at different charge and discharge rates (C-rate = 1C–5C) over time in a 4S5P battery pack.

During the discharging mode, it is evident that heat generation steadily increases after the battery discharge begins. However, as time passes, the rate of heat generation decreases, eventually reaching a point where the changes in heat generation become negligible. On the other hand, during the charging phase, there are two periods of maximum heat generation within the first ten minutes, followed by a consistent level of heat generation. The two peaks in the heat generation rate during charging are attributed to different thermal effects within the battery: the first peak occurs due to the internal resistance (IR drop) at the start of charging, while the second peak appears near the end of charging due to concentration polarization and slower lithium-ion diffusion, leading to increased energy dissipation Figure 8 compares the cell voltage during both the discharge and charge processes with a 3C rate. As can be seen, the voltage increases dramatically as soon as the charging process starts, which causes high heat generation in the cell. This phenomenon justifies the peak heat generation in charging cells. The initial voltage jump observed during charging is attributed to the internal resistance (IR drop), which causes a sudden increase when current is applied, whereas during discharge, the voltage drop is smaller due to the absence of charge accumulation, a more gradual electron flow, and different reaction kinetics. This behavior aligns with expected lithium-ion battery characteristics and has been validated against experimental data.

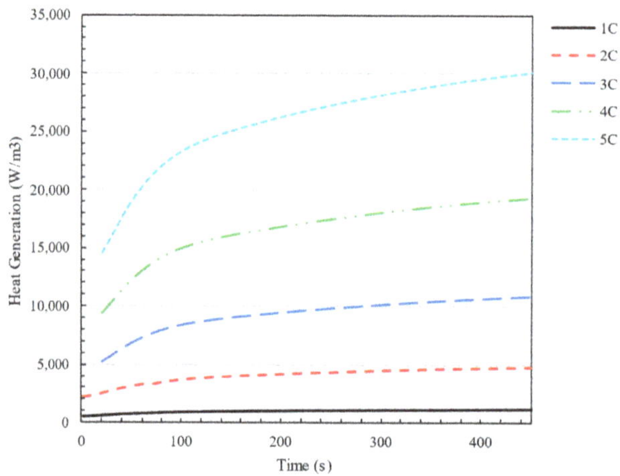

(a)

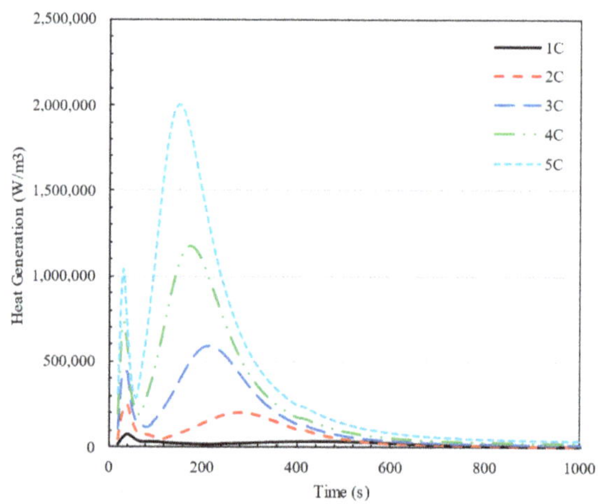

(b)

Figure 7. Changes in heat generation in a cell at C-rate = 1C–5C over time in 4S5P configuration in (**a**) discharging and (**b**) charging of the battery pack.

As is shown in Figure 7a, a rapid rise in heat generation occurs during the initial few minutes of charging the battery. The introduced relationship in Equation (20). is not followed in this section of the figure. Additionally, the peak time at which the battery's generated heat reaches its maximum varies depending on the charging rate (C-rate). As C-rate increases, the heat generates more and sooner, where at C-rate = 1, t_{peak} is 460 s; and at C-rate = 5, $t_{peak} = 150$ s.

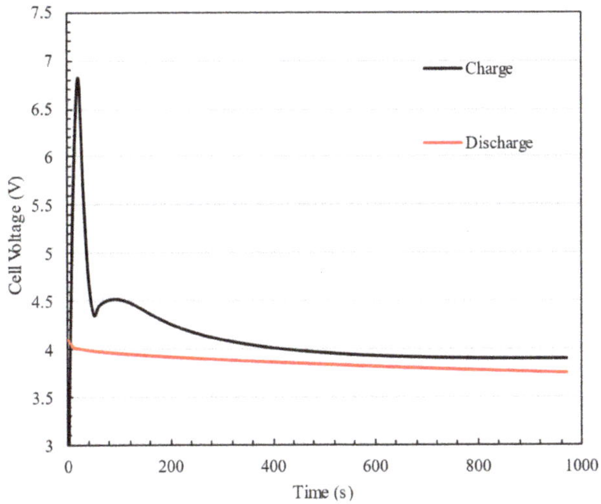

Figure 8. Cell voltage during the discharge and charge processes at a 3C rate.

Battery pack designers have many considerations when selecting the number of batteries in series or parallel. The energy and power of a battery pack can be calculated by the following equations [51]:

$$E_{pack} = n_s \cdot n_p \cdot Q_{cell} \cdot v_{cell} \tag{21}$$

$$P_{pack} = n_s \cdot n_p \cdot i_{cell} \cdot v_{cell} \tag{22}$$

where n_s is the number of batteries in series, and n_p is the number of batteries in parallel. So different battery pack configurations will not affect the energy and power of the battery pack, but economic and safety factors should be considered as the overall battery pack voltage should be less than 100 V, and the power electronics required for higher voltages will become more expensive. The generated heat by a battery is completely affected by the current and voltage on a specific battery. Thus, in this section, the amount of heat generated heat in a battery pack with different battery configurations at various charge and discharge rates is presented. The findings demonstrate that Equation (20) operates independently of battery design, including 2S10P, 4S5P, 5S4P, and 10S2P connections. Figure 9 depicts the effect of the number of batteries in series in a battery pack (n_s) on the heat generation of each cell. This figure also shows that heat generation in the battery pack is completely related to the square of the number of series lines in the battery pack. The trends highlight that, as more cells are added in series, the cumulative heat generation escalates rapidly. This quadratic relationship emphasizes the importance of thermal management strategies in large-scale battery packs, where increasing the number of cells in series can lead to excessive heat buildup, potentially impacting performance and safety. The yellow, grey, and blue dotted lines represent the corresponding fitted curves for each dataset, illustrating the trend of heat generation as a function of n

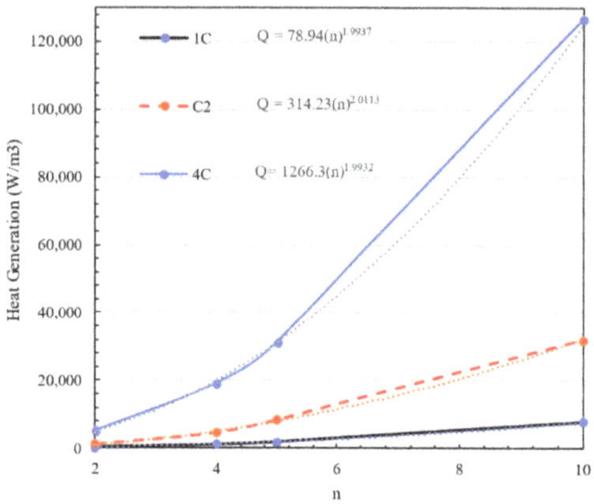

Figure 9. The effect of number of batteries in series (n_s) in a battery pack on heat generation.

6.2. Effectiveness of Incorporating Porous Materials

This study uses a hybrid BTMS to control temperature and temperature difference through a battery pack using air and a cold plate. In this system, air enters the battery pack from the upside, passes through the gaps between batteries, and exits from the side walls. The gaps between batteries define a porous medium. Porous materials provide a great passive cooling option for BTMS due to their availability, different properties, and cheap production costs [52]. In this section, the effect of porous material on the maximum temperature and temperature difference in a battery pack is investigated. In this section, five different porous materials, such as aluminum foam, copper foam, silicon carbide, aluminum oxide, and graphite, are considered. The properties of these materials are presented in Table 4. In this section, in all cases, the ambient temperature and cold plate temperature are 300 K.

Table 4. Properties of different porous materials [52].

Material	ρ(kg/m^3)	c_p(j/kg·K)	k(w/m·K)	Permeability	ϵ
Aluminum foam	2719	871	202.4	1.42×10^{-6}	0.958
Copper foam	8978	386	398	9.24×10^{-8}	0.944
Silicon Carbide	3210	700	430	1.93×10^{-8}	0.890
Aluminum oxide	3891	765	35	9.00×10^{-7}	0.878
Graphite	2130	700	1700	7.69×10^{-10}	0.728

Figure 10 compares and illustrates the maximum temperature and temperature difference in the battery pack for various porous materials when the battery charges and discharges at 5C. It can be seen from the obtained results in Figure 10a that graphite performs better in maintaining the battery pack's maximum temperature. The results show that using graphite improves the battery pack's temperature distribution when comparing temperature difference values. In discharge, all porous materials show good performance where the temperature difference in all cases is under 1 K and the maximum temperature is about 302.61 K for aluminum oxide and 300.17 K for graphite.

As shown in Figure 10b, in charging, heat generation increases to a peak amount, and after that, it decreases to a stabilized amount. This heat generation trend affects the battery temperature, so, in battery charging, cells experience a peak temperature, and then the temperature stabilizes when the cooling system acts in the system. The battery used in this study has an upper working temperature of 318 K, as shown in Figure 10b. Also, it shows that the only material that can keep the battery pack's peak temperature below 318

K in 5C charging is graphite, where the peak temperature is 316.98 K. Graphite is the only porous material that can keep cells' temperatures in the safety range (273 K < T < 318 K) in 5C charging.

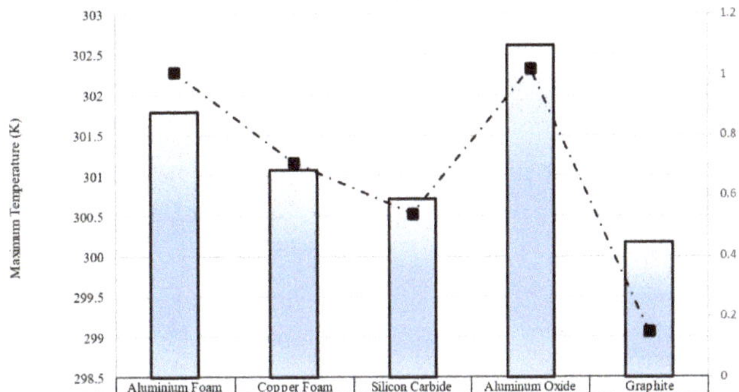

(a)

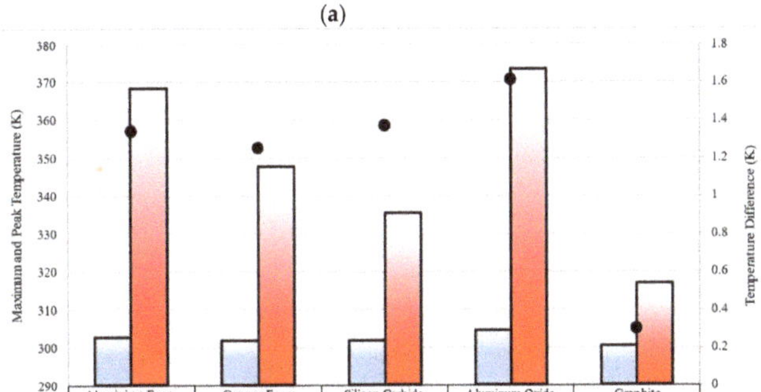

(b)

	Aluminium Foam	Copper Foam	Silicon Carbide	Aluminum Oxide	Graphite
Maximum Temperature	301.79	301.06	300.71	302.61	300.17
Temperature Difference	1.01	0.71	0.54	1.02	0.15

	Aluminium Foam	Copper Foam	Silicon Carbide	Aluminum Oxide	Graphite
Balanced Temperature	302.61	301.8803	301.81	304.37	300.35
Peak Temperature	368.5	347.75	335.44	373.31	316.98
Temperature Difference	1.34	1.252324	1.37	1.61	0.3

Figure 10. The performance of aluminum foam, copper foam, silicon carbide, aluminum oxide, and graphite in 5C (**a**) discharging and (**b**) charging of the battery pack.

Figure 11 demonstrates that charging the battery pack using graphite between the batteries results in the battery temperature remaining below 318 K, which is within the acceptable limit, while charging at a rate of 1C. The battery temperature exceeds the permissible range significantly when charging for 3C and 5C batteries. On the other hand, if graphite fills the gap between the batteries, the battery temperature remains below 318 K, demonstrating graphite's effectiveness in regulating the battery pack temperature when charging at a rate of 5C. Comparing the two plots shows that when the graphite is placed between the batteries, the temperature stabilizes more quickly. The battery pack reaches a steady temperature after 10 min of charging at 5C. Without using graphite, the time is over one hour. In the nonporous media configuration, there is no additional material between the battery cells, meaning that only air occupies the gaps, and cooling relies solely on natural or forced convection rather than enhanced heat dissipation through porous structures.

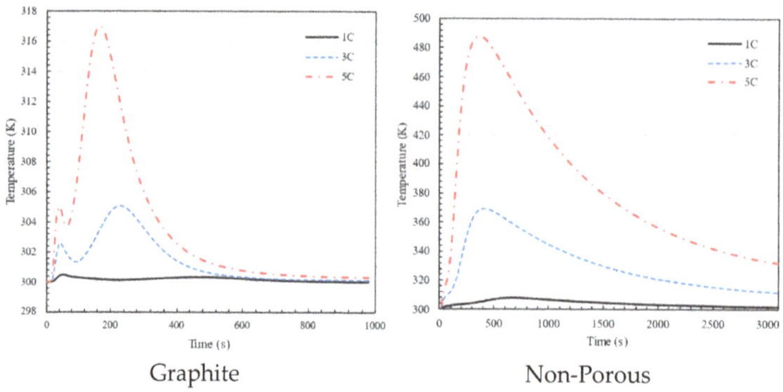

Figure 11. Maximum temperature in 1C, 3C and 5C charging of battery for the battery pack using graphite and non-porous media.

Figure 12 illustrates temperature contours on the surface of a battery pack cell, comparing the utilization of pure medium, aluminum foam, and graphite in the space between cells. Graphite has enhanced thermal performance by substantially reducing both the maximum temperature and temperature differences over the surface of the cell in comparison to aluminum foam. This indicates the efficacy of graphite in dissipating heat, which is essential for preventing overheating and ensuring equitable distribution of heat within the battery pack. The results highlight the potential of graphite as a preferable material for improving the dependability, safety, and lifespan of battery systems by boosting thermal control.

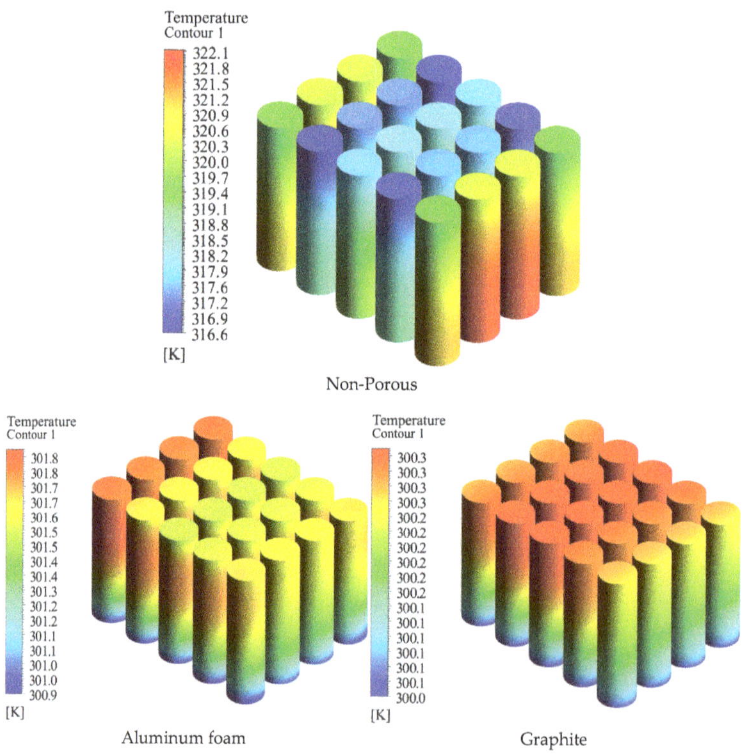

Figure 12. Temperature contour on the surface of the battery when pure medium, aluminum foam, and graphite are utilized in the gap between the batteries.

Figure 13 displays the temperature variation in the battery during charging with and without porous material. The allowable temperature variation in the battery is 5 K. The method described in this study exceeds the limit only in 5C mode, when the temperature differential between the two batteries drops below 5 K after 6 min. Without porous material, while charging at rates of 3C and 5C for one hour, the battery's temperature exceeds the acceptable limit, highlighting the improved performance of the developed system compared to the porous condition.

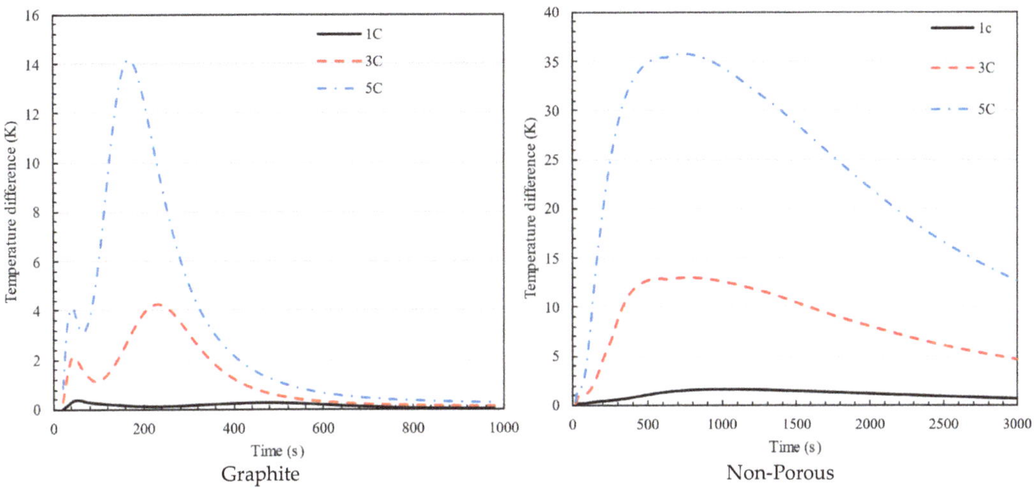

Figure 13. Temperature difference in 1C, 3C, and 5C charging of battery for the battery pack using graphite and non-porous media.

The absence of the first peak in the non-porous plot compared to the graphite plot is due to differences in heat dissipation mechanisms. In the graphite configuration, the high thermal conductivity of the graphite plate allows for rapid heat spreading, resulting in an initial peak as the system quickly redistributes heat before stabilizing. Conversely, in the non-porous case, heat dissipation relies primarily on natural convection, which is significantly slower and less effective at distributing heat. This leads to a more gradual temperature rise without a distinct initial peak, as heat accumulates more uniformly over time.

Figures 14 and 15 show the highest temperature and temperature variation for the discharge mode, respectively. Even if the temperature and temperature difference are within the acceptable range in all scenarios during discharge, the presence of a graphite causes the temperature to stabilize at a lower value compared to when there is no porous medium. Furthermore, comparing the highest temperature values reveals that the temperature stabilizes rapidly when using graphite, while the maximum temperature remains constant in the battery. Comparing temperature differences allows us to highlight graphite's outstanding ability to achieve an appropriate temperature distribution.

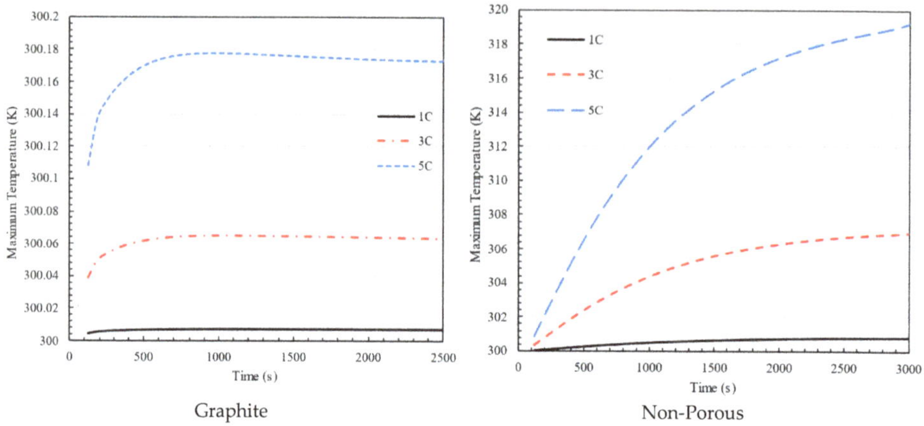

Figure 14. Maximum temperature in 1C, 3C, and 5C discharging of battery for the battery pack using graphite and non-porous media.

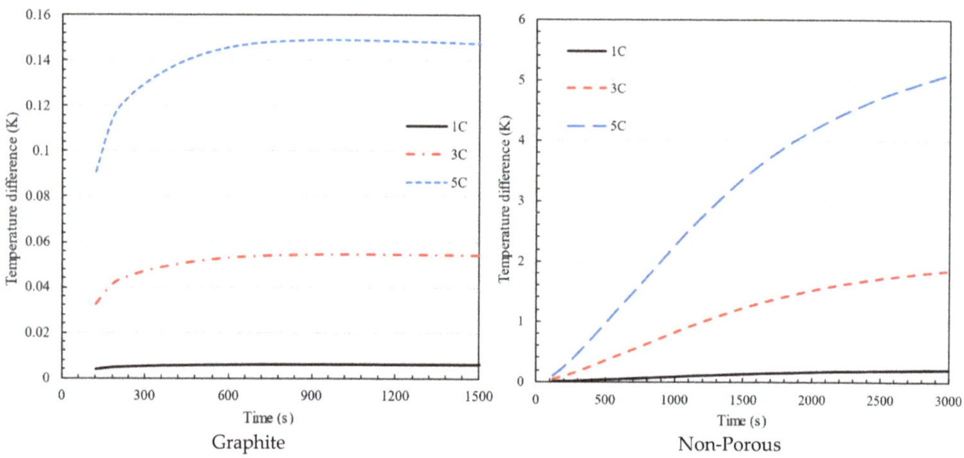

Figure 15. Temperature difference in 1C, 3C, and 5C discharging of battery for the battery pack using graphite and non-porous media.

Temperature variations inside battery packs have a substantial effect on their performance, safety, durability, and effectiveness. Uniform temperature across cells promotes peak performance and effectiveness, minimizing the chances of thermal runaway and other safety concerns. It also prolongs the battery's lifetime by minimizing damage from severe temperatures.

Figure 16 shows the temperature distribution on a horizontal wall (y = 35 mm) for batteries with and without the presence of graphite at 5C charging and discharging of the battery pack. As shown in both the charging and discharging of the battery pack, using graphite can help to distribute heat more uniformly throughout the battery pack. For non-porous battery packs, a temperature difference of about 5 K and 7 K for discharging and charging, respectively, is observed between cells in the side lines and central lines. This difference is less than 0.5 K for a battery pack containing graphite. Furthermore, Figure 16 shows that, when graphite is used in the battery pack, the thermal concentration will not be on cells, and the gap between them will be in a uniform thermal condition.

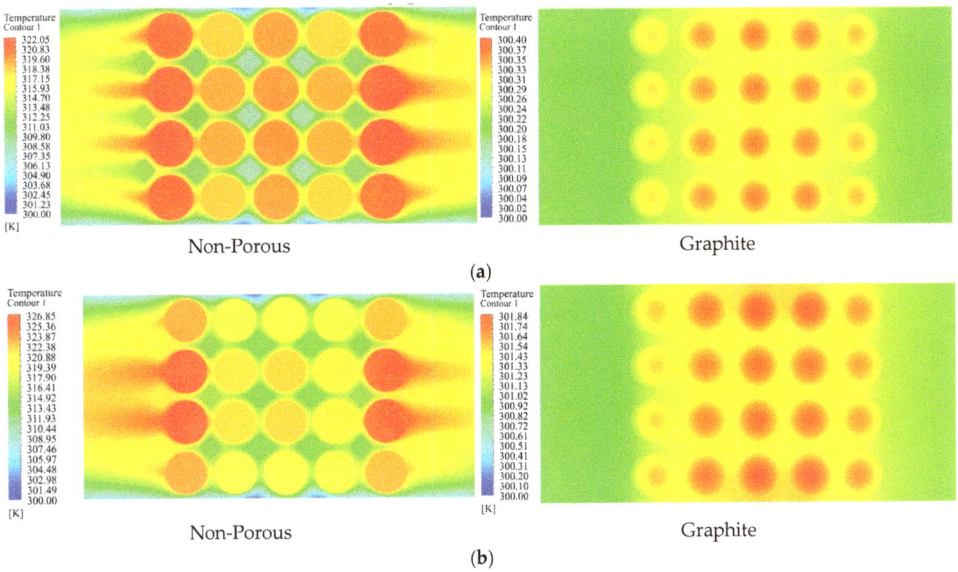

Figure 16. Temperature distribution in horizontal plan (y = 35 mm) for graphite and non-porous battery packs at 5C charging and discharging process. (**a**) Discharge. (**b**) Charge.

Figure 17 illustrates the impact of graphite integration on the thermal behavior of the battery pack under 5C charging conditions, comparing thermal abuse and normal conditions for both graphite and non-porous cases. The non-porous case under thermal abuse (black line) shows a rapid temperature rise, exceeding 700 K, indicating a high risk of thermal runaway and insufficient heat dissipation. In contrast, the porous configuration with graphite limits the maximum temperature to approximately 350 K, demonstrating superior thermal management by effectively spreading and dissipating heat. Additionally, under normal conditions with graphite, the temperature remains stable near 300 K, confirming that the designed cooling system effectively manages heat generation during standard operation.

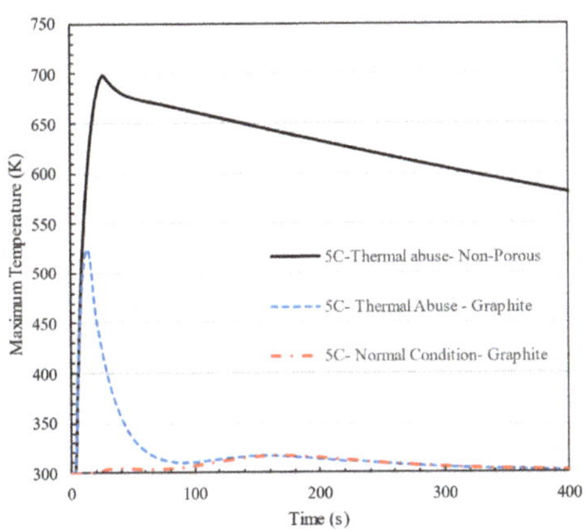

Figure 17. Effect of graphite on the thermal stability of the designed battery pack under 5C charging conditions, comparing thermal abuse and normal conditions.

This comparison highlights the effectiveness of the porous graphite cooling strategy in preventing extreme temperature rises and enhancing the safety of the battery pack under high C-rate and thermal abuse scenarios.

6.3. Battery Pack Performance at Different Conditions

ANSYS Fluent software (Release 2023 R1) was employed in this study to evaluate the thermal equilibrium of battery temperatures under varying system design parameters. These parameters include ambient temperature, air velocity entering the battery pack, and the cold plate surface temperature. A parametric study was conducted to analyze the influence of these factors on the maximum temperature of the battery pack and the temperature differential between its maximum and minimum values. The results demonstrate the proposed system's capability to regulate the maximum temperature and minimize temperature variation across different operating conditions. In this configuration, aluminum foam is used as filler material between the batteries, and all simulations were conducted at a discharge rate of 5C.

Figure 18a illustrates the effect of air inlet velocity on the maximum battery pack temperature at various ambient temperatures, with the cold plate temperature fixed at 280 K. The trend shows that the maximum temperature increases with the rise in ambient temperature. Even with a temperature difference of up to 40 K between the cold plate and the ambient temperature, the battery thermal management system (BTMS) successfully maintains the maximum temperature within acceptable limits.

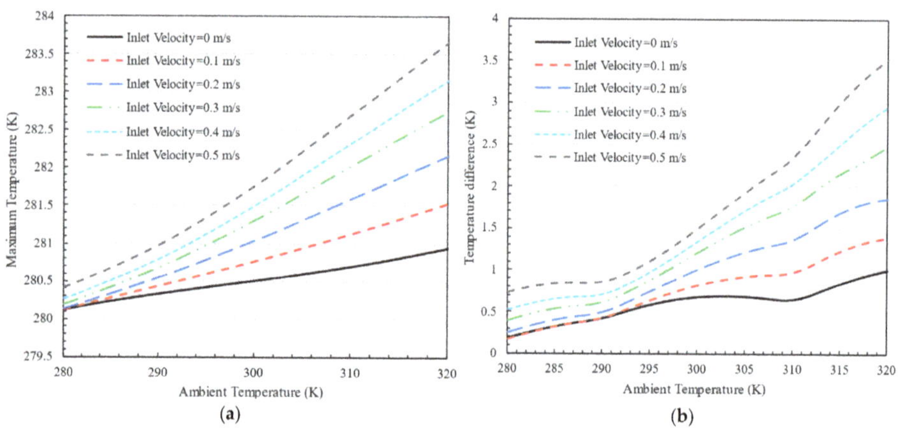

Figure 18. The effect of air inlet velocity on the (**a**) maximum temperature and (**b**) temperature difference in the battery pack at different ambient temperatures for $T_{coldplate} = 280$ K.

Similarly, Figure 18b depicts the temperature difference within the battery pack as a function of air inlet velocity. As the air inlet velocity increases, so does the temperature differential. However, even at higher ambient temperatures, the system keeps this difference below 3.5 K, indicating the system's ability to maintain a temperature difference of less than 3 K in most cases. Notably, as the ambient temperature exceeds the cold plate temperature, increasing the air velocity results in higher overall battery temperatures and a greater temperature difference within the pack.

Figure 19a illustrates the impact of air inlet velocity on the maximum battery pack temperature at different ambient temperatures, with the cold plate temperature set to 300 K. The graph indicates that as the ambient temperature rises, the maximum battery temperature also increases. Notably, as the air inlet velocity increases, the rate of temperature increase becomes more pronounced. However, in all scenarios, the designed battery thermal management system (BTMS) successfully maintains the battery pack's temperature within an acceptable range. Additionally, the graph shows that the temperature difference between the battery pack and the cold plate remains under 2 K in all conditions.

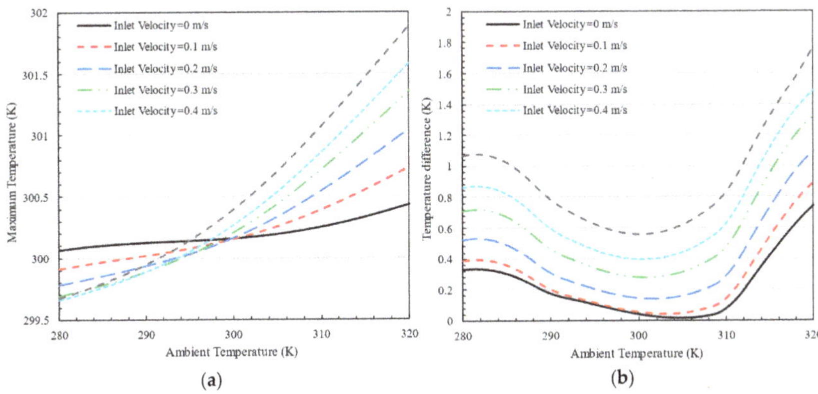

Figure 19. The effect of air inlet velocity on the (**a**) maximum temperature and (**b**) temperature difference in the battery pack at different ambient temperatures for $T_{coldplate} = 300$ K.

Figure 19b highlights the temperature difference within the battery pack when the cold plate is set to 300 K. The temperature difference is minimal when the ambient temperature is between 300 K and 310 K, with a differential of less than 2 K in all cases. This indicates improved thermal performance compared to the scenario with a cold plate temperature of 280 K, ensuring better temperature uniformity within the battery pack. As shown in Figure 19, the temperature rise is more rapid as the inlet velocity increases due to enhanced convective heat transfer. Higher air velocities facilitate greater heat dissipation from the battery surface, but they also increase heat transfer rates within the battery pack, causing a sharper initial temperature to rise before stabilization. At a room temperature of 320 K, the zero-inlet velocity condition exhibits a lower maximum temperature compared to others because, in the absence of forced convection, the dominant heat transfer mechanism is natural convection, which leads to slower but more uniform heat dissipation, preventing localized overheating.

7. Conclusions

A hybrid thermal management system was developed in the study, incorporating an air-cooling system, a cold plate, and porous materials. An extensive investigation was carried out to study heat generation in the battery pack during charging and discharging. A model was developed to forecast heat transfer in the battery pack with varying forms and charging rates. A study was conducted to examine the impact of five different types of porous materials on the maximum temperature and temperature difference in a battery pack. This study evaluated these factors with and without the presence of porous materials. The battery pack design was subjected to testing in various weather conditions. The main points of the present study are as follows:

- The heat produced during charging and discharging is directly proportional to the square of the C-rate, applicable to all battery pack configurations.
- The heat produced in a battery pack increases in direct proportion to the square of the number of batteries connected in series.
- Five distinct porous materials, including aluminum foam, copper foam, silicon carbide, aluminum oxide, and graphite, are analyzed. The results indicate that graphite

demonstrates superior performance in maintaining the maximum cell temperature and temperature difference within an acceptable range throughout both the charging and discharging of the battery pack.
- The suggested battery pack can maintain cells within the permissible range during charging and discharging at rates up to 5C.
- Porous materials help maintain a stable temperature in the battery pack within 10 min of the battery beginning to charge or discharge. In the scenario where a porous material is not utilized, the temperature stabilizes after one hour.
- Analyzing temperature contours of a battery pack reveal improved temperature dispersion during both the charging and draining processes.
- The suggested BTMS is evaluated under various conditions. BTMS performance is consistently satisfactory in all scenarios during the 5C charging of the battery pack.

These findings state that using graphite can obviously increase the efficiency of a hybrid battery pack while decreasing energy costs in active cooling systems such as air-cooling and liquid-cooling thermal management systems.

Author Contributions: Conceptualization, A.R.; investigation, A.R.; writing—original draft preparation, A.R.; writing—review and editing, M.A.; visualization, A.R., supervision, M.A. and M.D.; project administration, A.R. All authors have read and agreed to the published version of the manuscript.

Funding: This research received no external funding.

Data Availability Statement: The original contributions presented in this study are included in the article. Further inquiries can be directed to the corresponding author.

Conflicts of Interest: The authors declare no conflict of interest.

Nomenclature

	Unit	Description
A	m^2	Area
C	F	Capacitor
C_p	j/kg·K	Specific heat capacity
d_m	m	Pore diameter
I	ampere	Current
$\dot{j}$	m^3/s	Volumetric current transfer rate
k	W/m·K	PCM thermal conductivity
K_{TEC}	W/m·K	TEC material thermal conductivity
L_f	J/kg	Latent heat
$\dot{q}$	W	Heat generation
Q_c	W	Heat absorption
Q_H	W	Heat release
R	ohm	Resistor
R_S	ohm	Internal resistance
R_{TEC}	$S{\cdot}m^{-1}$	Electrical resistance
t	s	Time
T	K	Temperature
T_c	K	Cold temperature
T_h	K	Hot temperature
v	m/s	Velocity
V	volt	Voltage
V_{OCV}	volt	Open-circuit voltage
α	$\mu V{\cdot}K^{-1}$	Seebeck coefficient
β	-	Fluid fraction

ε	-	Porosity
σ_+	Ohm·m	Effective electric conductivity for positive electrode
σ_-	Ohm·m	Effective electric conductivity for negative electrode
φ_+	Joule/kg	Phase potential for positive electrode
φ_-	Joule/kg	Phase potential for negative electrode
ρ	kg/m^3	density
BTMS	Battery thermal management system	
CFD	Computational fluid dynamics	
COP	Coefficient of performance	
SOC	State of charge	

References

1. Ravi, S.S.; Brace, C.; Larkin, C.; Aziz, M.; Leach, F.; Turner, J.W. On the pursuit of emissions-free clean mobility—Electric vehicles versus e-fuels. *Sci. Total Environ.* **2023**, *875*, 162688. [CrossRef] [PubMed]
2. Hu, H.; Xu, X.; Sun, X.; Li, R.; Zhang, Y.; Fu, J. Numerical study on the inhibition control of lithium-ion battery thermal runaway. *ACS Omega* **2020**, *5*, 18254–18261. [CrossRef]
3. Zhao, J.; Song, C.; Li, G. Fast-charging strategies for lithium-ion batteries: Advances and perspectives. *ChemPlusChem* **2022**, *87*, e202200155. [CrossRef] [PubMed]
4. Su, Y.D.; Preger, Y.; Burroughs, H.; Sun, C.; Ohodnicki, P.R. Fiber optic sensing technologies for battery management systems and energy storage applications. *Sensors* **2021**, *21*, 1397. [CrossRef]
5. Rahmani, A.; Dibaj, M.; Akrami, M. Recent advancements in battery thermal management systems for enhanced performance of Li-ion batteries: A comprehensive review. *Batteries* **2024**, *10*, 265. [CrossRef]
6. Saavedra, J.; Gonzalez Cuadrado, D. Thermal Sensor Allocation for Effective and Efficient Heat Transfer Measurements in Transportation Systems. *Sensors* **2023**, *23*, 2803. [CrossRef] [PubMed]
7. Afzal, A.; Abdul Razak, R.K.; Mohammed Samee, A.D.; Kumar, R.; Ağbulut, Ü.; Park, S.G. A critical review on renewable battery thermal management system using heat pipes. *J. Therm. Anal. Calorim.* **2023**, *148*, 8403–8442. [CrossRef]
8. Rahmani, A.; Dibaj, M.; Akrami, M. A study on a battery pack in a hybrid battery thermal management system integrating with thermoelectric cooling. *Case Stud. Therm. Eng.* **2024**, *61*, 104856. [CrossRef]
9. Mariam, E.; Ramasubramanian, B.; Reddy, V.S.; Dalapati, G.K.; Ghosh, S.; Sherin, T.; Chakrabortty, S.; Motapothula, M.R.; Kumar, A.; Ramakrishna, S.; et al. Emerging trends in cooling technologies for photovoltaic systems. *Renew. Sustain. Energy Rev.* **2024**, *192*, 114203. [CrossRef]
10. Sharif, M.; Seker, H. Smart EV charging with context-awareness: Enhancing resource utilization via deep reinforcement learning. *IEEE Access* **2024**, *12*, 7009–7027. [CrossRef]
11. Wu, Y.; Huang, Z.; Li, D.; Li, H.; Peng, J.; Stroe, D.; Song, Z. Optimal battery thermal management for electric vehicles with battery degradation minimization. *Appl. Energy* **2024**, *353*, 122090. [CrossRef]
12. Lin, X.W.; Li, Y.B.; Wu, W.T.; Zhou, Z.F.; Chen, B. Advances on two-phase heat transfer for lithium-ion battery thermal management. *Renew. Sustain. Energy Rev.* **2024**, *189*, 114052. [CrossRef]
13. Shahzad, K.; Cheema, I.I. Low-carbon technologies in automotive industry and decarbonizing transport. *J. Power Sources* **2024**, *591*, 233888. [CrossRef]
14. Zhang, S.B.; Nie, F.; Cheng, J.P.; Yang, H.; Gao, Q. Optimizing the air flow pattern to improve the performance of the air-cooling lithium-ion battery pack. *Appl. Therm. Eng.* **2024**, *236*, 121486. [CrossRef]
15. Kumar, R.; Singh, L.K.; Gupta, A.K. Evaluating air-cooling performance of lithium-ion-battery module with various cell arrangements. *Energy Technol.* **2024**, *12*, 2301061. [CrossRef]
16. Chen, K.; Zhang, Z.; Wu, B.; Song, M.; Wu, X. An air-cooled system with a control strategy for efficient battery thermal management. *Appl. Therm. Eng.* **2024**, *236*, 121578. [CrossRef]
17. Moosavi, A.; Ljung, A.L.; Lundström, T.S. A study on the effect of cell spacing in large-scale air-cooled battery thermal management systems using a novel modeling approach. *J. Energy Storage* **2023**, *72*, 108418. [CrossRef]
18. Oyewola, O.M.; Idowu, E.T. Effects of step-like plenum, flow pattern and inlet flow regime on thermal management system. *Appl. Therm. Eng.* **2024**, *243*, 122637. [CrossRef]
19. Lan, X.; Li, X.; Ji, S.; Gao, C.; He, Z. Design and optimization of a novel reverse layered air-cooling battery management system using U and Z type flow patterns. *Int. J. Energy Res.* **2022**, *46*, 14206–14226. [CrossRef]
20. Wang, Y.; Yu, Y.; Jing, Z.; Wang, C.; Zhou, G.; Zhao, W. Thermal performance of lithium-ion batteries applying forced air cooling with an improved aluminium foam heat sink design. *Int. J. Heat Mass Transf.* **2021**, *167*, 120827. [CrossRef]

21. Zhang, S.B.; He, X.; Long, N.C.; Shen, Y.J.; Gao, Q. Improving the air-cooling performance for lithium-ion battery packs by changing the air flow pattern. *Appl. Therm. Eng.* **2023**, *221*, 119825. [CrossRef]
22. Zhang, F.; Liu, P.; He, Y.; Li, S. Cooling performance optimization of air cooling lithium-ion battery thermal management system based on multiple secondary outlets and baffle. *J. Energy Storage* **2022**, *52*, 104678. [CrossRef]
23. Kang, D.; Lee, P.Y.; Yoo, K.; Kim, J. Internal thermal network model-based inner temperature distribution of high-power lithium-ion battery packs with different shapes for thermal management. *J. Energy Storage* **2020**, *27*, 101017. [CrossRef]
24. Shahid, S.; Agelin-Chaab, M. Development and analysis of a technique to improve air-cooling and temperature uniformity in a battery pack for cylindrical batteries. *Therm. Sci. Eng. Prog.* **2018**, *5*, 351–363. [CrossRef]
25. Oyewola, O.M.; Awonusi, A.A.; Ismail, O.S. Design optimization of Air-Cooled Li-ion battery thermal management system with Step-like divergence plenum for electric vehicles. *Alex. Eng. J.* **2023**, *71*, 631–644. [CrossRef]
26. Jiang, W.; Feng, G.; Wang, H.; Chang, Z.; Tan, X.; Ji, J.; Zhang, Y.; Zhou, J. Analysis of hybrid active-passive prismatic Li-ion battery thermal management system using phase change materials with porous-filled mini-channels. *J. Energy Storage* **2024**, *80*, 110144. [CrossRef]
27. Verma, S.P.; Saraswati, S. Numerical and experimental analysis of air-cooled Lithium-ion battery pack for the evaluation of the thermal performance enhancement. *J. Energy Storage* **2023**, *73*, 108983. [CrossRef]
28. Zhao, G.; Wang, X.; Negnevitsky, M.; Li, C. An up-to-date review on the design improvement and optimization of the liquid-cooling battery thermal management system for electric vehicles. *Appl. Therm. Eng.* **2023**, *219*, 119626. [CrossRef]
29. Maknikar, S.K.; Pawar, A.M. Application of phase change material (PCM) in battery thermal management system (BTMS): A critical review. *Mater. Today Proc.* 2023; *in press*. [CrossRef]
30. Kavasoğullari, B.; Karagöz, M.E.; Yildiz, A.S.; Biçer, E. Numerical investigation of the performance of a hybrid battery thermal management system at high discharge rates. *J. Energy Storage* **2023**, *73*, 108982. [CrossRef]
31. Jiang, Z.; Li, H.; Sun, Z.; Qu, Z. Experimental study on 18650 lithium-ion battery-pack cooling system composed of heat pipe and reciprocating air flow with water mist. *Int. J. Heat Mass Transf.* **2024**, *222*, 125171. [CrossRef]
32. Sharma, D.K.; Agarwal, P.; Prabhakar, A. Effect of fin design and continuous cycling on thermal performance of PCM-HP hybrid BTMS for high ambient temperature applications. *J. Energy Storage* **2023**, *74*, 109360. [CrossRef]
33. Asli, M.; König, P.; Sharma, D.; Pontika, E.; Huete, J.; Konda, K.R.; Mathiazhagan, A.; Xie, T.; Höschler, K.; Laskaridis, P. Thermal management challenges in hybrid-electric propulsion aircraft. *Prog. Aerosp. Sci.* **2024**, *144*, 100967. [CrossRef]
34. Moaveni, A.; Siavashi, M.; Mousavi, S. Passive and hybrid battery thermal management system by cooling flow control, employing nano-PCM, fins, and metal foam. *Energy* **2024**, *288*, 129809. [CrossRef]
35. Rahmani, A.; Dibaj, M.; Akrami, M. Computational investigation of magnetohydrodynamic flow and melting process of phase change material in a battery pack using the lattice Boltzmann method. *J. Energy Storage* **2024**, *78*, 110046. [CrossRef]
36. Pakrouh, R.; Hosseini, M.J.; Ranjbar, A.A.; Rahimi, M. A novel liquid-based battery thermal management system coupling with phase change material and thermoelectric cooling. *J. Energy Storage* **2023**, *64*, 107098. [CrossRef]
37. Yang, M.; Mathew, G.; Nemati, H.; Moghimi, M.A. A novel approach for active cooling of a battery at cell level: Air-cooled mini-channel heat sink, enhanced with intermittent metal foam. *J. Energy Storage* **2024**, *81*, 110374. [CrossRef]
38. Sutheesh, P.M.; Atul, A.P.; Rohinikumar, B. Numerical and experimental investigations of thermal performance of lithium-ion battery with hybrid cooling system under dry-out condition. *J. Energy Storage* **2024**, *84*, 110889. [CrossRef]
39. Rahmani, A.; Dibaj, M.; Akrami, M. Enhancing Heat Storage Cooling Systems via the Implementation of Honeycomb-Inspired Design: Investigating Efficiency and Performance. *Energies* **2024**, *17*, 351. [CrossRef]
40. Bargal, M.H.; Abdelkareem, M.A.; Tao, Q.; Li, J.; Shi, J.; Wang, Y. Liquid cooling techniques in proton exchange membrane fuel cell stacks: A detailed survey. *Alex. Eng. J.* **2020**, *59*, 635–655. [CrossRef]
41. Kwon, K.H.; Shin, C.B.; Kang, T.H.; Kim, C.S. A two-dimensional modeling of a lithium-polymer battery. *J. Power Sources* **2006**, *163*, 151–157. [CrossRef]
42. Vanaclocha Hervas, C. Comparative Study of Three Electrochemical Cell Models for the CFD Simulation of a Battery Module. Ph.D. Thesis, Universitat Politècnica de València, Valencia, Spain, 2021.
43. Yetik, O.; Karakoc, T.H. A numerical study on the thermal performance of prismatic Li-ion batteries for hibrid electric aircraft. *Energy* **2020**, *195*, 117009. [CrossRef]
44. Yetik, O.; Karakoc, T.H. Estimation of thermal effect of different busbars materials on prismatic Li-ion batteries based on artificial neural networks. *J. Energy Storage* **2021**, *38*, 102543. [CrossRef]
45. Yetik, O.; Karakoc, T.H. Thermal and electrical analysis of batteries in electric aircraft using nanofluids. *J. Energy Storage* **2022**, *52*, 104853. [CrossRef]
46. Chen, H.; Zhang, T.; Hua, Y.; Gao, Q.; Han, Z.; Yang, K.; Xu, Y.; Liu, X.; Xu, X.; Wang, S. Simulation study on the interaction between the battery module and busbar under typical driving conditions of electric vehicles. *Case Stud. Therm. Eng.* **2023**, *45*, 103006. [CrossRef]

47. He, H.; Zhang, X.; Xiong, R.; Xu, Y.; Guo, H. Online model-based estimation of state-of-charge and open-circuit voltage of lithium-ion batteries in electric vehicles. *Energy* **2012**, *39*, 310–318. [CrossRef]
48. Hossein Bashiri, A.; Sangtarash, A.; Zamani, M. The effect of the porous media on thermal management of lithium-ion battery pack; a comparative and numerical study. *Therm. Sci. Eng. Prog.* **2022**, *34*, 101427. [CrossRef]
49. ANSYS Inc. *ANSYS Fluent Theory Guide, Release 2023 R1*; ANSYS Inc.: Canonsburg, PA, USA, 2023; Available online: https://www.ansys.com (accessed on 20 December 2024).
50. Li, C.; Cui, N.; Chang, L.; Cui, Z.; Yuan, H.; Zhang, C. Effect of parallel connection topology on air-cooled lithium-ion battery module: Inconsistency analysis and comprehensive evaluation. *Appl. Energy* **2022**, *313*, 118758. [CrossRef]
51. Plett, G.L. *Battery Management Systems, Volume II: Equivalent-Circuit Methods*; Artech House: London, UK, 2015.
52. Rashidi, S.; Esfahani, J.A.; Karimi, N. Porous materials in building energy technologies—A review of the applications, modelling and experiments. *Renew. Sustain. Energy Rev.* **2018**, *91*, 229–247. [CrossRef]

batteries

MDPI

Article

Online Core Temperature Estimation for Lithium-Ion Batteries via an Aging-Integrated ECM-1D Coupled Model-Based Algorithm

Yiqi Jia, Lorenzo Brancato, Marco Giglio and Francesco Cadini *

Dipartimento di Meccanica, Politecnico di Milano, Via La Masa, 1, 20156 Milano, Italy; yiqi.jia@polimi.it (Y.J.); lorenzo.brancato@polimi.it (L.B.); marco.giglio@polimi.it (M.G.)
* Correspondence: francesco.cadini@polimi.it

Abstract: Thermal management is pivotal for ensuring the safe and efficient operation of LIBs under dynamic conditions. Accurate core temperature monitoring remains a key BTMS challenge for predicting thermal distributions and mitigating TR risks. This study proposes a real-time core temperature estimation framework integrating joint EKF with an electro-thermal-aging model (ECM-1D). Using only surface temperature and voltage measurements, it simultaneously estimates core temperature, SOC, and capacity with bidirectional electro-thermal coupling. The hybrid approach pre-calibrates temperature/SOC/SOH-dependent parameters offline while updating capacity online. Validation under extreme conditions (high-rate cycling, aging, and ISCs) demonstrates 60% lower core temperature RMSE during high-rate cycling, a maximum estimation error below 0.9 K, and 58.9% reduction in SOC estimation error under aging conditions versus existing methods. The framework reliably tracks core temperature trends despite parasitic heat and signal noise, enabling earlier critical temperature warnings. This provides a foundation for TR prevention, advancing battery safety for EV and grid storage applications. Future extensions could integrate physical aging mechanisms and enhance fault detection capabilities.

Keywords: battery thermal management system; model-based algorithm; lithium-ion battery; core temperature estimation; thermal runaway

1. Introduction

The accelerating depletion of fossil fuels and the relentless rise in CO_2 emissions have intensified global efforts to transition toward sustainable energy systems in recent decades [1]. Among emerging energy storage technologies, LIBs have emerged as a cornerstone due to their high energy density, decreasing costs, and long life cycle [2]. These attributes have enabled transformative applications across portable electronics, EVs, and grid-scale renewable energy storage systems [3]. However, the growing demand for high-performance LIBs—particularly in EVs requiring rapid charging (high C-rate operation), extended environmental tolerance, and enhanced energy density—can promote higher temperature variations which create thermal management complexities and have exposed critical limitations in conventional thermal management systems [4,5].

Central to these limitations is the strong temperature dependence of LIBs. Extensive studies have demonstrated that their electrochemical performance is optimal only within a narrow thermal window (288.15–308.15 K) [6–8]. Deviation from this range initiates multiscale degradation processes, such as SEI decomposition, lithium plating, and capacity

https://doi.org/10.3390/batteries11040160

fade, which collectively escalate the risk of TR [9,10]. More critically, spatial temperature gradients within cells—driven by heterogeneous heat generation under extreme operating conditions—exacerbate these degradation mechanisms [11,12]. For example, mild thermal gradients (3 K) within the battery's active region can accelerate battery degradation by 300% [13]. Also, it has been reported that the radial temperature differentials exceed 10 K in cylindrical cells during high-rate discharging [14–16], while similar gradients in large-format cells exacerbate three-dimensional heat accumulation and local hotspots [12]. Such localized thermal variations promote uneven aging patterns that critically compromise both performance metrics and safety parameters [17,18].

The development of BTMS and BMS capable of monitoring the battery SOT is critical for ensuring LIB's operational reliability and thermal stability, while effective thermal regulation safeguards against electrochemical degradation driven by thermal variations, conventional surface-mounted sensors cannot directly measure internal thermal dynamics due to structural encapsulation [19]. This limitation is particularly significant because core temperature—a more direct indicator of SOT than surface temperature—is essential for optimizing BMS/BTMS control strategies [20].

To address this challenge, accurate multiphysics analysis of internal thermal gradients and electrothermal coupling mechanisms is imperative. Such analysis enables online estimation of core temperatures, which has emerged as a crucial research priority [20,21]. Enhanced SOT estimation facilitates intelligent thermal management through adaptive control strategies, improving operational efficiency and mitigating degradation induced by thermal gradients. Furthermore, integrating multiphysics electrothermal coupling characteristics supports battery health prognostics and early failure detection, while precise remaining useful life predictions enhance both safety and economic viability—key to sustainable battery lifecycle management.

Current methodologies for estimating or predicting battery core temperature are broadly categorized into data-driven and model-based approaches. Data-driven methods leverage machine learning algorithms trained on historical datasets to infer thermal and electrical states. Representative examples include the 2D-CNN developed by Li et al. for simultaneous core temperature and SOC estimation [22], and the spatiotemporal neural network by Liu et al., which integrates RGB neural networks with EKF [23]. EIS-based methods also have been extensively researched and combined with data-driven approaches in recent years for battery temperature prediction [24]. For example, Yuan et al. developed an advanced hybrid framework that integrates EIS-based volume-averaged temperature estimation with LSTM networks for thermodynamic parameter analysis, significantly improving core temperature prediction accuracy [25]. Additionally, comparative studies have systematically evaluated the performance of various data-driven models for this application [26]. Despite the high predictive accuracy, data-driven approaches face inherent limitations: their reliance on battery-specific historical datasets hinders generalization across diverse configurations and operating conditions, while their black-box nature complicates integration with functional safety-critical BMS requirements.

The evolution of model-based approaches for battery thermal management reflects a progressive integration of thermal and electrical dynamics to enhance estimation accuracy and computational efficiency. Initially, Sun et al. pioneered a KF-based framework that integrated Forgez's reduced-order thermal model [15], achieving internal temperature estimation with a maximum error of 1 K [27]. This work laid the foundation for thermal-centric state estimation but did not account for temperature-dependent electrical behaviors. Subsequently, Ma et al. advanced this paradigm by incorporating temperature-dependent electrical resistance into a ROM, enabling more accurate heat generation calculations and improving temperature estimation precision [28]. Building on these insights, researchers recognized the necessity of capturing bidirectional electrothermal coupling—where electrical parameters influence heat generation, and temperature variations conversely alter electrical dynamics [29]. For instance, Zhang et al. developed a simplified thermoelectric model that combined a 1RC ECM with a lumped 1D thermal model, explicitly deriving heat generation from ohmic losses and achieving real-time temperature estimation with an RMSE of 1 K [30]. These iterative advancements underscore the transition from standalone thermal models to tightly coupled electrothermal frameworks, balancing accuracy with computational tractability for real-time BMS applications.

Despite advancements in online core temperature estimation, existing approaches rarely address aging-induced variations in electrical–thermal parameters under dynamic operational conditions. Chen et al. [31] proposed an AFFRLS-dual EKF algorithm to concurrently update all ECM and thermal model parameters alongside SOC, SOH, and SOT across the battery's lifecycle. Although their method demonstrated robust temperature estimation under varying SOH levels and ambient temperatures, several limitations persist: simultaneous estimation of multiple parameters complicates filter-parameter adjustments and introduces computational complexity for real-time implementation; validation is restricted to constant current discharging profiles rather than real-world dynamic profiles; and critical scenarios such as high-rate charge–discharge cycles or fault-induced thermal transients—significantly impacting temperature distribution and model stability—remain unexamined. Additionally, parameter dependencies were indirectly derived from SOH-dependent OCV relationships instead of direct correlations.

To address these challenges, this study develops a model-based framework for online battery core temperature estimation by integrating SOC and capacity tracking through a coupled electrothermal model. To ensure computational efficiency without sacrificing accuracy, a hybrid parameterization strategy is implemented: temperature- and SOC-dependent parameters (e.g., entropy coefficient) are pre-calibrated offline through designed experiments, while temperature-, SOC-, and SOH-dependent ECM parameters are pre-calibrated offline via high-fidelity simulations. Both sets of parameters are stored as lookup tables. Only battery capacity—a key parameter linking SOC, SOH, and heat generation—is dynamically updated online. Simultaneously, the core temperature and SOC, and capacity are estimated online through a joint EKF algorithm. This proposed method is tested under extreme scenarios—including high rate charge–discharge cycles, ISCs, and accelerated aging protocols—the framework achieves RMSE < 0.9 K for core temperature estimation. Notably, it enables real-time capacity fade tracking under continuous aging stress and offers relatively good performance even under battery ISC faults. Compared to prior work, this approach advances the field as follows: (i) Enhancing core temperature estimation accuracy through a coupled thermal-electrical–aging model that dynamically captures parameter interactions. (ii) Achieving online multi-state joint estimation within a lean computational framework. (iii) Demonstrating robustness across diverse extreme dynamic profiles built by a high-fidelity battery model. By explicitly embedding aging effects into the estimation architecture, this method provides a physically interpretable and computationally efficient solution for adaptive thermal management in real-world applications.

The remainder of this paper is structured as follows: Section 2 first presents the proposed methodology, detailing the coupled electro-thermal model formulation, parameter estimation framework, and algorithm implementation. Building upon this foundation, Section 3 systematically analyzes the core temperature estimation accuracy under diverse operating conditions through comprehensive simulation studies. Finally, Section 4 concludes with key findings and discusses future work.

2. Methodology

The proposed methodology establishes a closed-loop framework for online core temperature estimation, integrating offline parameter characterization with real-time state observation. As illustrated in Figure 1, the framework operates through two sequential phases: offline parameter identification and online joint state estimation.

In the offline phase, critical model parameters are systematically determined through simulated characterization studies under controlled variations of thermal conditions, SOC levels, and aging states. These parameters form the foundation for a reduced-order coupled electro-thermal model that explicitly incorporates multi-domain dynamics, including temperature-dependent electrical responses and aging-induced capacity fade.

The online phase employs an EKF algorithm to simultaneously track three interdependent states: (i) SOC through voltage-current dynamics, (ii) core temperature via thermal state propagation, and (iii) capacity degradation through incremental parameter adaptation. The algorithm embeds the reduced-order ECM-1D thermal battery model to maintain computational efficiency while preserving essential electro-thermal couplings. Subsequent sections provide detailed technical implementation of three core components: reduced-order model development; offline parameter estimation process; and joint EKF algorithm implementation.

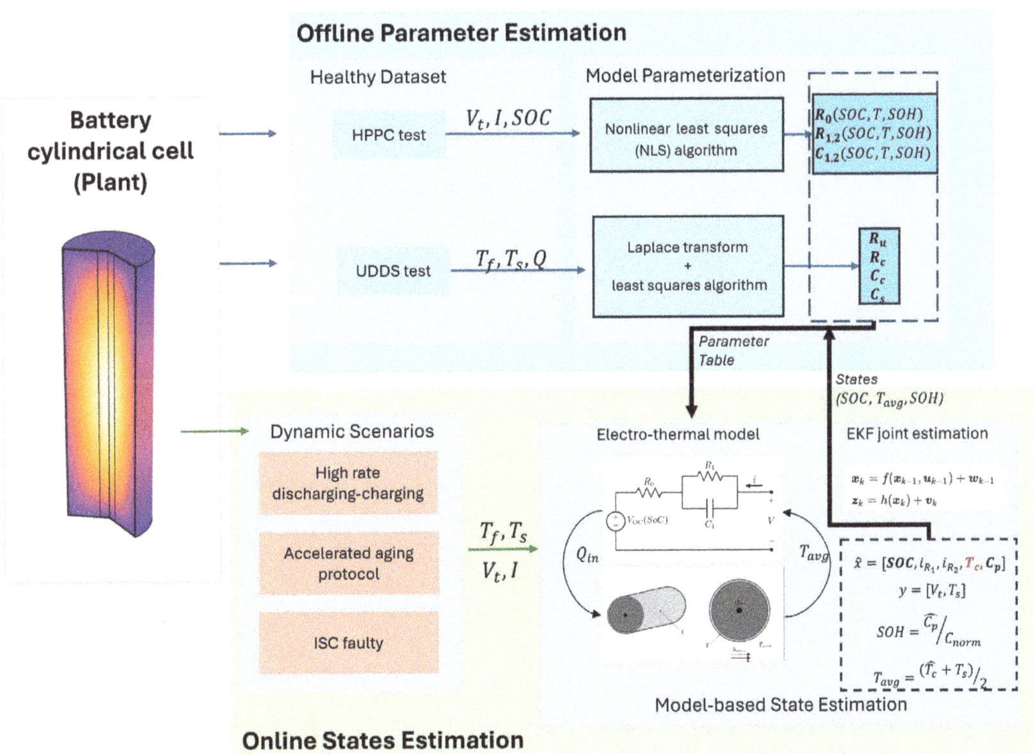

Figure 1. Overall online core temperature estimation framework.

2.1. Model Development

The proposed electro-thermal model architecture extends the framework by Lin et al. [32] through enhanced parameter coupling mechanisms, as illustrated in Figure 2.

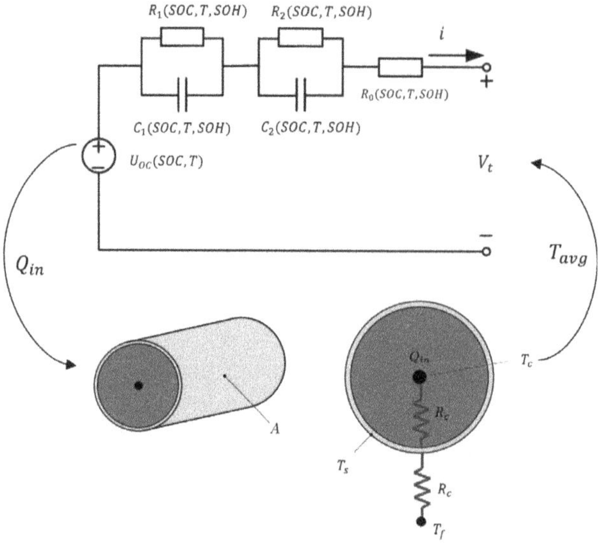

Figure 2. Reduced-order battery model.

This control-oriented formulation synergistically integrates electrical and thermal dynamics via two coupled subsystems.

The electrical subsystem employs a second-order ECM with temperature- and state-dependent parameters, as defined by Equation (1).

$$V_t = U_{oc}(SOC,T) - R_0(T,SOC,SOH)i - \sum_{k=1}^{2} R_k(T,SOC,SOH)i_{R_k} \tag{1}$$

where

- $U_{oc}(SOC,T)$ is the OCV, which incorporates temperature and SOC compensation as reported in Equation (2);
- R_0 and R_k (for $k = 1,2$) denote the ohmic and polarization resistances, respectively, each dependent on temperature (T), SOC, and SOH;
- In this coupled model, the electrical subsystem's temperature T is represented by the average temperature T_{avg}, which is derived from the thermal model.

The OCV with temperature compensation is defined by Equation (2).

$$U_{oc}(SOC,T) = U_{oc,ref}(SOC) + (T - T_{ref})\frac{\partial U_{oc}}{\partial T} \tag{2}$$

with the reference temperature T_{ref} defined as 293.15 K.

Here, $U_{oc,ref}(SOC)$ is the reference OCV at T_{ref}, and $\frac{\partial U_{oc}}{\partial T}$ is the entropic (temperature sensitivity) coefficient, typically provided as a function of SOC.

The state-of-charge evolution is described via coulomb counting, as defined by Equation (3).

$$SOC(t) = SOC_0 - \frac{\eta}{C_p}\int_0^t i\,dt \tag{3}$$

where

- SOC_0 is the initial SOC;
- C_p is the battery capacity, which may degrade over time;
- The integration is performed with a 1 s time step.

The polarization currents are modeled by first-order dynamics, as defined by Equation (4).

$$\frac{di_{R_k}}{dt} = \frac{1}{\tau_k}(-i_{R_k} + i), \quad k = 1,2 \tag{4}$$

with the time constants defined as $\tau_k = R_k C_k$. All parameters in the electrical model are summarized in Table 1.

The thermal behavior of the battery is captured using a two-state lumped parameter model, which divides the battery into the core and the surface regions through the radial direction of the cell. This model provides a balance between model accuracy and computational efficiency by capturing essential temperature gradients without resorting to a full spatial model. It is built based on the following considerations:

- Longitudinal simplification: The high thermal conductivity along the longitudinal axis ensures a uniform temperature distribution, allowing the neglect of temperature variations along the axial direction of the cell.
- Radial resolution: Significant temperature gradients in the radial direction of the cell motivate a core-surface discretization.

The governing equations for radial heat transfer are as follows:

$$C_c \frac{dT_c}{dt} = Q_{in} + \frac{T_s - T_c}{R_c} \tag{5}$$

$$C_s \frac{dT_s}{dt} = \frac{T_f - T_s}{R_u} - \frac{T_s - T_c}{R_c} \tag{6}$$

where T_c and T_s denote the core and surface temperatures, respectively.

The heat generation combines irreversible and reversible (entropic) components, as defined in Equation (7).

$$Q_{in} = -\left[\underbrace{(V_t - U_{oc}(SOC,T))i}_{\text{Irreversible heating}} + \underbrace{T_{avg}\, i \frac{\partial U_{oc}}{\partial T}(SOC)}_{\text{Entropic heating}}\right] \tag{7}$$

Here, T_{avg} is defined later as the average battery temperature. Natural air convection between the cell surface and air is considered as the cooling type in this research, as defined in Equation (6), which can be extended to other cooling types via the tunable thermal parameter R_u. Other parameters in the thermal model are summarized in Table 1.

The proposed model incorporates three essential coupling mechanisms:

- Thermal feedback: The average battery temperature is dynamically updated as Equation (8).

  $$T_{avg} = \frac{1}{2}(T_c + T_s), \tag{8}$$

  which in turn updates the temperature-dependent ECM parameters.
- Entropic heating: Reversible thermodynamic effects are captured via the entropic heating term in Equation (7).
- Aging mechanisms: Battery degradation is characterized at each time step and dynamically updates the SOH-dependent ECM parameters. The state-of-health is defined as follows:

  $$SOH = \frac{C_p}{C_{norm}} \tag{9}$$

  where C_{norm} denotes the nominal battery capacity and SOH modifies ECM parameters through pre-calibrated aging maps.

The SOC-OCV curve and the entropy coefficient curve, both as functions of SOC, are typically provided in battery data sheets or determined experimentally.

Table 1. Model nomenclature

Symbol	Physical Meaning	Value	Unit
V_t	Terminal voltage	-	V
U_{oc}	OCV	-	V
R_0	Ohmic resistance	-	Ω
R_1, R_2	Polarization resistances	-	Ω
C_1, C_2	Polarization capacitances	-	F
τ_1	Fast time constant (charge transfer)	-	s
τ_2	Slow time constant (diffusion)	-	min
i	Applied current	-	A
i_{R_k}	Polarization currents ($k = 1, 2$)	-	A
η	Coulombic efficiency	1	-
C_p	Battery capacity	-	Ah
C_{norm}	Nominal capacity	1.258	Ah
R_c	Core-to-surface thermal resistance	1.00	K/W
R_u	Surface-to-coolant thermal resistance	11.93	K/W
C_c	Core heat capacity	33.45	J/K
C_s	Surface heat capacity	3.35	J/K
T_f	Coolant/ambient temperature	-	K
T_{avg}	Average temperature	-	K
T_{ref}	Reference temperature	293.15	K
Q_{in}	Total heat generation rate	-	W
$\frac{\partial U_{oc}}{\partial T}$	Entropic coefficient	-	V/K
w	Process noise	-	-
n	Measurement noise	-	-
$\hat{T}_c$	Estimated core temperature	-	K
Δt	Sampling time	1	s

2.2. *Offline Parameter Estimation*

The parameter identification framework employs a hybrid approach combining physical characterization and data-driven optimization. The study focuses on a commercial 18650 LIB (1.258 Ah nominal capacity), with simulated datasets generated through a validated COMSOL 6.3 Multiphysics® air-cooled battery model. This model couples a lumped-parameter simplification of the SPM with a 2D axisymmetric thermal model, where the thermal-electrical coupling is achieved through joule heating terms calculated from ECM parameters, referred as 'plant model'. Detailed specifications can be found in [33].

The overall framework for electrical and thermal parameter identification is summarized in Table 2. Both approaches utilize hybrid methodologies combining physical insights with gray-box parameter optimization, enabling systematic analysis of battery behaviors through distinct identification procedures. This offline parameter calibration process can be extended to other LIB types by leveraging either existing model parameters from the literature or experimental characterization data, ensuring broader applicability of the proposed approach. For the battery states or working conditions beyond the range mentioned in Table 2, a re-calibration will be required.

Table 2. Comparative framework for parameter identification.

Criteria	Electrical Model (ECM)	Thermal Model
Test Conditions	- SOC: 0.2–1 every 0.1 - T: 288.15–358.15 K every 10 K - SOH: 0.65–1 every 0.05	- Air natural convection ($h = 20$ W/(m^2·K)) - Casing material constraints
Working Profile	HPPC test	UDDS profile
Measured Parameters	- Terminal voltage (V_t) - Current (i) - Ambient temperature (T_f)	- Surface temperature (T_s) - Ambient temperature (T_f) - Current (i)
Identification Algorithm	NLS Levenberg-Marquardt optimization	Laplace Transform analysis Grey-box NLS estimation with physical constraints
Key Parameters	$\theta = [R_0, R_1, R_2, C_1, C_2]$	$\theta = [R_u, R_c, C_c, C_s]$

The parameter estimation results revealed distinct characteristics among the ECM components. It can be observed that the resistance parameters exhibited marked variations across different SOH, which evidenced the importance of integrating the SOH-dependency into the model. The complete parameterization results are presented in Figure 3, with complementary thermal characteristics detailed in Table 1.

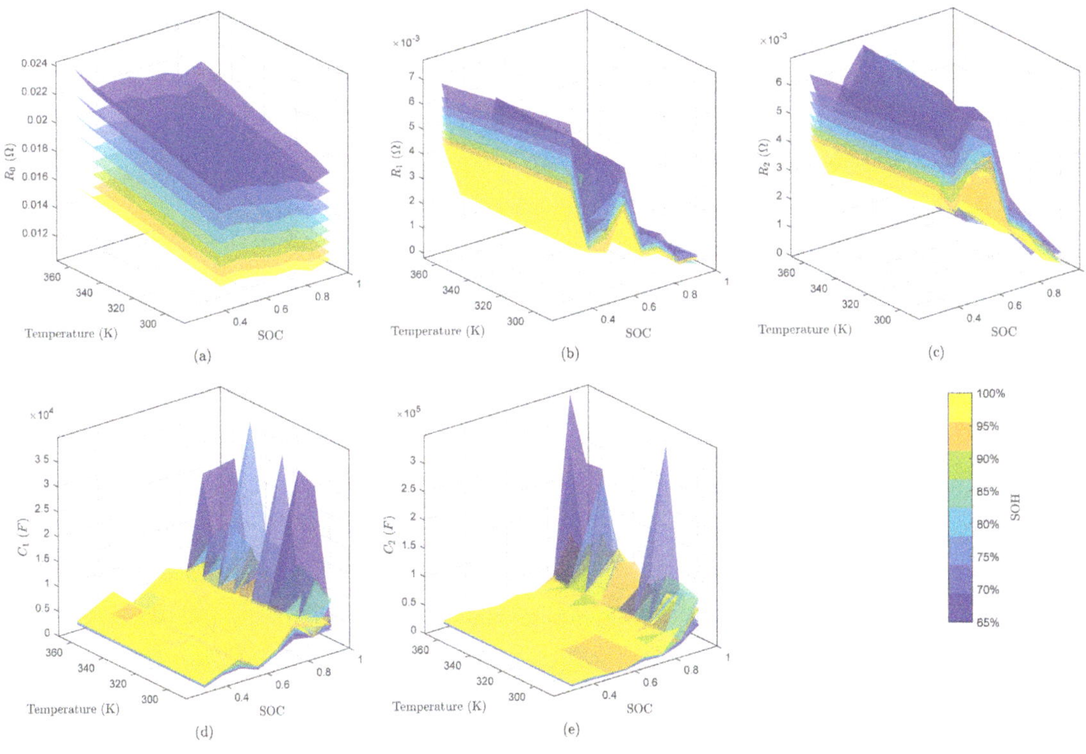

Figure 3. ECM parameters: (**a**) R_0, (**b**) R_1, (**c**) R_2, (**d**) C_1, (**e**) C_2.

2.3. *Joint EKF Algorithm Implementation*

Joint EKF enables simultaneous tracking of the battery cell's dynamic states and unknown parameters by augmenting them into a unified state vector, which is recursively updated using the EKF framework [34]. In this study, the algorithm is adapted to enhance the estimation accuracy of the SOT while accounting for battery aging effects through capacity estimation integration. The joint estimation process addresses the coupled dynamics between the thermal behavior and battery health, where aging-induced capacity loss alters both the heat generation rates and the thermal response. The algorithm implementation follows the phased structure detailed in Table 3. In the kth step, the augmented state vector is denoted as follows:

$$x_k = [SOC_k,\, i_{R_1,k},\, i_{R_2,k},\, T_{c,k},\, C_{p,k}]^T, \tag{10}$$

where SOC_k represents the battery's state-of-charge, $i_{R_1,k}$ and $i_{R_2,k}$ denote the polarization currents of the first and second RC branches (capturing fast and slow dynamic effects, respectively), $T_{c,k}$ is the core temperature, and $C_{p,k}$ corresponds to the battery capacity that reflects aging effects. The input vector is given as follows:

$$u_k = [i_k,\, V_k,\, T_{f,k}]^T, \tag{11}$$

where i_k is the battery current input, V_k is the terminal voltage, and $T_{f,k}$ represents the coolant or ambient temperature for thermal management. The measurement vector is defined as follows:

$$y_k = [V_{t,k},\, T_{s,k}]^T. \tag{12}$$

with $V_{t,k}$ being the measured terminal voltage and $T_{s,k}$ the measured surface temperature of the battery.

According to the coupled model introduced in Section 2.1, the discrete-time electro-thermal dynamics are described and observed by the following process and measurement equations: Process equations:

$$\mathrm{SOC}_{k+1} = \mathrm{SOC}_k - \frac{\Delta t}{C_{\mathrm{p},k}} i_k + w_{\mathrm{SOC}} \tag{13}$$

$$i_{\mathrm{R1},k+1} = i_{\mathrm{R1},k} + \frac{\Delta t}{\tau_{1,k}}\left(-i_{\mathrm{R1},k} + i_k\right) + w_{i_{\mathrm{R1}}} \tag{14}$$

$$i_{\mathrm{R2},k+1} = i_{\mathrm{R2},k} + \frac{\Delta t}{\tau_{2,k}}\left(-i_{\mathrm{R2},k} + i_k\right) + w_{i_{\mathrm{R2}}} \tag{15}$$

$$T_{\mathrm{c},k+1} = T_{\mathrm{c},k} + \frac{\Delta t}{C_\mathrm{c}}\left[Q_{\mathrm{in},k} + \frac{T_{\mathrm{s},k} - T_{\mathrm{c},k}}{R_\mathrm{c}}\right] + w_{T_c} \tag{16}$$

$$C_{\mathrm{p},k+1} = C_{\mathrm{p},k} + w_{C_p} \tag{17}$$

where Δt is the sampling time interval. The $w_{\mathrm{SOC}}, w_{i_{\mathrm{R}_1}}, w_{i_{\mathrm{R}_2}}, w_{T_c}, w_{C_p}$ denote the process noise terms associated with SOC, polarization currents, core temperature, and battery capacity, respectively. Other parameters has been explained in previous Section 2.1. Measurement equations:

$$V_{\mathrm{t},k+1} = U_{\mathrm{oc},k} - R_{1,k}\, i_{\mathrm{R1},k} - R_{2,k}\, i_{\mathrm{R2},k} - R_{0,k}\, i_k + n_{V_t} \tag{18}$$

$$T_{\mathrm{s},k+1} = T_{\mathrm{s},k} + \frac{\Delta t}{C_\mathrm{s}}\left[\frac{T_{\mathrm{f},k} - T_{\mathrm{s},k}}{R_\mathrm{u}} + \frac{T_{\mathrm{s},k} - T_{\mathrm{c},k}}{R_\mathrm{c}}\right] + n_{T_k} \tag{19}$$

where n_{V_t} and n_{T_k} represent the measurement noise associated with the terminal voltage and the surface temperature, respectively.

As demonstrated above in Section 2.2, key model parameters exhibit coupled dependencies on SOC, SOH, and thermal conditions. These relationships are characterized by the following: (i) temperature/SOC-dependent polarization components: $\{C_1, C_2\} = f(T_{avg}, SOC, SOH))$; (ii) aging-sensitive ohmic resistance: $\{R_0, R_1, R_2\} = f(T_{\text{avg}}, SOC, SOH)$; (iii) capacity degradation metric: $SOH = C_p/C_{\text{norm}}$.

This proposed online battery core temperature estimation method is tested through diverse dynamic battery operating conditions, with the corresponding results presented and discussed in the subsequent chapter.

Table 3. Joint extended Kalman filter algorithm steps.

Step	Mathematical Formulation
Augmentation	$z_k = \begin{bmatrix} x_k \\ \theta_k \end{bmatrix}$
Initialization	$\hat{z}_0^+ = \mathbb{E}[z_0]$ $P_0^+ = \mathbb{E}\left[(z_0 - \hat{z}_0^+)(z_0 - \hat{z}_0^+)^\top\right]$
Prediction	$\hat{z}_k^- = f(\hat{z}_{k-1}^+, u_{k-1})$ $P_k^- = F_{k-1}P_{k-1}^+F_{k-1}^\top + Q_k$ $F_{k-1} = \left.\frac{\partial f}{\partial z}\right\|_{\hat{z}_{k-1}^+}$
Update	$K_k = P_k^- H_k^\top (H_k P_k^- H_k^\top + R_k)^{-1}$ $\hat{z}_k^+ = \hat{z}_k^- + K_k(y_k - h(\hat{z}_k^-))$ $P_k^+ = (I - K_k H_k)P_k^-$ $H_k = \left.\frac{\partial h}{\partial z}\right\|_{\hat{z}_k^-}$

3. Results and Discussion

This chapter evaluates the performances of the proposed algorithm under three dynamic operational scenarios characterized by significant temperature variations: high-rate current charge–discharge cycles, accelerated aging simulations, and ISC fault conditions. The ambient temperature T_f under simulations is set as 298.15 K. Validation is performed via coupled SPM-2D thermal simulations (adapted from [33]), addressing the technical constraints associated with non-disruptive experimental validation [35], while the model has certain limitations, including the absence of explicating aging mechanisms and the assumption of homogeneous material distribution, it effectively overcomes experimental challenges associated with core temperature measurement while simulating various dynamical operation conditions. The simulations employ a standard ('Normal') mesh element size in 'plant model', which has been verified through grid independence studies to provide sufficient accuracy while maintaining computational efficiency.

High-rate charge–discharge cycles are first examined to assess the algorithm's capability in capturing thermal dynamics through comparative analysis with the conventional standalone thermal model applied by Sun et al. [27], as well as to evaluate the stability of the aging-related parameter estimates. In accelerated aging simulations, the method adapts to capacity fade and parameter drift, leading to enhanced core temperature and SOC tracking compared with the prior work proposed by [36], while enabling real-time SOH monitoring. Finally, under ISC fault conditions, the algorithm maintains reliable core temperature estimation despite electrical–thermal disturbances, thereby facilitating early detection of TR. These comparative analyses underscore the algorithm's advantages in dynamic performance, aging adaptability, and fault robustness. Notably, on a laptop equipped with an AMD Ryzen 7 5800H processor (AMD, Santa Clara, CA, USA), each estimation step was completed within 0.06 s, outperforming most existing approaches reported in the literature [37] and demonstrating compatibility with real-time BMS implementation.

3.1. Dynamic Performance Evaluation

This subsection evaluates the algorithm's capability to capture rapid thermal dynamics under extreme power conditions through high-rate charge–discharge cycle testing. A custom high-rate current profile (Figure 4) is designed to rigorously assess dynamic response performance. Under peak current rates exceeding 20 C, substantial heat accumulation induces pronounced core-to-surface temperature differentials (up to 6 K) and elevates the overall battery temperature to approximately 365 K within minutes.

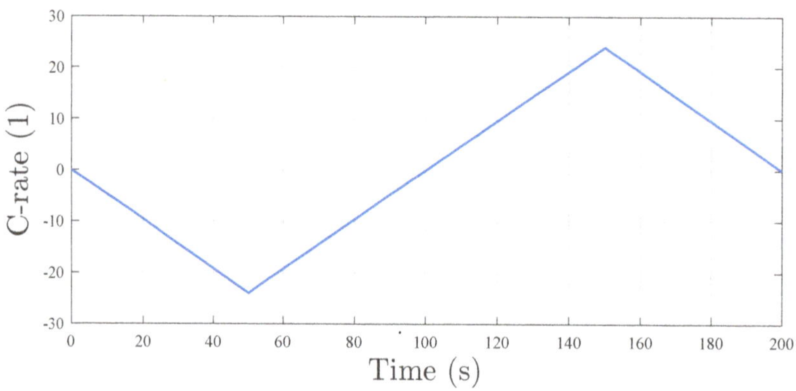

Figure 4. High rate discharging/charging current profile.

A comparative analysis with a standalone thermal model (referred to as 'thermal model' in subsequent sections) highlights the critical role of electrothermal coupling in the proposed framework. Specifically, the compared method proposed by Lin et al. [38] integrates a two-state thermal model into an EKF but excludes electrical measurements and neglects electrothermal interactions. This conventional approach assumes simplified Joule heating with a fixed internal resistance ($R_0 = 0.01\,\Omega$), derived from average parameter estimates (typically sourced from datasheet in practical applications).

As demonstrated in Figure 5, the proposed methodology achieves superior estimation accuracy throughout the operational cycles. A 62.5% reduction in RMSE quantifies the model's enhanced performance. Crucially, under healthy battery conditions (SOH = 1), the estimated capacity exhibits minimal variations (relative error < 0.16%), as shown in Figure 5c, confirming immunity to false degradation indications in non-aging scenarios.

The performance enhancement primarily stems from a bidirectional electro-thermal coupling mechanism that operates through two critical pathways: (1) dynamic quantification of reversible heat generation during phase transitions via SOC-dependent entropy coefficients, and (2) continuous compensation of RC network parameters through real-time thermal adaptation. This dual-path architecture enables simultaneous thermal–electrical state co-estimation while maintaining parameter consistency across operational temperature gradients.

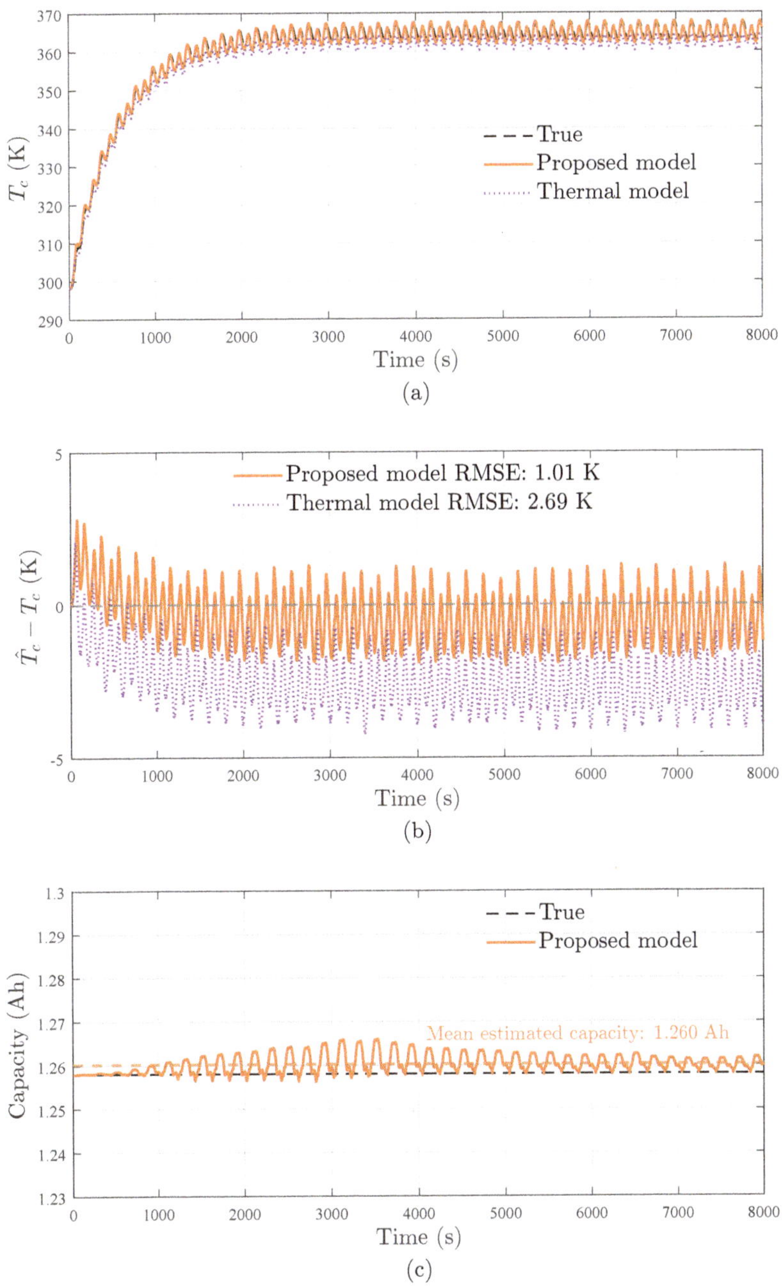

Figure 5. Core temperature and capacity estimation results under high-rate discharging/charging profiles: (**a**) the estimated core temperature (T_c), (**b**) temperature estimation error of core temperature with RMSE comparison, (**c**) the estimated battery capacity (C_p).

3.2. Health-Adaptive Capability Analysis

This subsection evaluates the algorithm's ability to adapt to battery aging by simultaneously tracking SOH, SOC, and core temperature under dynamic operational conditions. An accelerated aging protocol is designed using UDDS cycles combined with a controlled capacity fade profile (Figure 6d). The capacity loss (Q_{loss}) is governed by calendar aging dynamics, where the degradation rate (I_{loss}) depends on the nominal capacity (C_{norm}) and aging time constant (τ_{loss}) through the linear relationship defined in Equations (20)–(22).

$$\frac{\partial Q_{\text{loss}}}{\partial t} = I_{\text{loss}}, \quad Q_{\text{loss}}|_{t=0} = 0 \tag{20}$$

$$I_{\text{loss}} = \frac{C_{\text{norm}}}{\tau_{\text{loss}}} \tag{21}$$

$$C_{\text{p}} = C_{\text{norm}} - Q_{\text{loss}} \tag{22}$$

Unlike constant current profiles, the UDDS cycles impose dynamic and variable loads, adding significant electro-thermal complexity that closely mimics real driving conditions. This protocol utilized to test two interconnected functionalities of the proposed method: (1) core temperature estimation accuracy under aging, and (2) joint SOH-SOC tracking performance.

The proposed method is compared with Pang et al.'s electrothermal framework [36] (referred as 'electro-thermal model'), which integrates a simplified thermal model with temperature-dependent ECM parameters and irreversible heating assumptions (Equation (7)). Unlike the proposed aging-adaptive architecture, Pang's approach lacks explicit capacity loss. Which means, it assumes a fixed battery capacity and does not update parameters as the cell ages. This comparison highlights the improvements gained through our SOH-adaptive framework.

Figure 6 shows the estimation results of battery states of the proposed method and the compared method. Key results are summarized as follows: First, with the SOH-adaptive framework, our proposed method reduces maximum core temperature error from 1.15 K to 0.89 K (RMSE decreased by 11%). Second, real-time capacity estimation enables significant SOC tracking enhancement, yielding a 58.9% reduction in RMSE. In contrast, Pang's model exhibits accumulated SOC errors (Figure 6c) due to unaccounted capacity loss, which further degrade temperature estimation accuracy (since incorrect SOC affects predicted voltage and heat generation). Third, The proposed framework reliably tracks the battery's capacity fade over time (Figure 6d), confirming its ability to estimate SOH in real time alongside SOC and temperature. This joint estimation capability not only enhances the model's accuracy in reflecting aging but also provides a foundation for advanced health monitoring strategies for the battery management.

Comparative analysis confirms the framework's superiority in sustaining temperature estimation accuracy below 1 K throughout the battery lifespan—a critical advancement over conventional electrothermal models without capacity adaptation. This performance gap highlights the necessity of online co-estimating electrical, thermal, and degradation states of the battery, a capability absent in prior approaches.

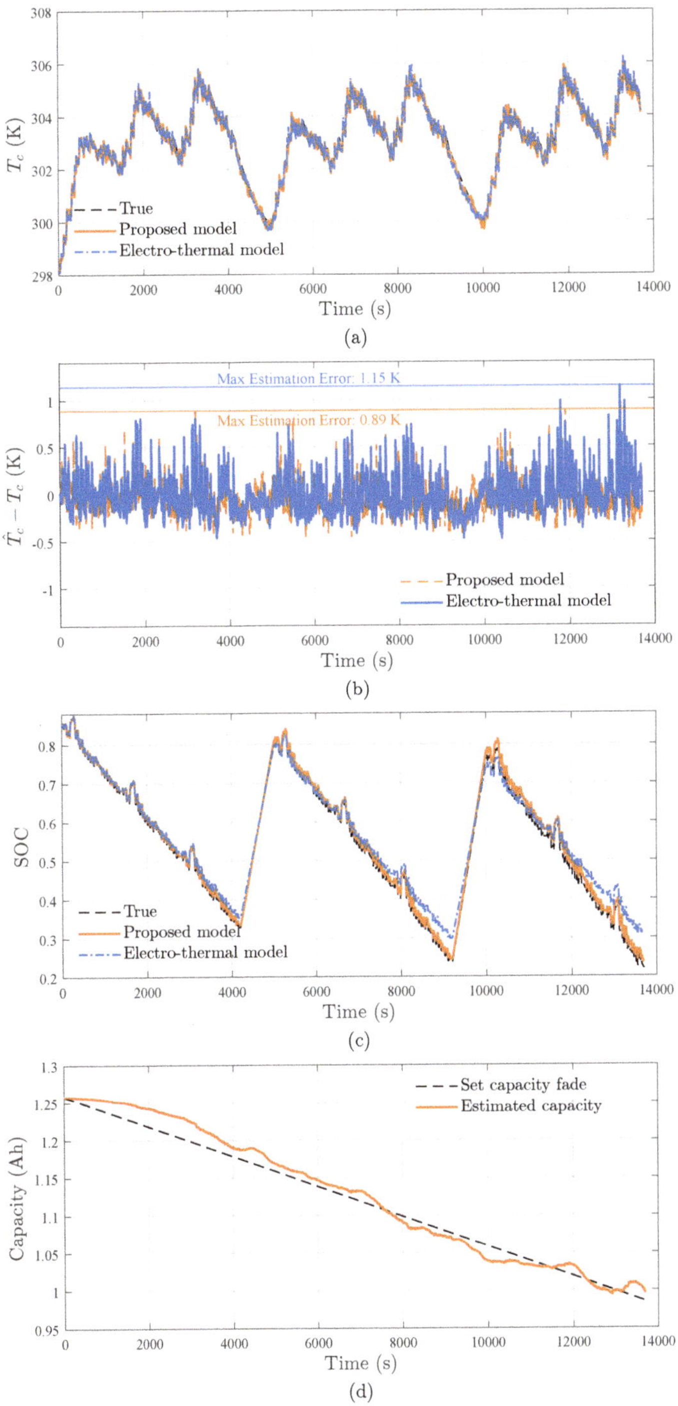

Figure 6. Online state estimation performance during accelerated aging tests demonstrating: (**a**) core temperature (T_c) estimation, (**b**) core temperature estimation error, (**c**) SOC estimation, and (**d**) capacity (C_p) estimation.

3.3. Fault Scenario Robustness Analysis

The accurate estimation of battery SOT becomes increasingly critical under ISC faults, as elevated internal temperature gradients exacerbate TR risks [39]. Simultaneously, ISC-induced electrical–thermal disturbances—including abrupt voltage collapse, abnormal heat generation, and accelerated SOC depletion—severely distort observable signals [10], thereby degrading temperature estimation reliability. To evaluate the robustness of the proposed algorithm, three ISC severity levels (soft: 30 Ω, moderate: 10 Ω, severe: 5 Ω) as classified in related studies and a progressive ISC scenario (30 Ω to 5 Ω over 70 min) are simulated within UDDS cycles [40,41]. The ISC heat source is modeled as uniformly distributed across electrode layers, while equivalent resistances (R_{ISC}) emulate fault dynamics, consistent with our prior ISC detection framework [42].

Figure 7a demonstrates core temperature estimation performance under abrupt ISC levels. Despite increased RMSE (>1 K) compared to healthy conditions introduced in the previous subchapter, the algorithm relatively reliably tracks core temperature trends across all fault levels. This resilience stems from real-time reconciliation of SOC and thermal feedback, which mitigates error propagation caused by abnormal charge depletion. Additionally, the capacity estimation exhibits a gradual decline during ISC operation—a misinterpretation of actual capacity loss but a potential indicator of SOH degradation induced by ISC-induced damage. This phenomenon and its implications for SOT accuracy will be further investigated in future work.

Figure 7b illustrates the evolution of estimated core temperature ($\hat{T}_c$), true core temperature (T_c), and surface temperature (T_s) under the progressive ISC event. A significant thermal gradient (up to 9 K) between the core and surface is observed during fault progression, highlighting the inadequacy of surface temperature monitoring for internal risk detection.

To quantify the safety improvements, a critical temperature threshold (383.15 K, corresponding to separator melting [39]) is established. Beyond this threshold, separator melting initiates more micro ISCs within the battery cell, which rapidly progress to macroscopic ISCs as the separator structural integrity fails. This cascade effect leads to accelerated temperature rise rates, ultimately culminating in thermal runaway and fire incidents. In Figure 7b, the red zone marks the hazardous temperature range ($\geq$383.15 K) where TR becomes imminent. The proposed framework detects when the core hotspot enters this critical zone, enabling earlier alarm triggering and higher TR prevention likelihood compared with relying solely on T_s as the direct input signal. The proposed method triggers the critical temperature alarm 612.4 s earlier via real-time core temperature estimation ($\hat{T}_c$) compared to surface temperature-based alerts (T_s). During this critical interval, the core temperature rises by 7.84 K, while surface measurements lag significantly, failing to reflect internal hotspot development. This delayed surface response risks catastrophic failure, as localized core temperatures may exceed safety limits long before surface sensors react. Overall, relying solely on surface temperature (Figure 7b) critically underestimates internal risks during fault evolution. Bridging this gap, the proposed framework can potentially help BTMS to deploy preventive measures earlier, decisively enhancing safety in real-world fault scenarios.

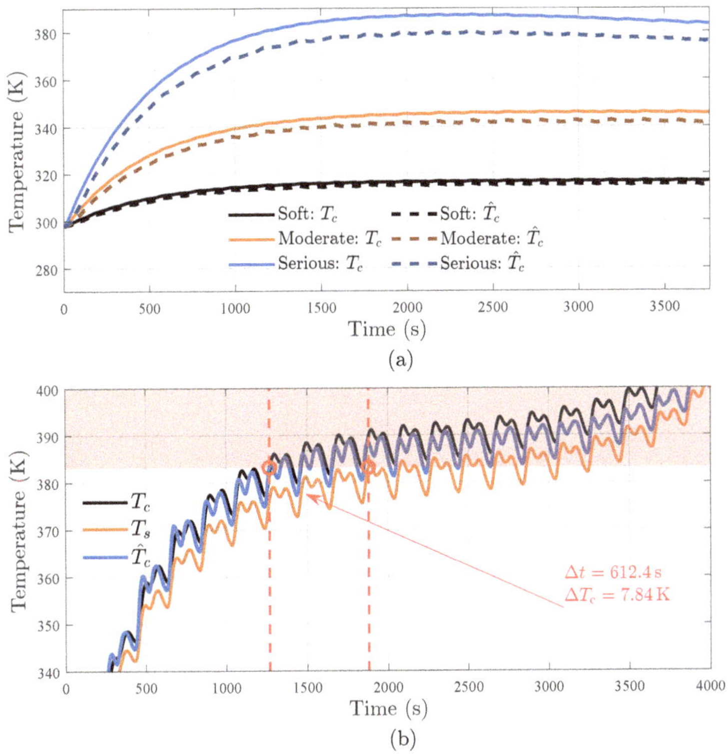

Figure 7. Core temperature estimation results under ISC conditions: (**a**) abrupt ISCs, and (**b**) SOT under evolving ISC event.

4. Conclusions

This study presents an online framework for estimating the core temperature of LIBs by integrating the EKF algorithm with an electro-thermal–aging coupled model (ECM-1D). The framework enables real-time joint estimation of core temperature, SOC, and capacity (indicative of SOH) through dynamic updates of temperature-dependent electrical parameters and aging-induced capacity degradation. By explicitly capturing bidirectional electro-thermal interactions, the method achieves enhanced accuracy in battery state estimation compared to prior approaches under extreme conditions, including a 62.5% reduction in RMSE during high-rate cycling, a maximum core temperature estimation error of <0.9 K, and a 58.9% RMSE reduction in SOC estimation under accelerated aging. Furthermore, the proposed approach demonstrates robust performance under challenging operating conditions, such as ISC scenarios.

The framework offers practical value for BTMS and BMS by improving battery core temperature estimation and enabling simultaneous monitoring of SOC, SOH. Its computational efficiency, achieved through hybrid offline calibration and online parameter adaptation, ensures compatibility with embedded BMS implementations. Future work will focus on integrating the physical correlations between complex aging mechanisms and electro-thermal measurements to further enhance multi-state estimation accuracy of the algorithm. Also, to enhance the framework's generality, future work may integrate physics-based aging mechanisms (e.g., lithium plating, SEI growth) into the 'plant model', enabling systematic robustness analysis. Further validation through experimental tests or high-fidelity simulations under real-world operational scenarios would strengthen its practical relevance. Additionally, combining the framework with advanced fault detection modules will establish a comprehensive health monitoring system. These advancements contribute to safer and more reliable LIB operation in EVs and grid-scale energy storage applications.

Author Contributions: Conceptualization, Y.J. and F.C.; methodology, Y.J., L.B. and F.C.; software, Y.J. and L.B.; validation, Y.J.; formal analysis, Y.J.; investigation, Y.J. and F.C.; resources, M.G. and F.C.; data curation, Y.J.; writing—original draft preparation, Y.J.; writing—review and editing, Y.J., L.B. and F.C.; visualization, Y.J.; supervision, F.C.; project administration, M.G. and F.C.; funding acquisition, Y.J. and M.G. All authors have read and agreed to the published version of the manuscript.

Funding: This research was funded by China Scholarship Council grant number No.202108320086.

Data Availability Statement: The original contributions presented in this study are included in the article. Further inquiries can be directed to the corresponding author.

Acknowledgments: The author Yiqi Jia would like to thank the China Scholarship Council for the financial support (CSC, No.202108320086).

Conflicts of Interest: The authors declare no conflicts of interest.

Abbreviations

The following abbreviations are used in this manuscript:

LIB	Lithium-Ion Battery
TR	Thermal Runaway
BTMS	Battery Thermal Management System
EKF	Extended Kalman Filter
ECM	Equivalent Circuit Model
SOC	State of Charge
SOH	State of Health
ISC	Internal Short Circuit
RMSE	Root Mean Square Error
EV	Electric Vehicle
SEI	solid-electrolyte interphase
BMS	Battery Management System
SOT	State of Temperature
2D-CNN	Two-Dimensional Convolutional Neural Network
EIS	Electrochemical Impedance Spectroscopy
LSTM	Long Short-Term Memory
KF	Kalman Filter
ROM	Reduced-Order Model
RC	Resistor-Capacitor
AFFRLS	Adaptive Forgetting Factor Recursive Least Squares
OCV	Open-Circuit Voltage
SPM	Single Particle Model
HPPC	Hybrid Pulse Power Characterization
UDDS	Urban Dynamometer Driving Schedule
NLS	Nonlinear Least Squares

References

1. Abas , N.; Kalair, A.; Khan, N. Review of fossil fuels and future energy technologies. *Futures* **2015**, *69*, 31–49. [CrossRef]
2. Ren, D.; Feng, X.; Lu, L.; Ouyang, M.; Zheng, S.; Li, J.; He, X. An electrochemical-thermal coupled overcharge-to-thermal-runaway model for lithium ion battery. *J. Power Sources* **2017**, *364*, 328–340. [CrossRef]
3. Goodenough, J.; Kim, Y. Challenges for rechargeable batteries. *J. Power Sources* **2011**, *196*, 6688–6694. [CrossRef]
4. Xiao, H.; Zhao, J.; Gao, Q.; Zhang, W.; Cheng, X.; Song, C.; Li, G. Recent Advances in Fast-Charging Lithium-Ion Batteries: Mechanism, Materials, and Future Opportunities. *Chem. Eng. J.* **2025**, *506*, 159927. [CrossRef]

5. Gerssen-Gondelach, S.J.; Faaij, A.P. Performance of Batteries for Electric Vehicles on Short and Longer Term. *J. Power Sources* **2012**, *212*, 111–129. [CrossRef]
6. Ma, S.; Jiang, M.; Tao, P.; Song, C.; Wu, J.; Wang, J.; Deng, T.; Shang, W. Temperature Effect and Thermal Impact in Lithium-Ion Batteries: A Review. *Prog. Nat. Sci. Mater. Int.* **2018**, *28*, 653–666. [CrossRef]
7. Ouyang, D.; Weng, J.; Chen, M.; Wang, J. Impact of High-Temperature Environment on the Optimal Cycle Rate of Lithium-Ion Battery. *J. Energy Storage* **2020**, *28*, 101242. [CrossRef]
8. Zhu, G.; Wen, K.; Lv, W.; Zhou, X.; Liang, Y.; Yang, F.; Chen, Z.; Zou, M.; Li, J.; Zhang, Y.; et al. Materials Insights into Low-Temperature Performances of Lithium-Ion Batteries. *J. Power Sources* **2015**, *300*, 29–40. [CrossRef]
9. Yang, S.; Hua, Y.; Qiao, D.; Lian, Y.; Pan, Y.; He, Y. A Coupled Electrochemical-Thermal-Mechanical Degradation Modelling Approach for Lifetime Assessment of Lithium-Ion Batteries. *Electrochim. Acta* **2019**, *326*, 134928. [CrossRef]
10. Feng, X.; Ouyang, M.; Liu, X.; Lu, L.; Xia, Y.; He, X. Thermal Runaway Mechanism of Lithium Ion Battery for Electric Vehicles: A Review. *Energy Stor. Mater.* **2018**, *10*, 246–267. [CrossRef]
11. Troxler, Y.; Wu, B.; Marinescu, M.; Yufit, V.; Patel, Y.; Marquis, A.J.; Brandon, N.P.; Offer, G.J. The effect of thermal gradients on the performance of lithium-ion batteries. *J. Power Sources* **2014**, *247*, 1018–1025. [CrossRef]
12. Xia, G.; Cao, L.; Bi, G. A Review on Battery Thermal Management in Electric Vehicle Application. *J. Power Sources* **2017**, *367*, 90–105. [CrossRef]
13. Li, S.; Zhang, C.; Zhao, Y.; Offer, G.J.; Marinescu, M. Effect of thermal gradients on inhomogeneous degradation in lithium-ion batteries. *Commun. Eng.* **2023**, *2*, 74. [CrossRef]
14. Zhang, G.; Cao, L.; Ge, S.; Wang, C.Y.; Shaffer, C.E.; Rahn, C.D. In Situ Measurement of Radial Temperature Distributions in Cylindrical Li-Ion Cells. *J. Electrochem. Soc.* **2014**, *161*, A1499. [CrossRef]
15. Forgez, C.; Do, D.V.; Friedrich, G.; Morcrette, M.; Delacourt, C. Thermal Modeling of a Cylindrical LiFePO4/Graphite Lithium-Ion Battery. *J. Power Sources* **2010**, *195*, 2961–2968. [CrossRef]
16. Yu, Y.; Vergori, E.; Worwood, D.; Tripathy, Y.; Guo, Y.; Somá, A.; Greenwood, D.; Marco, J. Distributed Thermal Monitoring of Lithium Ion Batteries with Optical Fibre Sensors. *J. Energy Storage* **2021**, *39*, 102560. [CrossRef]
17. Carter, R.; Kingston, T.A.; Atkinson, R.W.; Parmananda, M.; Dubarry, M.; Fear, C.; Mukherjee, P.P.; Love, C.T. Directionality of thermal gradients in lithium-ion batteries dictates diverging degradation modes. *Cell Rep. Phys. Sci.* **2021**, *2*, 100351. [CrossRef]
18. Olabi, A.; Maghrabie, H.M.; Adhari, O.H.K.; Sayed, E.T.; Yousef, B.A.; Salameh, T.; Kamil, M.; Abdelkareem, M.A. Battery Thermal Management Systems: Recent Progress and Challenges. *Int. J. Thermofluids* **2022**, *15*, 100171. [CrossRef]
19. Bandhauer, T.M.; Garimella, S.; Fuller, T.F. A Critical Review of Thermal Issues in Lithium-Ion Batteries. *J. Electrochem. Soc.* **2011**, *158*, R1. [CrossRef]
20. Zheng, Y.; Che, Y.; Hu, X.; Sui, X.; Stroe, D.I.; Teodorescu, R. Thermal state monitoring of lithium-ion batteries: Progress, challenges, and opportunities. *Prog. Energy Combust. Sci.* **2024**, *100*, 101120. [CrossRef]
21. Li, S.; Patel, A.N.; Zhang, C.; Amietszajew, T.; Kirkaldy, N.; Offer, G.J.; Marinescu, M. Internal temperature estimation for lithium-ion batteries through distributed equivalent circuit network model. *J. Power Sources* **2024**, *611*, 234701. [CrossRef]
22. Li, Y.; Ma, C.; Liu, K.; Chang, L.; Zhang, C.; Duan, B. A Novel Joint Estimation for Core Temperature and State of Charge of Lithium-Ion Battery Based on Classification Approach and Convolutional Neural Network. *Energy* **2024**, *308*, 132721. [CrossRef]
23. Liu, K.; Li, K.; Peng, Q.; Guo, Y.; Zhang, L. Data-Driven Hybrid Internal Temperature Estimation Approach for Battery Thermal Management. *Complexity* **2018**, *2018*, 9642892. [CrossRef]
24. Li, D.; Wang, L.; Duan, C.; Li, Q.; Wang, K. Temperature prediction of lithium-ion batteries based on electrochemical impedance spectrum: A review. *Int. J. Energy Res.* **2022**, *46*, 10372–10388. [CrossRef]
25. Yuan, A.; Cai, T.; Luo, H.; Song, Z.; Wei, B. Core Temperature Estimation of Lithium-Ion Battery Based on Numerical Model Fusion Deep Learning. *J. Energy Storage* **2024**, *102*, 114148. [CrossRef]
26. Arafat, K.A.A.; Yousuf Bhuiyan, S.M.; Mahamud, R.; Parvez, I. Investigating the Performance of Different Machine Learning Models for Forecasting Li-ion Battery Core Temperature Under Dynamic Loading Conditions. In Proceedings of the eIT, Eau Claire, WI, USA, 30 May–1 June 2024; pp. 1–7. [CrossRef]
27. Sun, J.; Wei, G.; Pei, L.; Lu, R.; Song, K.; Wu, C.; Zhu, C. Online Internal Temperature Estimation for Lithium-Ion Batteries Based on Kalman Filter. *Energies* **2015**, *8*, 4400–4415. [CrossRef]
28. Ma, Y.; Cui, Y.; Mou, H.; Gao, J.; Chen, H. Core Temperature Estimation of Lithium-Ion Battery for EVs Using Kalman Filter. *Appl. Therm. Eng.* **2020**, *168*, 114816. [CrossRef]
29. Li, J.; Cheng, Y.; Jia, M.; Tang, Y.; Lin, Y.; Zhang, Z.; Liu, Y. An Electrochemical-Thermal Model Based on Dynamic Responses for Lithium Iron Phosphate Battery. *J. Power Sources* **2014**, *255*, 130–143. [CrossRef]
30. Zhang, C.; Li, K.; Deng, J. Real-Time Estimation of Battery Internal Temperature Based on a Simplified Thermoelectric Model. *J. Power Sources* **2016**, *302*, 146–154. [CrossRef]
31. Chen, S.; Wang, Z.; Zhang, P.; Yu, Y.; Liu, X.; Li, L.; Sun, J. Online core temperature estimation method for lithium-ion batteries over the entire lifecycle. *J. Energy Storage* **2025**, *107*, 115033. [CrossRef]

32. Lin, X.; Perez, H.E.; Mohan, S.; Siegel, J.B.; Stefanopoulou, A.G.; Ding, Y.; Castanier, M.P. A lumped-parameter electro-thermal model for cylindrical batteries. *J. Power Sources* **2014**, *257*, 1–11. [CrossRef]
33. Multiphysics, C. Thermal Modeling of a Cylindrical Lithium-Ion Battery in 2D. COMSOL Application Library: Battery_Design_Module/Thermal_Management/li_battery_thermal_2d_axi. 2023. Available online: https://www.comsol.com/model/thermal-modeling-of-a-cylindrical-lithium-ion-battery-in-2d-10221 (accessed on 10 March 2025).
34. Plett, G.L. Dual and Joint EKF for Simultaneous SOC and SOH Estimation. In Proceedings of the 21st International Battery, Hybrid and Fuel Cell Electric Vehicle Symposium & Exhibition (EVS21), Monte Carlo, Monaco, 2–6 April 2005; pp. 1–12.
35. Sun, L.; Sun, W.; You, F. Core temperature modelling and monitoring of lithium-ion battery in the presence of sensor bias. *Appl. Energy* **2020**, *271*, 115243. [CrossRef]
36. Pang, H.; Guo, L.; Wu, L.; Jin, J.; Zhang, F.; Liu, K. A novel extended Kalman filter-based battery internal and surface temperature estimation based on an improved electro-thermal model. *J. Energy Storage* **2021**, *41*, 102854. [CrossRef]
37. Hossain Lipu, M.; Abd Rahman, M.; Mansor, M.; Ansari, S.; Meraj, S.T.; Hannan, M. Hybrid and combined states estimation approaches for lithium-ion battery management system: Advancement, challenges and future directions. *J. Energy Storage* **2024**, *92*, 112107. [CrossRef]
38. Lin, X.; Perez, H.E.; Siegel, J.B.; Stefanopoulou, A.G.; Li, Y.; Anderson, R.D.; Ding, Y.; Castanier, M.P. Online Parameterization of Lumped Thermal Dynamics in Cylindrical Lithium Ion Batteries for Core Temperature Estimation and Health Monitoring. *IEEE Trans. Control Syst. Technol.* **2013**, *21*, 1745–1755. [CrossRef]
39. Feng, X.; Fang, M.; He, X.; Ouyang, M.; Lu, L.; Wang, H.; Zhang, M. Thermal runaway features of large format prismatic lithium-ion battery using extended volume accelerating rate calorimetry. *J. Power Sources* **2014**, *255*, 294–301. [CrossRef]
40. Lai, X.; Jin, C.; Yi, W.; Han, X.; Feng, X.; Zheng, Y.; Ouyang, M. Mechanism, modeling, detection, and prevention of the internal short circuit in lithium-ion batteries: Recent advances and perspectives. *Energy Stor. Mater.* **2021**, *35*, 470–499. [CrossRef]
41. Zhang, G.; Wei, X.; Tang, X.; Zhu, J.; Chen, S.; Dai, H. Internal short circuit mechanisms, experimental approaches and detection methods of lithium-ion batteries for electric vehicles: A review. *Renew. Sustain. Energy Rev.* **2021**, *141*. [CrossRef]
42. Jia, Y.; Brancato, L.; Giglio, M.; Cadini, F. Temperature enhanced early detection of internal short circuits in lithium-ion batteries using an extended Kalman filter. *J. Power Sources* **2024**, *591*, 233874. [CrossRef]

batteries

MDPI

Review

Advances in the Battery Thermal Management Systems of Electric Vehicles for Thermal Runaway Prevention and Suppression

Le Duc Tai and Moo-Yeon Lee *

Department of Mechanical Engineering, Dong-A University, 37 Nakdong-Daero 550, Saha-gu, Busan 49315, Republic of Korea; 2377988@donga.ac.kr
* Correspondence: mylee@dau.ac.kr; Tel.: +82-51-200-7642

Abstract: In response to the global imperative to reduce greenhouse gas emissions and fossil fuel dependency, electric vehicles (EVs) have emerged as a sustainable transportation alternative, primarily utilizing lithium-ion batteries (LIBs) due to their high energy density and efficiency. However, LIBs are highly sensitive to temperature fluctuations, significantly affecting their performance, lifespan, and safety. One of the most critical threats to the safe operation of LIBs is thermal runaway (TR), an uncontrollable exothermic process that can lead to catastrophic failure under abusive conditions. Moreover, thermal runaway propagation (TRP) can rapidly spread failures across battery cells, intensifying safety threats. To address these challenges, developing advanced battery thermal management systems (BTMS) is essential to ensure optimal temperature control and suppress TR and TRP within LIB modules. This review systematically evaluates advanced cooling strategies, including indirect liquid cooling, water mist cooling, immersion cooling, phase change material (PCM) cooling, and hybrid cooling based on the latest studies published between 2020 and 2025. The review highlights their mechanisms, effectiveness, and practical considerations for preventing TR initiation and suppressing TRP in battery modules. Finally, key findings and future directions for designing next-generation BTMS are proposed, contributing valuable insights for enhancing the safety and reliability of LIB applications.

Keywords: batteries; electric vehicles; thermal management; thermal runaway

1. Introduction

In recent decades, environmental pollution and global warming have become important challenges for societies around the world. Many human activities that have greenhouse gas emissions have contributed significantly to climate change [1]. One of the largest sources of these emissions is the transportation sector, where fossil fuel-powered internal combustion engine (ICE) vehicles are a major contributor to air pollution and carbon dioxide emissions [2,3]. In response to growing environmental concerns, electric vehicles (EVs) have emerged as a promising alternative to traditional ICE vehicles, offering a cleaner and more sustainable transportation option [4,5].

Lithium-ion batteries (LIBs) are the core components of EVs, serving as their primary energy storage systems. LIBs are favored for their high energy density, long lifecycle, low self-discharge rate, and high conversion rate, making them ideal for use in EVs [6–8]. However, the performance, lifetime, and safety of LIBs are significantly dependent on their operating temperature range. Studies have shown that LIBs perform best within a temperature range of 25–40 °C [9–11]. In addition, the temperature difference between cells in the battery module should be kept below 5 °C to ensure uniform performance and prevent rapid aging [12–14]. Operation at temperatures outside this range can have adverse effects. Specifically, LIBs operating at low temperatures reduce battery capacity and accelerate aging, while high temperatures can lead to loss of performance and loss of thermal control, posing significant safety risks in extreme cases [15–17].

Additionally, thermal runaway (TR) in LIBs is a significant safety concern, primarily triggered by abusive conditions such as thermal, electrical, and mechanical stress [18–20]. As the temperature of a battery cell increases, exothermic reactions within the cell become more intense, leading to an uncontrollable rise in temperature. Once a critical temperature threshold is surpassed, typically between 130 and 200 °C, the battery can enter TR, where the temperature can escalate to 600–1000 °C, often accompanied by the emission of flammable gases, smoldering, and combustion [21,22]. The heat generated from a single cell undergoing TR can appear limited. However, in densely packed battery systems, the proximity of cells can result in thermal runaway propagation (TRP). This chain reaction occurs when the heat from a failing cell triggers TR in adjacent cells, causing a rapid spread of TR throughout the battery pack. The intense heat generated by TRP can lead to severe consequences, including widespread fires, explosions, and damage to surrounding systems, posing a significant threat to public safety [23]. Notable fire incidents, such as those involving EVs, such as the BMW i3, Tesla Model X, and Porsche Panamera, highlight the danger of TRP, where a single cell failure triggered catastrophic fires [22]. The risk of TR and TRP is exacerbated by common causes such as overheating, internal short circuits (ISC), overcharging, and manufacturing defects, which initiate the internal exothermic reactions within the cell. Furthermore, the spread of TR can be accelerated by the design of the battery, with heat transferring rapidly through the battery shell to neighboring cells. Consequently, effective battery thermal management systems (BTMS) are critical in preventing the onset of TR and suppressing its propagation, ensuring the safety and reliability of the LIB application systems [24,25].

Liquid cooling and phase change material (PCM) cooling are two advanced battery thermal management techniques that play a crucial role in preventing and suppressing TR in LIBs. Liquid cooling systems are widely recognized for their high heat transfer efficiency as they utilize a circulating coolant to absorb and dissipate heat from the battery cells. This method allows for precise temperature control and uniform cooling, which is vital for maintaining the battery within the optimal operating temperature range. Liquid cooling also offers the flexibility to scale for various battery sizes and configurations, making it suitable for applications ranging from EVs to large-scale energy storage systems [26,27]. On the other hand, PCM cooling systems provide the solution to manage battery temperature effectively by utilizing materials that absorb and store excess heat during phase transitions from solid to liquid. PCM cooling can effectively reduce the risk of TR by preventing overheating and maintaining the battery's temperature within safe limits. Additionally, the large latent heat of the PCM can significantly delay the TRP time from the TR-triggered cell to other cells in the battery pack [28–31]. When integrated into battery packs, both liquid cooling and PCM cooling systems offer outstanding benefits, ensuring thermal stability, enhancing battery lifespan, and improving overall safety by reducing the likelihood of TR incidents.

Recently, several review articles have been published addressing thermal management strategies to prevent TR in LIBs, as summarized in Table 1. Unlike previous reviews, this study provides an in-depth and systematic analysis on thermal runaway prevention/suppression using advanced cooling strategies, including indirect liquid cooling, water mist cooling, immersion cooling, PCM cooling, and hybrid cooling systems. The novelty of this review distinctly lies in its comprehensive and specific focus on evaluating advanced BTMS tailored explicitly for both the TR prevention and TRP suppression in LIB modules used in EVs, while most previous reviews have been limited to a TR prevention perspective. Additionally, this review uniquely incorporates and synthesizes the most recent advancements published between 2020 and 2025, addressing the significant gap identified in the prior literature that frequently overlooked the latest developments or offered limited insights into advanced BTMS for TR prevention and TRP suppression. Furthermore, this review explicitly discusses the effectiveness and limitations of each advanced cooling approach, offering clear evaluations and critical comparisons that were insufficiently addressed in earlier studies. Thus, the present review distinctly advances the current state-of-the-art by providing practical insights and specific directions to guide ongoing and future research in battery safety and thermal management. The structure of this review is as follows: Section 2 outlines the mechanisms leading to TR and the severe impacts of the heat generated during TR events; Section 3 discusses advanced cooling strategies for TR prevention and TRP suppression in LIBs; and Section 4 summarizes the key findings and presents recommendations for future research directions.

Table 1. Recent reviews on battery thermal management strategies to prevent TR in LIBs.

Review	Focus of Review	Highlights	Missing Topics/Future Perspectives
Jiang et al. (2022) [32]	• Comprehensive review of BTMS for LIBs across normal, sub-zero, and abusive conditions. • Discusses battery cooling (air cooling, liquid cooling, PCM cooling, and hybrid cooling) and battery heating methods. • Covers TR origins and prevention strategies, but is not exclusively focused on TR prevention, and addresses overall thermal management, including general operation, low-temperature performance, and abusive conditions.	• Reviews heat generation mechanisms during both normal use and abusive events. • Summarizes BTMS methods, including air cooling, liquid cooling, PCM cooling, and hybrid cooling. • Discusses battery heating methods such as alternating current (AC) heating, self-heating, and external heating under sub-zero temperatures. • Addresses TR prevention via both battery material modification and thermal management design. • Highlights research challenges, especially in developing BTMS capable of handling wide temperature ranges and multiple abusive scenarios.	• Does not specifically address thermal runaway propagation (TRP) suppression. • Focuses on both traditional cooling, such as air cooling. Lacking in-depth analysis and evaluation of advanced cooling strategies, namely liquid cooling and PCM cooling. • Omits recent developments in advanced BTMS for TR prevention and TRP suppression.
Shahid et al. (2022) [19]	• Comprehensive review of the entire TR process in LIBs, including initiation, propagation, gas venting, prediction, prevention, and mitigation. • Covers both BTMS for TR prevention and mitigation strategies (insulation, fire barriers, and fire suppression) after TR occurs. • BTMS is an important part, but not the sole focus.	• Summarizes thermal abuse, mechanical abuse, and electrical abuse mechanisms that cause TR. • Details experimental and numerical studies on TRP and gas emission during TR events. • Reviews BTMS technologies (air cooling, liquid cooling, PCM cooling, and hybrid cooling systems) as key preventive measures. • Critically discusses mitigation strategies post-TR, including fire suppression, explosion barriers, and thermal insulation methods. • Provides future research directions, including the development of better TR prediction models, lightweight BTMS, and integrated safety designs for EV batteries.	• Provides a limited performance evaluation of advanced cooling methods specifically for EV battery modules. • Insufficient coverage of recent advancements in BTMS for TR prevention and TRP suppression. • Lacks clear identification of key future research directions for the development of advanced BTMS.

Table 1. *Cont.*

Review	Focus of Review	Highlights	Missing Topics/Future Perspectives
Chavan et al. (2023) [20]	• Reviews thermal management systems for LIBs with an emphasis on TR mitigation using battery cooling methods. • Focused primarily on cooling technologies (air cooling, liquid cooling, PCM cooling, heat pipes, and hybrid cooling) as a strategy to prevent and suppress TR in EV batteries. • Discusses TR mechanisms, TR modeling, and TRP prevention techniques.	• Provides a detailed classification and evaluation of active and passive battery cooling techniques specifically for TR prevention. • Highlights the importance of hybrid cooling (PCM and liquid cooling) to balance normal operation efficiency and TR suppression capability. • Reviews the necessity of thermal insulation layers between cells to enhance TR propagation resistance. • Stresses that cooling alone is insufficient for complete TR suppression, recommending combined strategies (cooling and insulation). • Suggests future research directions in improving thermal conductivity, enhancing system compactness, and integration with battery management systems (BMS).	• Limited in-depth analysis of advanced battery thermal management strategies specifically aimed at preventing TRP. • Does not incorporate the most recent advancements in BTMS published in the past few years.
Zhi et al. (2024) [33]	• Comprehensive review of technologies for the prevention, early detection, and mitigation of TR in LIBs. • Covers intrinsic safety technologies (material modifications), thermal management systems (air, liquid, PCM, and hybrid cooling), early warning and detection methods, and fire prevention/extinguishing techniques. • Focuses on TR broadly, not specifically on BTMS alone.	• Summarizes thermal management technologies (air, liquid, PCM, and hybrid cooling) as part of TR prevention strategies, but thermal management is not the sole focus. • Includes extensive discussion on material innovation (flame-retardant electrolytes, electrodes, separators) and early warning technologies (sensor systems). • Reviews fire prevention and fire suppression technologies of TR. • Proposes challenges and future directions for comprehensive TR risk mitigation rather than focusing exclusively on advanced BTMS designs.	• While the review includes an evaluation of conventional cooling methods, it lacks a focused and in-depth analysis of advanced BTMS tailored for LIB applications. • Insufficient research has addressed the effectiveness of advanced cooling technologies in suppressing TRP, limiting the practical insights for improving battery safety. • The review does not provide specific or detailed recommendations for the future development of advanced BTMS aimed at preventing TR and suppressing TRP, which are critical for high-energy battery applications.
Yang et al. (2024) [34]	• Reviews TR mechanisms and thermal management technologies specifically for LIBs in electric aircraft. • Focuses on TR prevention and mitigation strategies tailored for aviation applications. • Covers both enhanced heat removal techniques (air cooling, liquid cooling, and PCM cooling) and heat/fire propagation prevention methods (thermal insulation, fire-retardant technologies).	• Evaluates various thermal management strategies (air cooling, liquid cooling, PCM cooling, and thermochemical cooling) for their suitability in electric aircraft, based on weight, power, and effectiveness. • Highlights the potential of PCM combined with thermochemical materials and bio-inspired insulation materials as promising lightweight solutions. • Recommends passive fire-retardant methods combined with PCM.	• The review primarily focuses on advanced BTMS for TR prevention in aviation applications, limiting its relevance to EV contexts. • It lacks a detailed and systematic evaluation of advanced BTMS strategies specifically designed to prevent TRP in LIB systems. • There is an absence of clearly defined recommendations for the future development and optimization of advanced BTMS aimed at enhancing both TR prevention and TRP suppression under realistic operating conditions.

2. TR Characterization

2.1. TR Mechanism

The basic mechanisms that lead to TR in LIBs include mechanical abuse, electrical abuse, and thermal abuse [18]. These mechanisms can initiate a series of chain reactions that escalate the temperature within the battery and propagate failures, potentially causing fires or explosions. The abusive conditions associated with the TR mechanism of LIBs are shown in Figure 1.

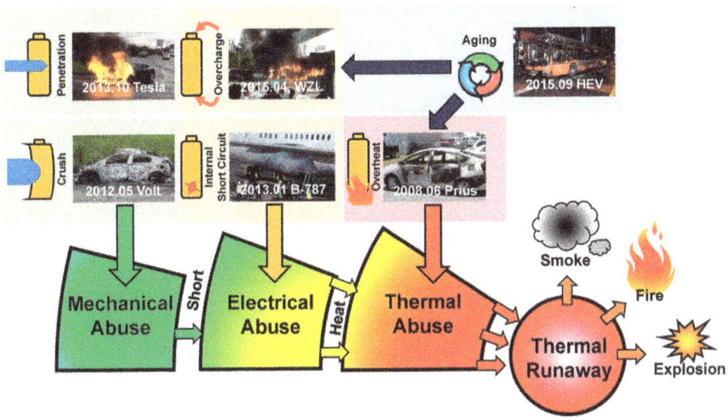

Figure 1. Abusive conditions are related to the TR mechanisms of LIBs [18].

Mechanical abuse typically arises from impacts such as vehicle crashes or external forces that deform the battery pack. Such deformation can lead to the rupture of the separator between the anode and cathode, which, in turn, causes an ISC [19]. The ISC generates significant heat and leads to the initiation of TR. Additionally, penetration by an external object (e.g., a nail) can trigger a short circuit and lead to rapid heat generation as the battery discharges. The extent of damage depends on the location of the penetration. In particular, low heat dissipation areas are more susceptible to TR [20]. As the mechanical forces increase, the battery pack's deformation can become non-uniform, concentrating stress and elevating the risk of failure.

Electrical abuse in LIBs primarily involves external short circuits (ESC) and overcharging or overdischarging. ESC occurs when electrodes with different voltages come into contact, often due to external forces such as water immersion, contamination, deformation during mechanical abuse, or electric shock [18]. When an ESC occurs, the current increases rapidly to its peak value and then decreases to a plateau before dissipating as the battery discharges. This process generates heat within the circuit. However, in typical ESC conditions, this heat is often insufficient to trigger TR unless certain conditions are met, such as an increase in the mass transfer coefficient for lithium ions at the anode or a larger surface area of the anode, which leads to greater current flow and faster heat generation [35]. Overcharging is another form of electrical abuse and a significant contributor to TR. This condition arises when the battery is charged beyond its recommended voltage limit, typically due to a failure in the battery management system (BMS) that prevents the battery from stopping the charging process [18]. Overcharging results in excessive energy being introduced into the battery, leading to increased heat generation from both ohmic heating and side reactions at the electrodes [35,36]. These reactions are particularly severe at the anode, where lithium plating can occur due to excessive lithium intercalation, causing an increase in temperature. In addition, overcharging can lead to the collapse of the cathode material structure and the release of oxygen, which further accelerates electrolyte decomposition and produces additional gases [37,38]. Similarly, overdischarging occurs when the battery is drained beyond its capacity, causing a negative voltage in certain cells, leading to abnormal heat generation and potential capacity degradation [18]. Overdischarging can also cause the dissolution of copper collectors in the battery, resulting in ISC and TR [39–41].

Thermal abuse occurs when the battery experiences overheating, often due to mechanical or electrical failures, poor battery connections, and external fires. Thermal abuse can escalate quickly, as the rising internal temperature accelerates the breakdown of the separator and the electrolyte, leading to ISC and further increasing heat generation. This can create a series of chain reactions where the temperature rise causes additional breakdowns, culminating in full-scale TR [19,20,42].

2.2. TRP

TRP refers to the spread of the TR phenomenon from the triggered TR cell to neighboring cells within a battery pack. Once TR is initiated in a single cell, it generates significant heat that can trigger the same event in adjacent cells and potentially cause a TRP throughout the entire pack [19]. Feng et al. conducted an experimental investigation on penetration-induced TRP within a large-format LIB pack. In the six-battery module test, the study indicated that a maximum temperature up to 791.8 °C occurred within a single battery during TRP. The analysis showed that approximately 12% of the total heat released by the initiating battery is sufficient to trigger TR in adjacent cells, with the majority of heat transferring through the battery shell rather than the pole connector [43]. These temperatures and heat transfer characteristics were recorded without active cooling and represent the severity of uncontrolled TRP in densely packed battery modules. Chen et al. explored TRP in cylindrical battery packs and found that larger battery packs exhibited more serious TRP. Specifically, the study showed that 32 battery cells simultaneously ignited during the TR event in a battery pack consisting of 99 cells [44]. The intensity and speed of TRP are also influenced by the state of charge (SOC) of the batteries with higher SOCs, leading to faster and more intense TRP [45].

2.3. Heat Generation During TR

During TR in LIBs, a series of exothermic reactions generate significant amounts of heat, rapidly escalating the internal temperature of the battery. The heat generation occurs in multiple stages, each corresponding to specific reactions triggered by elevated temperatures. These reactions include the decomposition of the solid electrolyte interphase (SEI), reactions at the negative and positive electrode–electrolyte interfaces, and the decomposition of the electrolyte. Additionally, heat can also be generated due to an ISC, which occurs when the separator partially melts as the temperature rises, allowing for the anode and cathode materials to come into direct contact. This ISC further contributes to heat release. The total heat generation is calculated by summing the heat produced by each reaction, as defined by the SEI decomposition reaction (Q_{sei}), negative-electrolyte reaction (Q_{ne}), positive-electrolyte reaction (Q_{pe}), electrolyte decomposition (Q_e), and the heat generated by ISC (Q_{ISC}). The total heat generation during TR is calculated using Equation (1) [46],

$$Q_{TR} = Q_{sei} + Q_{ne} + Q_{pe} + Q_e + Q_{ISC} \tag{1}$$

The ranges of the starting reaction temperatures and the formulas for calculating the heat released for each reaction process are presented in Table 2. The physical and kinetic parameters used in the heat generation equations are presented in Table 3.

The immense heat generation during TR poses significant safety risks in LIBs. As the battery undergoes TR, exothermic reactions lead to rapid and uncontrollable temperature increases, often exceeding 1000 °C in the absence of cooling strategies [22]. This excessive heat can result in the rupture of the battery casing, release of toxic gases, fires, and explosions. Such catastrophic events can be triggered by relatively minor incidents, such as overcharging or physical damage, highlighting the critical need for preventive measures. Advanced BTMS are essential to mitigate these risks by controlling the temperature within safe operating limits, preventing TR initiation and suppressing TRP. By incorporating advanced cooling strategies such as liquid cooling, PCM cooling, and hybrid cooling, BTMS can prevent TR and significantly reduce the chances of a domino effect in multi-cell configurations, enhancing the safety and reliability of battery-powered systems.

Therefore, reviewing robust thermal management solutions is paramount to ensuring the safe operation of LIBs, particularly in high-energy applications like EVs.

Table 2. Heat generation equations of TR reactions [22,46,47].

TR Reactions	Temperature Range for the Initiation of Reaction	Reaction Rate Equation	Heat Generation Equation
SEI decomposition reaction	60–130 °C	$R_{sei} = \frac{dc_{sei}}{dt} = -A_{sei} \cdot c_{sei} \cdot exp\left[-\frac{E_{a,sei}}{RT}\right]$	$Q_{sei} = -H_{sei} \cdot W_{sei} \cdot R_{sei}$
negative-electrolyte reaction	100–130 °C	$R_{ne} = \frac{dt_{sei}}{dt} = -\frac{dc_{ne}}{dt} = -A_{ne} \cdot c_{ne} \cdot exp\left[-\frac{t_{sei}}{t_{sei,ref}}\right] \cdot exp\left[-\frac{E_{a,ne}}{RT}\right]$	$Q_{ne} = -H_{ne} \cdot W_{ne} \cdot R_{ne}$
positive-electrolyte reaction	150–300 °C	$R_{pe} = \frac{dc_{pe}}{dt} = A_{pe} \cdot \alpha \cdot (1-\alpha) \cdot exp\left[-\frac{E_{a,pe}}{RT}\right]$	$Q_{pe} = H_{pe} \cdot W_{pe} \cdot R_{pe}$
electrolyte decomposition	Around 200 °C	$R_e = \frac{dc_e}{dt} = -A_e \cdot c_e \cdot exp\left[-\frac{E_{a,e}}{RT}\right]$	$Q_e = -H_e \cdot W_e \cdot R_e$
ISC	Around 130 °C	$R_{ISC} = -\frac{dSOC}{dt} = ISC_{cond} \cdot SOC \cdot A_{ISC} \cdot exp\left[-\frac{E_{a,ISC}}{RT}\right]$	$Q_{ISC} = H_{ISC} \cdot R_{ISC} = V \cdot C \cdot 3600 \cdot \eta \cdot R_{ISC}$

Table 3. Parameters for the heat generation equations of TR reactions [22,46,47].

Parameters	Description	Unit
$A_{sei/ne/pe/e/ISC}$	Frequency factor of the reaction	s^{-1}
$E_{a,sei/ne/pe/e/ISC}$	Activation energy of the reaction	J/mol
$c_{sei/ne/e}$	Initial dimensionless content	-
α	Conversion degree of the positive electrode material	-
$t_{sei,ref}$	Initial dimensionless SEI layer thickness	-
$H_{sei/ne/pe/e/ISC}$	Heat of reaction	J/kg
$W_{sei/ne/pe/e}$	Material content	kg/m^3
V	Battery nominal voltage	V
C	Battery capacity	Ah
η	Efficiency factor	-
ISC_{cond}	Parameter to control heat generation by ISC: – $ISC_{cond} = 0$, $T < 130$ °C – $ISC_{cond} = 1$, $T \geq 130$ °C	-

2.4. Comparison Between Traditional and Advanced Cooling Strategies

Traditional BTMS primarily utilize air cooling, owing to their simplicity and low cost [48,49]. Air cooling relies on natural or forced convection to dissipate heat from battery modules [50,51]. However, the relatively low thermal conductivity and heat capacity of air limits its cooling effectiveness, especially in high-energy battery systems [52–54]. Consequently, air cooling often results in non-uniform temperature distributions and insufficient heat removal under abusive conditions, increasing the risk of TR.

In contrast, advanced cooling strategies, such as liquid cooling and PCM cooling, offer significantly higher heat transfer rates and more precise temperature control. These methods improve temperature uniformity, rapidly absorb and dissipate heat, and can suppress or delay TR and its propagation [55–57]. The limitations of traditional air cooling thus motivate the exploration and development of these advanced thermal management solutions, which are the focus of the subsequent sections.

3. Advanced Cooling Strategies for TR Prevention and Suppression in LIBs

This section presents a comprehensive review of advanced BTMS designed to prevent and suppress TR and TRP in LIB modules. With the increasing energy density and power demand in modern battery systems, traditional cooling methods are often insufficient to address the safety challenges posed by rapid heat generation and propagation. Therefore, advanced cooling strategies have emerged as critical components in enhancing thermal stability and operational safety. These advanced cooling strategies include liquid cooling methods, such as indirect liquid cooling, water mist cooling, and immersion cooling, which are discussed in Section 3.1. PCM cooling systems are detailed in Section 3.2. Finally, hybrid cooling approaches combining multiple mechanisms to maximize heat dissipation efficiency are presented in Section 3.3.

3.1. Liquid Cooling

3.1.1. Indirect Liquid Cooling

Indirect liquid cooling is a prevalent thermal management technique extensively studied for LIB modules, primarily due to its superior heat dissipation capabilities compared to traditional air cooling [58]. Numerous studies have investigated its potential for suppressing TRP events. Fu et al. demonstrated that using mini-channel cold plates based on indirect liquid cooling effectively delayed and even completely suppressed TRP, depending on the coolant velocity and channel width, as shown in Figure 2. Specifically, the results show that, by applying cold plates with mini-channels, the cooling system effectively delayed the occurrence of TRP at low coolant flow velocities of 0.2 m/s. At this flow rate, only battery 1 and battery 2 underwent TR, with maximum temperatures of approximately 360 °C and 200 °C, respectively, while battery 3 remained below the critical temperature threshold, reaching a maximum of 88 °C without experiencing TR. When the coolant velocity was further increased to 0.3 m/s, TR was confined solely to battery 1, the temperature of battery 2 remained below 88 °C, and battery 3 maintained a temperature below 55 °C, indicating complete suppression of TRP at this flow velocity. In addition, the mini-channel width significantly affected the cooling performance. At a fixed coolant flow velocity of 0.2 m/s, TRP was eliminated when the mini-channel width was $\geq$7 mm [59].

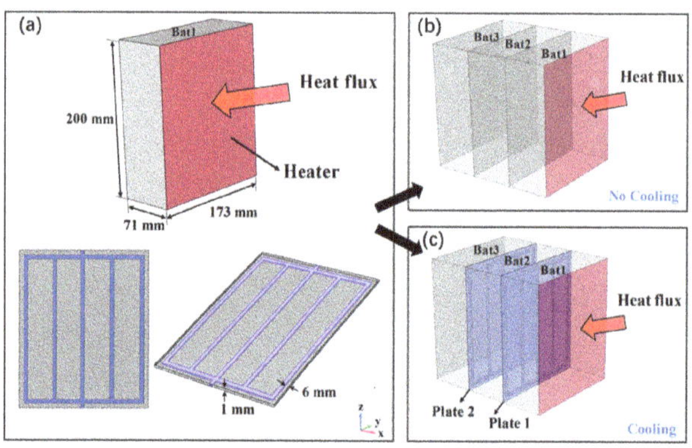

Figure 2. Geometric setup of the thermal management model: (**a**) individual cell with cooling plates, (**b**) battery pack without liquid cooling, and (**c**) battery pack featuring a liquid cooling system [59].

Similarly, Ke et al. highlighted that high coolant flow rates through serpentine-channel cooling significantly mitigated TRP, as presented in Figure 3. The results demonstrate that coolant flow rates of 96 L/h or higher effectively prevented TRP, maintaining the temperature of the triggered TR cell at around 450 °C while keeping neighboring cells below 80 °C. In contrast, lower flow rates of 0 L/h, 32 L/h, and 64 L/h were insufficient to suppress TRP, allowing for propagation to the adjacent cells. The study also showed that the expulsion of high-temperature electrolytes from the battery's positive side when TR is activated plays an important role in propagating TR to adjacent cells. This electrolyte expulsion mechanism was identified as the main contributor to TRP, while conduction and radiation played only minor roles [60].

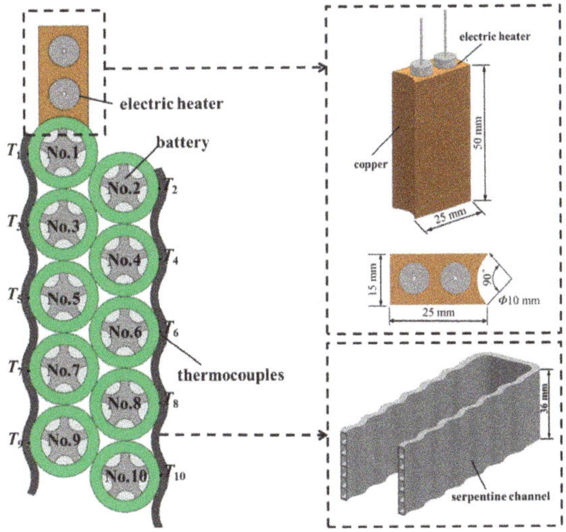

Figure 3. Layout of a battery module incorporating liquid cooling with a serpentine cooling channel [60].

However, despite their promising performance in TR prevention and TRP suppression, indirect liquid cooling systems present inherent challenges. Rui et al. pointed out that, while indirect liquid cooling reduced peak heat flux relatively, it was insufficient to fully prevent TRP without complementary thermal insulation. Aerogel insulation layers significantly enhanced suppression effectiveness, underlining the limitation of relying solely on indirect liquid cooling. Specifically, compared to the basic configuration, a configuration with indirect liquid cooling reduced the peak heat flux of a TR cell and adjacent cells from 885.7 W to 848.2 W. However, this decrease was not sufficient to prevent TRP. When a 1 mm thick aerogel insulation layer was introduced between cells in the battery module, the TRP was significantly slowed, with the average interval between TR events extending from 84.4 s to 384.2 s. Ultimately, the combined application of liquid cooling and thermal insulation layers successfully prevented TRP, maintaining the temperature of the triggered TR cell at around 500 °C while keeping adjacent cells below 80 °C [61].

Similarly, Liu et al. conducted a comprehensive study on the suppression of TRP in LIB packs using mini-channel cold plates and insulation layers. Simulations showed that adding insulation layers between battery cells significantly delayed the onset of TR in adjacent cells. Specifically, when 2 mm and 3 mm aerogel insulation layers were introduced, the time for TR to trigger in the adjacent cell was delayed by 312 s and 944 s, respectively. In addition, the TRP to the entire battery pack was delayed by 1293 s, which far exceeded safety standards. Furthermore, the proposed cooling strategy controlled the average temperature of adjacent cells not to exceed 80 °C, the LIB pack is protected and the propagation of TR is completely prevented [62]. These results emphasize the role of combining insulation layers with indirect liquid cooling systems for effective TR prevention and TRP suppression in LIB modules.

In addition, Yang et al. noted that, although indirect liquid cooling based on cold plates relatively effectively managed temperatures and suppressed TRP, coolant configurations could unintentionally transfer heat and trigger adjacent cell failures due to direct cell-to-cell conduction facilitated by coolant channels. Specifically, the study evaluated the TRP inhibiting effectiveness of BTMS using indirect liquid cooling with cold plates placed at various locations inside the battery pack, including bottom (BTMS-1a), side (BTMS-1b), and inter-cell (BTMS-1c) cold plates, as depicted in Figure 4. The research results show that BTMS-1a demonstrated the highest cooling rate, reaching 2.03 kW. Thus, maintaining the best temperature uniformity within the battery pack with 13.4 °C at a 5C discharge rate. However, BTMS-1a had the shortest propagation time of 74.4 s due to the direct contact between cells and the high-temperature coolant transferring heat to other cells, leading to TR triggering. When the coolant flow rate in each cold plate was increased to 0.75 L/min, BTMS-1b could further delay the occurrence of TR to 82.6 s. Notably, at a similar flow rate, BTMS-1c significantly enhanced cell cooling, completely inhibiting TRP. In this case, the maximum temperature of the triggered TR cell reached around 1000 °C, while the adjacent second cell remained near 100 °C and other neighboring cells were maintained at approximately 60 °C [63]. These research results demonstrate that optimization of cold plate layout and coolant flow plays an important role in TR prevention and TRP suppression in the LIB modules of indirect liquid cooling systems.

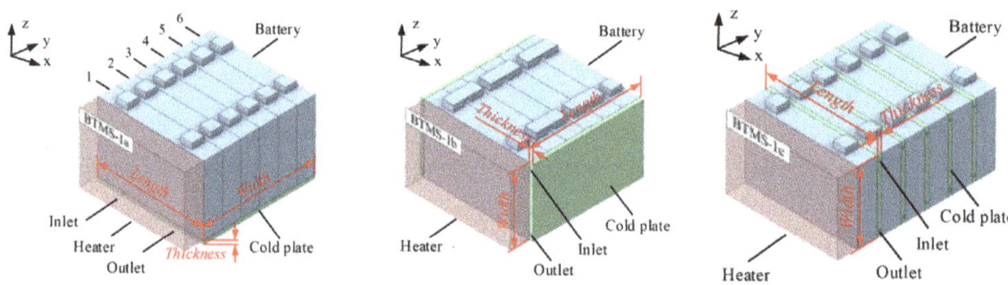

Figure 4. Battery thermal management configuration using liquid cooling with cold plates at different locations, including BTMS-1a: bottom cold plate, BTMS-1b: side cold plates, and BTMS-1c: inter-cell cold plates [63].

Recent advancements aim to address these shortcomings through novel structural designs. For example, Du et al. presented a novel BTMS that incorporates a triply periodic minimal surface (TPMS) porous structure in liquid cooling channels to enhance heat dissipation and reduce the risk of TR, according to Figure 5. The results show that the battery pack using the primitive porous liquid cooling structure showed a 12.43% reduction in maximum temperature and an 8.41% reduction in temperature difference compared to the straight-tube channel. In addition, under extreme operating conditions up to 50 °C, the primitive porous structure still maintained the maximum temperature of the battery pack below 54.66 °C, and the temperature difference did not exceed 3.76 °C, effectively preventing TR events [64].

Critically evaluating these findings reveals significant knowledge gaps regarding the long-term reliability of indirect liquid cooling, particularly concerning coolant leakage, corrosion, and material compatibility under prolonged operation and thermal abuse scenarios. The complexity and increased parasitic energy demands associated with pumping coolant through intricate channels also raise concerns regarding practical feasibility and economic viability in commercial EV battery applications. Addressing these gaps will require future research to optimize coolant materials, refine cooling channel architectures, integrate advanced insulation materials, and comprehensively test system performance under realistic operational conditions.

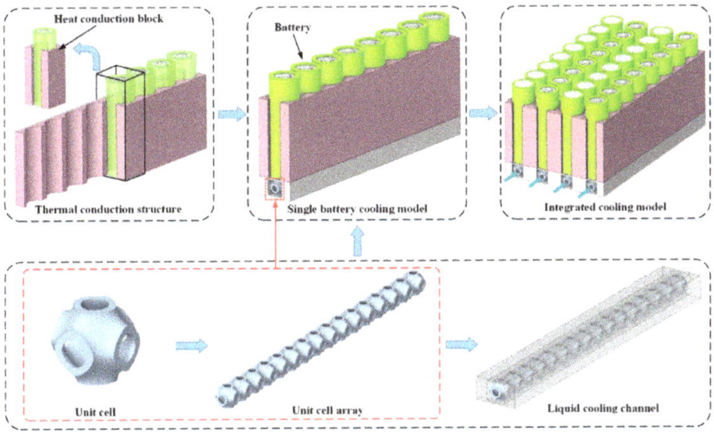

Figure 5. A schematic of liquid cooling using a primitive porous structure [64].

3.1.2. Water Mist

Water mist cooling has gained attention as a promising active cooling strategy for TR prevention and TRP suppression in LIB modules, attributed primarily to its high latent heat of vaporization and rapid cooling capacity. By using pure water with non-conductive properties, water mist cooling ensures the safety and reliability of LIBs cooling during TR events. Experimental studies have demonstrated notable cooling efficacy of water mist cooling under various TR conditions. Ma et al. investigated the cooling effect of water mist to suppress TRP in LIB modules under various conditions, as demonstrated in Figure 6. The study found that water mist applied effectively reduced the temperature rise in the cells during the TR process. When water mist was released 30 s before the onset of TR, it successfully suppressed the occurrence of TR. However, when the rate of temperature increase reached 2 °C/s, water mist could not completely suppress TR, but significantly reduced the temperature increase and surface temperature of the cells [65].

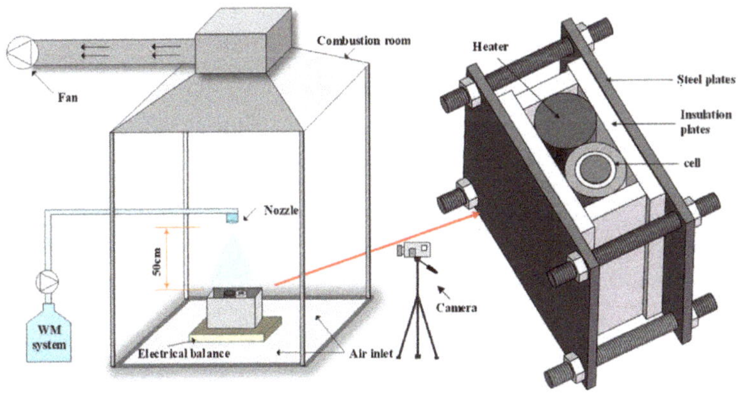

Figure 6. Experimental diagram of battery thermal management using a water mist cooling system [65].

Liu et al. conducted an experimental study on the cooling effect of fine water mist (FWM) to mitigate TR in LIBs. The results demonstrate that FWM significantly enhanced the cooling capacity, achieving a maximum improvement of 29.5% compared to air cooling. When the FWM flow rate increased from 0.1 mL/s to 0.8 mL/s, the cooling time to reduce the maximum temperature of the battery during TR to below 100 °C was reduced by 94.5 s, from 496.0 s to 401.5 s [66].

Despite these promising results, water mist cooling faces critical limitations, notably in densely packed battery modules, where the efficacy of water mist penetration is significantly compromised by flame-induced buoyancy and gas ejection phenomena. These limitations suggest that conventional water mist delivery systems might not provide sufficient penetration depth or uniform cooling under severe TR scenarios. Therefore, Liu et al. evaluated and improved the efficiency of water mist cooling by determining a critical water consumption threshold necessary for successful TRP suppression. The results demonstrate that water mist provided excellent cooling performance, achieving cooling rates exceeding 100 $°C \cdot s^{-1}$ and reducing the cell surface temperature to approximately 100 °C within a few seconds. At the same time, the research results also show that a critical water consumption rate of 1.95×10^{-4} $kg \cdot Wh^{-1}$ was sufficient to suppress TRP, even under extreme conditions where four out of five cells experienced TR [67]. The above research results show that identifying a critical water consumption threshold is effective for successful TRP suppression; however, this indicates that effective water mist cooling systems could demand substantial volumes of water, consequently increasing system complexity, cost, and weight.

Furthermore, recent innovations have focused on enhancing water mist cooling through intermittent spraying and hybrid approaches. Yao et al. demonstrated superior performance using intermittent mist spray cycles combined with extinguishing agents like $C_6F_{12}O$, markedly improving fire suppression effectiveness and significantly controlling battery temperature increases. This intermittent spraying approach not only enhanced cooling efficiency, but also optimized water consumption. Specifically, the study showed that integrating $C_6F_{12}O$ with water mist significantly enhanced the fire suppression ability compared to continuous water spray alone, effectively extinguishing the battery flame within 1 s. Notably, the intermittent water mist spray mode with a duty cycle (DC) of 0.5 and a cycle period (Pt) of 2 s showed superior cooling performance, limiting the maximum battery temperature to 115.9 °C and demonstrating the highest heat suppression ability. In addition, a longer intermittent cycle (DC = 0.1, Pt = 20 s) provided a prolonged cooling time, effectively reducing the battery temperature to below 50 °C, completely preventing the spread of uncontrolled TR [68].

Similarly, Xu et al. investigated the cooling effect of water mist with additives to suppress TR events in LIBs. The results show that water mist combined with additives such as Tween 20 (Polyoxyethylene (20) sorbitan monolaurate—$C_{18}H_{34}O_6(C_2H_4O)_{20}$) or 1-heptanol—significantly improved the cooling efficiency during TR events. In particular, water mist with Tween 20 provided the highest cooling efficiency, dissipating 12.46 ± 0.20 kJ of heat, an increase of 23.6% compared to pure water. This improvement was reflected in a reduction in temperature, with the surface temperature of the battery decreasing by 8.33 ± 0.28 °C/s. Furthermore, when water mist is applied, the cooling effect significantly reduces the fire extinguishing time, reducing the burn time from 92 s in the test without water mist to just 5 s [69].

Critically assessing the available studies, there remain significant gaps in understanding the optimal mist particle size, delivery frequency, and distribution mechanisms within battery packs. Additionally, water mist cooling systems must ensure robust electrical insulation to prevent short circuits or leakage currents that could compromise safety. Future research should prioritize developing tailored mist delivery systems that ensure deep penetration and uniform coverage, as well as investigating integrated cooling strategies that combine water mist cooling with other reliable extinguishing agents. Addressing these gaps is crucial for enhancing the practical feasibility, effectiveness, and scalability of water mist cooling systems in commercial LIB applications.

3.1.3. Immersion Cooling

Immersion cooling, also known as direct liquid cooling, offers an effective thermal management solution by submerging battery cells in dielectric fluids to ensure rapid and uniform heat dissipation. This approach eliminates thermal interface resistance and facilitates high-efficiency heat removal, making it particularly effective for managing extreme thermal events such as TR and TRP. Ye et al. conducted an experimental investigation on immersion cooling to inhibit the TR behavior of LIBs and evaluate the suppression of TRP in battery modules. The findings indicate that a full 70 mm immersion significantly delayed the onset of TR, extending the safety valve rupture-to-TR trigger interval from 70 s (at 20 mm immersion) to 312 s and reducing the peak TR temperature to 306 °C. Additionally, cells in 100%, 75%, and 50% SOC designate greater battery capacity, which elevates the risk of TR and thermal hazards. Furthermore, immersion cooling effectively inhibited the inter-cell TRP with peak temperature reductions of adjacent cells in 100%, 75%, and 50% SOC modules of 118.2 °C, 121.4 °C, and 111.5 °C, respectively [24].

In addition, several studies emphasize the role of fluid selection in optimizing the cooling efficacy of immersion cooling. Liu et al. investigated the use of immersion cooling with fluorinated liquids (HFE-7000 and SF33) to mitigate uncontrolled TRP in LIB packs, as shown in Figure 7. The results show that immersion cooling provided outstanding heat transfer rates and significantly reduced the surface temperature of the TR cell, with temperatures maintained at 518.8 °C and 448 °C for HFE-7000 and SF33, respectively. Furthermore, the immersion cooling system completely inhibited the propagation of TR, with the temperature maintained for neighboring cells at around 34 °C [25].

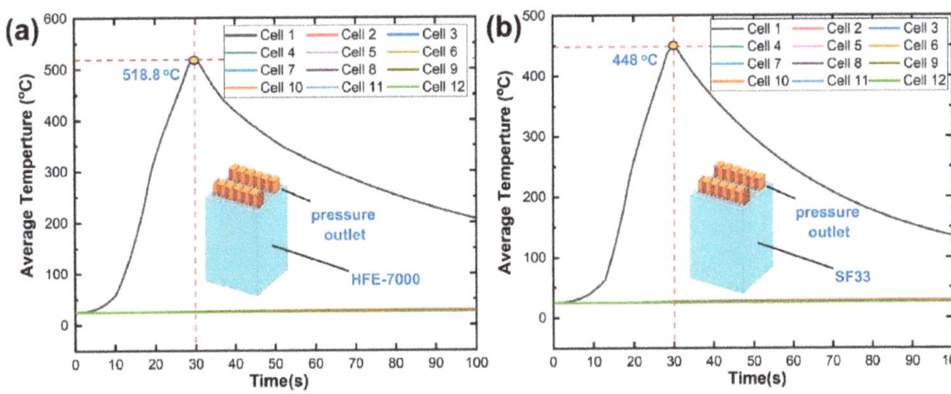

Figure 7. The average temperature variation of LIB cells in a battery pack with immersion cooling using (**a**) HFE-7000 and (**b**) SF33 [25].

Similarly, Sadar et al. investigated the use of Transformer Oil (TO) immersion cooling as a method to prevent TR in LIB packs under extreme conditions, as depicted in Figure 8. The results from TR testing show that the center cell temperature in the TO immersion setup reached a maximum of 222 °C, well below the critical temperature for TR, which typically exceeds 300 °C. Notably, TO prevented TR after 18,000 s, providing significant time for safe intervention. This extended time allows for the more effective identification of potential problems before catastrophic TR occurs [70].

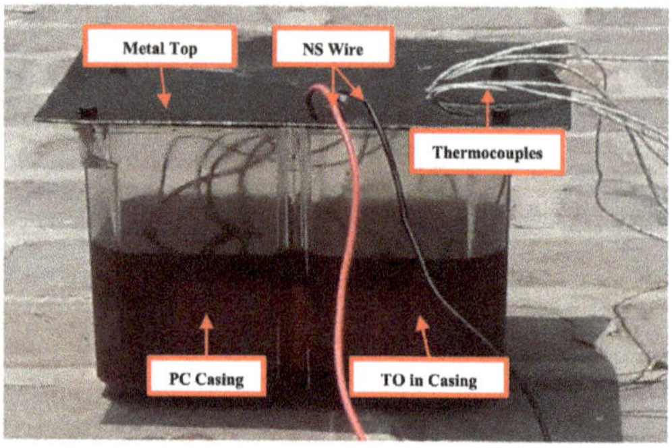

Figure 8. Transformer Oil (TO) immersion cooling setup for a battery pack in Polycarbonate (PC) casing [70].

Furthermore, some studies have further explored the flow configuration in immersion cooling setups, aiming to optimize the distribution of the fluid to achieve better temperature uniformity and lower maximum temperatures in LIB modules. Yang et al. examined the effectiveness of single-phase immersion cooling (SPIC) in managing the thermal behavior of 280 Ah $LiFePO_4$ batteries, focusing on the effects of different flow arrangements, including jet impingement, same sides and opposite sides, as demonstrated in Figure 9. The results show that jet impingement cooling provided the best temperature control, with a maximum temperature reduction of 1.29 °C compared to the same side arrangement. In addition, the temperature difference was 1.32 °C lower in the jet impingement system. The study also highlighted that no ignition or explosion occurred during the temperature spike during immersion cooling, with the maximum temperature during the temperature spike reaching 241.9 °C, demonstrating the safety and reliability of SPIC in preventing catastrophic TR events in large-scale battery systems [71].

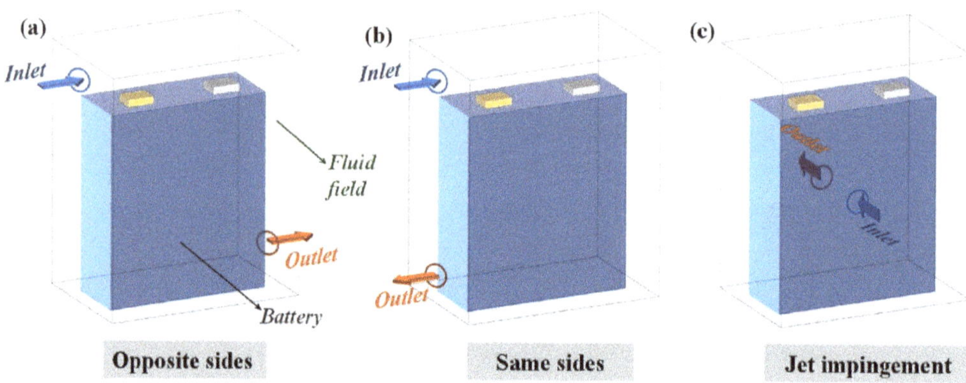

Figure 9. Different flow arrangement configurations in immersion cooling: (**a**) opposite sides, (**b**) same sides, and (**c**) jet impingement [71].

Similarly, Patil et al. examined the use of dielectric fluid immersion cooling for LIBs, focusing on optimizing the flow configuration and tab cooling to improve cooling efficiency and prevent TR, as illustrated in Figure 10. The study found that LIB cells immersed in a dielectric fluid flowing from bottom to top assisted with tab cooling demonstrated a 46.8% reduction in maximum temperature at the positive tab at a 3C discharge rate compared to natural convection. This method resulted in a significant improvement in cooling performance, keeping the battery temperature below 40 °C even at a high 5C discharge rate. Furthermore, when simulating thermal abuse conditions such as ISC, the peak temperature of the battery pack was limited to 341.7 °C, successfully preventing the propagation of TR throughout the cells [72]. These insights underscore the importance of optimized fluid dynamics in maximizing the effectiveness of immersion cooling systems.

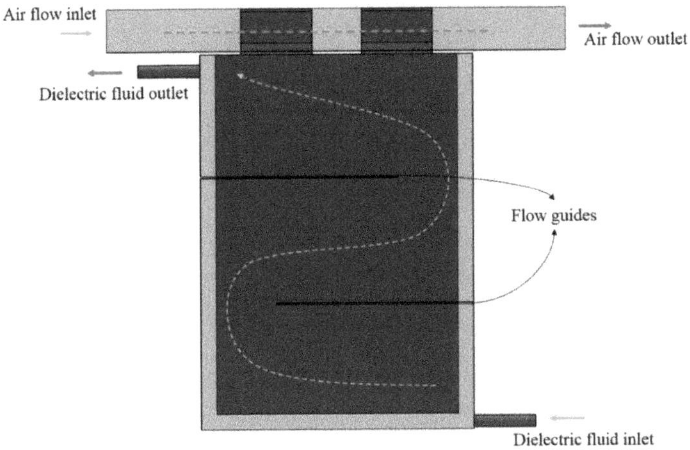

Figure 10. Immersion cooling configuration combined with tab cooling for the LIB pack [72].

While immersion cooling systems demonstrate exceptional performance in heat mitigation, several practical and safety challenges must be addressed. The high cost and limited environmental compatibility of many dielectric fluids, such as fluorinated liquids, restrict their widespread adoption. Chemical stability, potential toxicity, and disposal issues raise additional environmental and regulatory concerns. Moreover, ensuring structural sealing against leaks is vital, especially in two-phase systems where pressure fluctuations occur due to phase change. A critical safety consideration involves the dielectric strength and material compatibility of immersion fluids with cell packaging and electrical components. Any breach in insulation could compromise safety and lead to short circuits, particularly in systems exposed to thermal abuse. Therefore, the dielectric properties of fluids must be rigorously validated under real-world conditions, including long-duration cycling and extreme temperatures. Although current research demonstrates promising results, further studies are needed to evaluate the long-term reliability of immersion-cooled LIB packs, the environmental impact of different coolant formulations, and the trade-offs between performance, safety, and cost. Future directions should focus on developing biodegradable or low-toxicity dielectric fluids, enhancing sealing techniques, and exploring hybrid cooling strategies that incorporate immersion cooling with supplemental insulation or PCM for comprehensive TR prevention and TRP suppression.

3.2. PCM Cooling

PCM cooling has emerged as a promising passive thermal management strategy due to its ability to absorb large amounts of heat during phase transitions, thereby delaying or suppressing the rapid temperature escalation associated with TR and TRP in LIBs. Unlike active systems, PCM cooling does not require external power input, making it particularly attractive for compact energy-efficient applications. Liu et al. conducted numerical simulations to investigate the effectiveness of paraffin phase change material (PPCM) as a thermal safety measure against uncontrolled TRP in LIB packs, as shown in Figure 11. The results show a clear correlation between the thickness of PPCM and its ability to slow down TRP. Specifically, the TR onset in the adjacent cell was significantly delayed to 64 s, 266 s, and 414 s, corresponding to PPCM thickness increases of 1.8 mm, 3.6 mm, and 5.4 mm, respectively. Furthermore, combining PPCM with insulation layers provided additional protection and extended the TR onset delay in the adjacent cell to 798 s, corresponding to an approximately 43-fold improvement over the unprotected LIB pack [25].

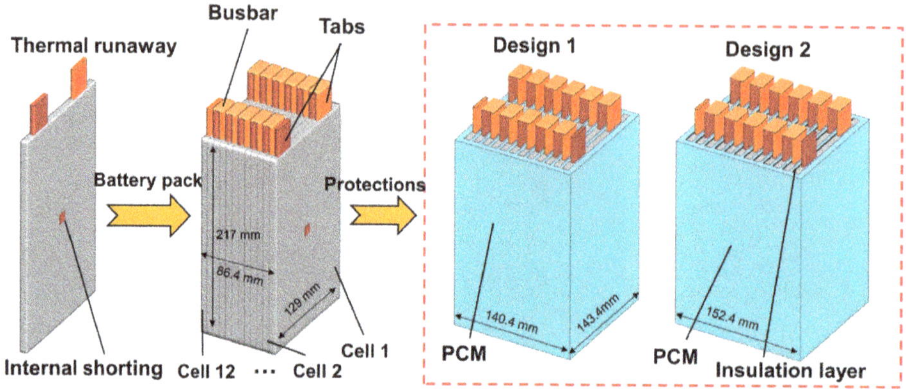

Figure 11. Battery thermal management model using PCM cooling (Design 1) and PCM cooling combined with an insulation layer (Design 2) [25].

Yang et al. evaluated the use of PCM cooling in BTMS to mitigate uncontrolled TRP, as presented in Figure 12. The results show that PCM cooling effectively absorbed and managed heat in TRP, reducing the maximum temperature to 913.5 °C, which was significantly lower than the temperature achieved with only using insulation materials of 1165.9 °C. In addition, PCM cooling could delay the propagation time of TR until 80.8 s due to the advantages of large latent heat and a high heat absorption rate [63].

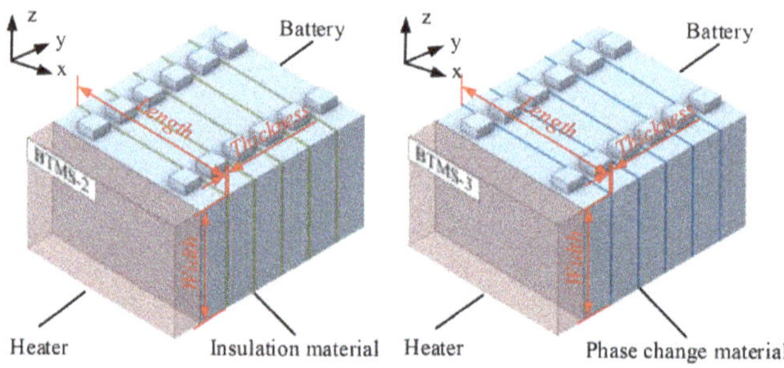

Figure 12. Battery thermal management configuration using insulation material (BTMS-2) and PCM cooling (BTMS-3) [63].

Sadar et al. explored the use of PCM cooling to prevent TR in LIB packs under extreme conditions, as demonstrated in Figure 13. During TR testing, the center cell temperature in the PCM cooling setup peaked at 434 °C, significantly lower than the 560 °C observed in the natural convection setup. Additionally, PCM cooling significantly delayed the time to the TR event by 1400 s, 500 s longer than the natural convection battery pack configuration [70].

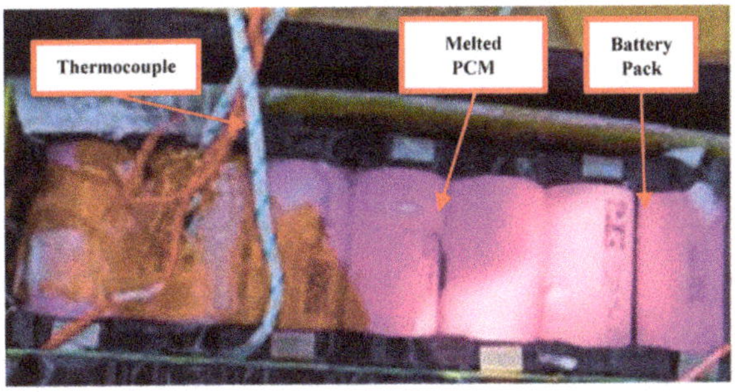

Figure 13. Setup for battery thermal management with PCM cooling [70].

Zhi et al. studied the application of a hydrated salt-based composite PCM consisting of a mixture of $Na_2SO_4 \cdot 10H_2O$ and $KAl(SO_4)_2 \cdot 12H_2O$ to the thermal management of LIBs and to mitigate TR, as shown in Figure 14. The results show that the maximum battery temperature decreased by 20.4%, from 65.95 °C, with air cooling to 52.47 °C with PCM cooling, and the temperature difference decreased by 59.7%, from 5.46 °C to 2.20 °C, respectively, during a 0.5C charge and a 1.5C discharge cycle. The PCM cooling also demonstrated its effectiveness in preventing TRP, with the peak temperature of the first cell during TR reduced from 427 °C (without PCM) to 383 °C (with PCM). Furthermore, the temperatures of the adjacent cells with the TR triggering cell were significantly reduced by the PCM barrier, completely preventing TR events [73]. These results demonstrate the feasibility of tailoring PCM compositions to enhance both thermal regulation and TRP suppression.

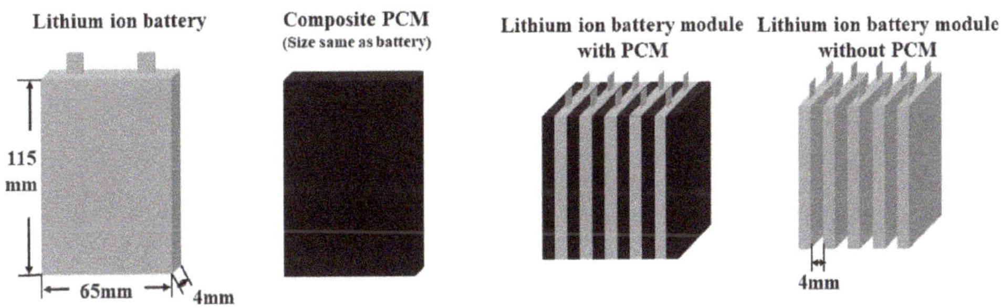

Figure 14. Battery thermal management strategy using PCM cooling [73].

Despite these advantages, PCM systems are constrained by inherently low thermal conductivity, which limits the rate of heat absorption and dissipation. This drawback becomes especially pronounced under high-power or prolonged abuse conditions where heat generation exceeds the phase change capacity of the PCM. While additives such as expanded graphite, metal foams, or thermally conductive fillers have been employed to enhance thermal conductivity, their integration must be balanced against material stability, cost, and added weight. In addition to performance concerns, safety and material compatibility issues must be considered. Paraffin-based PCMs, although being commonly used, may pose fire risks if containment is breached. Hydrated salts, while non-flammable, can exhibit phase separation or subcooling, which reduces reliability over long-term cycles. Therefore, selecting PCMs with stable thermal cycling properties, high latent heat, and minimal volumetric expansion is crucial for system durability and safety. Critically, PCM systems alone cannot maintain thermal equilibrium during extended thermal events without supplemental dissipation. As such, PCM cooling is best suited for transient heat suppression or as a component in hybrid systems. Future research should aim to optimize PCM formulations for enhanced conductivity and cycling stability, develop composite materials with integrated structural and thermal functions, and validate long-term performance under realistic operational scenarios. Additionally, standardization in testing methodologies for PCM-enhanced BTMS will be essential to accelerate their adoption in practical battery applications. Furthermore, research into combining PCM cooling with advanced cooling strategies such as liquid cooling will further improve the goal of temperature control and prevention of TR and TRP events in LIB modules.

3.3. Hybrid Cooling

Hybrid cooling systems, which integrate both active and passive cooling mechanisms, have gained prominence as highly effective strategies for suppressing TR and TRP in LIB modules. By combining the high latent heat absorption of PCM with the continuous heat removal capabilities of liquid cooling, hybrid systems aim to achieve optimal thermal regulation, temperature uniformity, and safety performance under both normal and abuse conditions. Xiao et al. investigated a hybrid cooling strategy that combined composite phase change material (CPCM) and liquid cooling to enhance the thermal safety of LIB packs under TRP conditions, as depicted in Figure 15. The research results show that increasing the coolant flow rate significantly delayed TRP, with the propagation time extended from 17.8 s at 0.009 m/s to 32.9 s at 0.029 m/s. Additionally, increasing battery spacing delayed propagation and altered its path, while decreasing the thermal conductivity of the CPCM substantially reduced TRP due to limited heat transfer. Conversely, increasing the number of cooling channels had a limited effect on TRP time, with the total TRP time changing from 25.5 s to 27 s when the number of cooling channels increased between 1 and 5 [46].

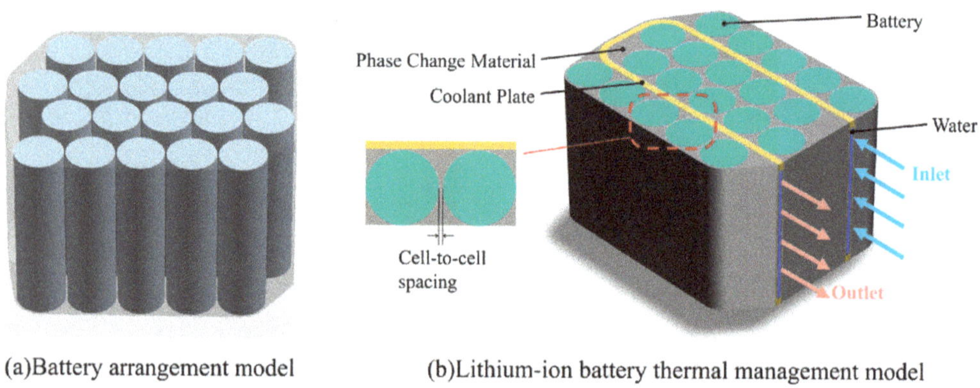

Figure 15. A hybrid cooling configuration combining PCM cooling and liquid cooling with cooling channels for the LIB pack [46].

Luo et al. proposed an adaptive hybrid cooling strategy to mitigate TR in LIB modules by integrating liquid cooling and PCM cooling with considerations of buoyancy-driven natural convection in PCM, as illustrated in Figure 16. The findings indicate that the proposed hybrid cooling approach extended the TRP interval by 60.6% (85.9 s) and reduced total coolant flow by 31.7%. In addition, the implementation of hybrid cooling improved the cooling performance without causing additional parasitic energy consumption or compromising the energy density of the battery pack [29].

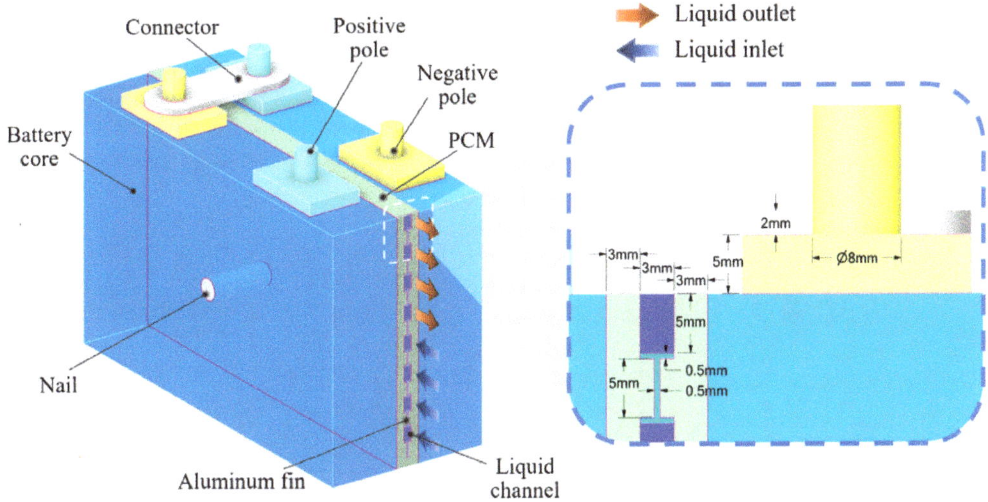

Figure 16. The geometry of hybrid cooling integrates PCM cooling with the microchannel cold plates [29].

Lu et al. proposed a thermal management system using a hybrid cooling strategy combining PCM cooling based on a paraffin/Al-foam composite and liquid microchannels to suppress TR in LIB packs, as depicted in Figure 17. The research results demonstrate that the proposed hybrid cooling system effectively suppressed TR, with the temperature of neighboring cells only increasing to about 80 °C in the first 600 s, which could not trigger a TR at a liquid flow rate of 0.01 L/min. The proposed hybrid cooling system outperformed individual PCM or liquid microchannel cooling methods and significantly improved battery safety by suppressing TR across multiple cells [74]. These findings underscore the hybrid approach's ability to balance energy efficiency, safety, and thermal performance.

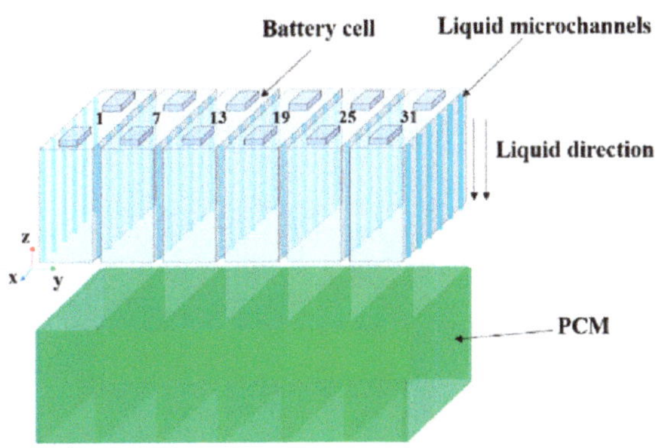

Figure 17. Proposed hybrid cooling system combining PCM cooling and liquid microchannel cooling [74].

Beyond thermal control, advanced hybrid systems have been designed with multi-functional components to improve system compactness and heat spreading. Ouyang et al. proposed a novel hybrid cooling system that integrates CPCM, aerogel for thermal insulation, and a cooling plate with nanofluid to enhance both heat dissipation and insulation, as demonstrated in Figure 18. The research showed that the proposed hybrid cooling system efficiently mitigates TRP and reduces the maximum battery temperature from 740.35 °C to

55.19 °C under normal heat dissipation conditions. Under extreme conditions, the maximum battery temperature was reduced from 707.36 °C to 107.19 °C. In addition, the cooling performance of the hybrid cooling system was further optimized with the best combination of factors, including a flow rate of nanofluid of 0.05 m/s, nanoparticle volume fraction of 1%, and a thickness of the PCM and insulating material of 3 mm, which resulted in a 23% reduction in maximum battery temperature and a 22% reduction in energy consumption compared to the baseline configuration [75].

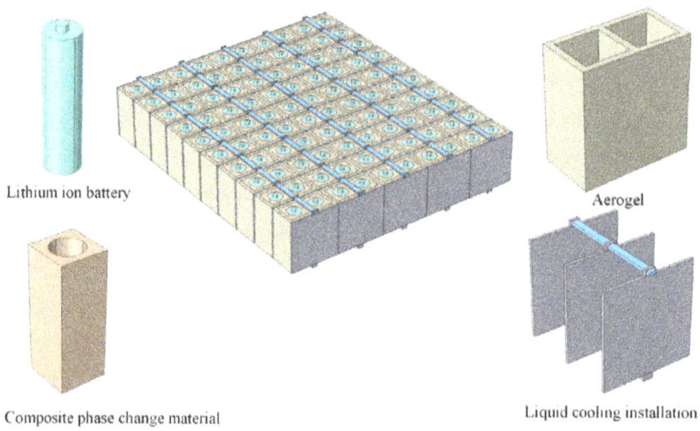

Figure 18. Structure of a hybrid cooling system combining PCM cooling, aerogel insulation material, and liquid cooling using cooling plates [75].

Similarly, Zhang et al. proposed a hybrid cooling strategy combining PCM and liquid cooling to mitigate uncontrolled TRP in LIB modules, as shown in Figure 19. The research results demonstrate that the proposed hybrid cooling system effectively controlled the battery temperature, with the maximum temperature during a 5C discharge cycle maintained at 39.8 °C, compared to 51.7 °C for the only liquid-cooled system. The hybrid cooling system also slowed down the uncontrolled TRP with a delayed time of 460 s between the two cells, while the liquid-cooled system had nearly simultaneous uncontrolled TR events. In addition, the higher PCM thermal conductivity (0.6 W/m·K) improved the conventional cooling performance, but increased the heat transfer rate, resulting in uncontrolled TRP in 5.6 min. Furthermore, increasing the water velocity in the cooling channel to 0.01 m/s successfully prevented uncontrolled TRP in the battery module [76].

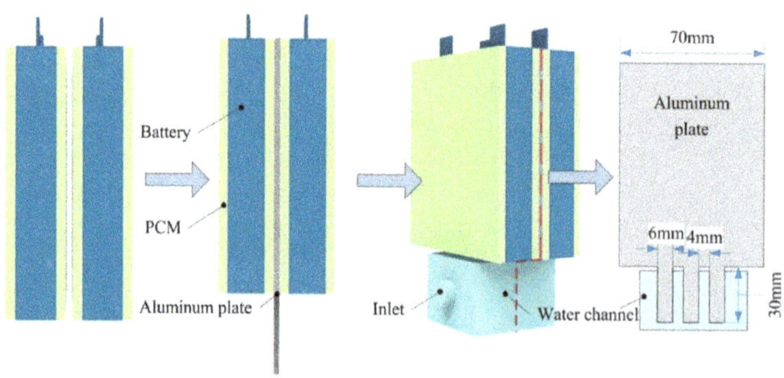

Figure 19. A hybrid cooling strategy combining PCM cooling and liquid cooling with an aluminum material cooling plate [76].

Wang et al. studied a hybrid cooling strategy that combines PCM cooling with wavy microchannel cooling plate (PCM-WMCP) to enhance cooling efficiency and temperature uniformity in LIBs, as presented in Figure 20. The numerical study found that the hybrid cooling system significantly reduced the maximum battery temperature compared to the air-cooled (AC) system and the PCM-only system. Specifically, at a discharge rate of 1.0C, the maximum battery temperature decreased from 59.49 °C (AC) to 43.65 °C (PCM-WMCP), and, at a discharge rate of 3.0C, the battery temperature decreased from 79.66 °C (AC) to 49.45 °C (PCM-WMCP). Additionally, when 10% graphite was added to the PCM to form a composite PCM (CPCM(B)), the system achieved a more uniform temperature distribution, reducing the temperature difference between the maximum and minimum battery temperatures from 2.14 °C to 1.61 °C. Furthermore, the hybrid CPCM(B)-WMCP system also demonstrated superior performance in preventing TR, with the time to reach the critical temperature for TR being 650 s slower than the PCM-only system [77].

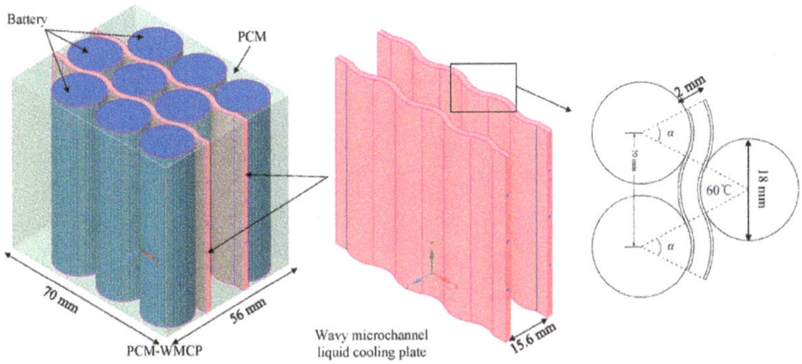

Figure 20. A hybrid cooling structure combining PCM cooling with wavy microchannel liquid cooling [77].

Gu et al. proposed a novel hybrid cooling that integrates CPCM (paraffin wax and boron nitride) and a variable wall liquid cooling plate (LCP), as depicted in Figure 21. The results show that the implementation of the variable wall LCP resulted in a significant reduction in the maximum battery temperature, which was reduced by 1.81 °C compared to the standard LCP. Furthermore, the study found that maintaining a flow rate above 0.89 g/s was sufficient to keep the maximum battery temperature below 40 °C during both the charge and discharge cycles, providing a practical hybrid cooling solution to prevent TR in high-performance batteries [78].

Zonouzi et al. explored an innovative hybrid cooling approach to the thermal management of LIBs by integrating helical tube liquid cooling with PCM cooling. The results demonstrate that the hybrid cooling system outperformed individual cooling methods, with the maximum battery temperature after 750 s being reduced to 34.45 °C in the hybrid cooling system, compared to 38.55 °C with only PCM cooling and 36.35 °C with only liquid cooling. The study also highlighted the positive impact of reducing the helical tube pitch, which significantly improved the cooling performance by providing a larger contact area for heat exchange and contributing to the effective prevention of potential TR events in LIB application systems [79].

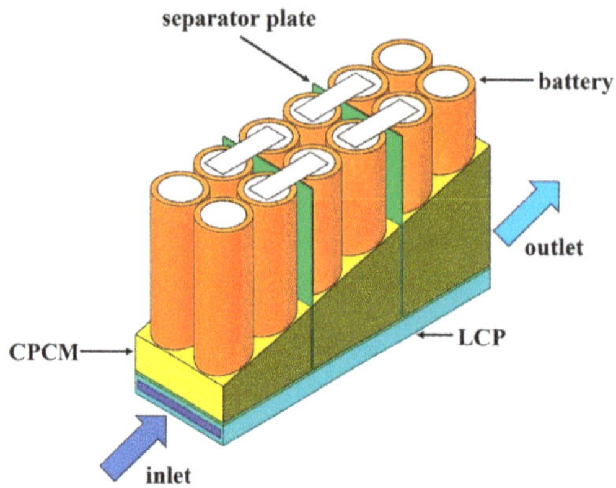

Figure 21. A hybrid cooling strategy using PCM cooling combined with a liquid cooling plate [78].

Liu et al. evaluated a hybrid cooling system combining a mini-channel cold plate (MCP) with PCM cooling for LIBs, especially under a high discharge rate and TR conditions. The results show that, during a high discharge rate of 2.5C, the optimized hybrid cooling structure maintained the battery temperature at 44.37 °C and a temperature difference between cells of no more than 2.01 °C, even at low coolant flow rates. Under external short-circuit conditions, the optimized structure maintained the temperature of the battery pack at 43.22 °C, providing effective thermal management. Most notably, when TR occurs due to an ISC, the optimized hybrid cooling configuration can delay TR propagation by up to 521 s, significantly improving safety by preventing TR from spreading to adjacent cells [80].

Ji et al. developed a novel hybrid cooling strategy combining phase change composite materials (PCCM) with a liquid cooling plate to suppress TRP in high-specific-energy ternary LIBs. The experimental results demonstrate that when a 2 mm thick PCCM was integrated with a liquid cooling plate, the TR in a 60 Ah battery module was successfully mitigated. Specifically, during the TR event, the TR trigger cell reached a temperature of approximately 650 °C; however, due to the heat absorption capacity of the PCCM and the cooling effect from the liquid cooling plate, the temperature of the adjacent cell remained under control and remaining below 100 °C throughout the process, effectively suppressing TRP [81].

Despite these advancements, hybrid cooling systems present several engineering and implementation challenges. The integration of multiple components increases system volume, design complexity, and manufacturing cost. Inefficient thermal coupling between PCM and liquid channels can limit their effectiveness during rapid thermal transients, while the saturation of PCM over extended events may diminish passive cooling capacity. Furthermore, additional structural materials may reduce energy density, a critical consideration in space-constrained EV applications. From a safety perspective, hybrid systems must be carefully designed to avoid coolant leakage, electrical shorting, and material degradation under thermal stress. Dielectric compatibility of coolants, structural sealing techniques, and insulation integrity are essential for maintaining electrical isolation and mechanical reliability over time. To address these issues, future research should prioritize the co-optimization of hybrid system geometries, including cooling channel layouts and PCM distributions to maximize thermal synergy and spatial efficiency. Investigating advanced composite PCMs with self-healing or thermally triggered reusability may provide further breakthroughs. Finally, comprehensive long-term performance validation under real-world duty cycles will be critical for enabling the safe and scalable deployment of hybrid BTMS in high-energy LIB applications.

4. Summary and Recommendations

TR events remain one of the most critical safety concerns in LIB applications. TR is often triggered by mechanical, electrical, or thermal abuse and is characterized by rapid heat generation, gas release, and potential fires or explosions. As LIBs are increasingly deployed in high-energy and high-power applications, the demand for advanced BTMS capable of preventing and suppressing TR events has become paramount. Advanced BTMS technologies offer promising solutions by enhancing heat dissipation, maintaining battery temperature uniformity, and responding flexibly to thermal anomalies. In this context, the present review focuses on the recent studies over the past 5 years on advanced BTMS for the prevention and suppression of TR events and TRP in LIB modules. The advanced cooling strategies summarized and evaluated in the present review include liquid cooling methods such as indirect liquid cooling, water mist cooling, and immersion cooling, followed by PCM cooling and, finally, hybrid cooling systems. The following summarizes the main conclusions and recommendations of the current review:

(a) Indirect liquid cooling is a widely adopted thermal management technique that offers superior heat dissipation compared to traditional air-cooled methods, primarily due to the higher thermal conductivity and specific heat capacity of liquid coolants such as water or water–glycol mixtures. This approach enables more effective temperature regulation and improved thermal uniformity across the battery module. However, the presence of multiple thermal interfaces between the battery cells and cold plates introduces significant thermal resistance, limiting the cooling effectiveness during critical (TR) events. Additionally, indirect liquid cooling systems often involve complex structural designs, increasing the risks related to coolant leakage, corrosion, and higher parasitic energy consumption. These factors contribute to increased system weight, maintenance demands, and overall cost, posing challenges for scaling to next-generation high-power LIB applications. Therefore, while indirect liquid cooling remains effective under standard operating conditions, its application for TR prevention and suppression in future high-capacity systems requires further optimization in terms of system design, coolant management, and safety validation.

(b) Water mist cooling has emerged as a promising strategy to mitigate TR and its propagation in LIBs due to its high latent heat of vaporization and rapid heat absorption through droplet evaporation. By cooling both the ambient environment and the battery surface, water mist can significantly reduce the temperature rise during TR events. Furthermore, water mist systems offer environmental advantages, being non-toxic, readily available and cost-effective. However, their effectiveness can be compromised in dense battery modules where flame-induced buoyancy and gas ejection limit mist penetration to inner cells. Additionally, large water volumes are often required to sustain sufficient cooling performance, which increases system complexity and weight. Despite these limitations, water mist cooling demonstrates strong potential for TR suppression, and further improvements in delivery methods, droplet sizing, hybrid agent use, and real-time thermal control are needed to enable its broader application in advanced battery thermal management systems.

(c) Immersion cooling, also known as direct liquid cooling, offers a highly effective approach for preventing and suppressing TR and TRP in high-energy LIB systems. By directly submerging battery cells in dielectric fluids, immersion cooling eliminates thermal interface resistance and achieves rapid, uniform heat dissipation. Single-phase immersion cooling provides flexible system design options, while two-phase immersion cooling leverages latent heat for enhanced cooling performance. However, challenges such as the high cost, environmental sensitivity, and chemical stability requirements of dielectric fluids, along with concerns over leakage, material compatibility, and system complexity restricted widespread adoption. Additionally, ensuring robust structural sealing is critical, particularly in two-phase systems where boiling-induced pressure variations occur. Despite these issues, innovative designs such as tab-assisted cooling, optimized coolant flows, and improved immersion architectures show strong potential to make immersion cooling a reliable and scalable solution for enhancing LIB safety against TR and TRP events.

(d) PCM cooling presents a promising passive thermal management strategy for LIBs, particularly in compact and safety-critical applications. PCMs absorb substantial amounts of heat through latent heat during phase transitions, effectively delaying or suppressing rapid temperature rises without external energy input. Paraffin- and hydrated salt-based PCMs have demonstrated significant benefits in retarding TRP by maintaining temperature uniformity within battery packs. Enhancements through composite formulations, such as adding expanded graphite or metallic fillers, further improve thermal conductivity and structural stability. However, the inherently low thermal conductivity of pure PCMs limits their standalone performance, especially under severe abuse conditions. Moreover, PCM systems lack active heat dissipation mechanisms, making them more suitable for low to moderate power densities or as supplementary layers in hybrid cooling architectures. Future developments in high-conductivity composite PCMs and integration with active cooling systems are critical to overcoming current limitations and unlocking the full potential of PCM cooling for advanced battery thermal management.

(e) Hybrid cooling systems that integrate passive and active thermal management techniques have emerged as highly effective solutions for mitigating TR and TRP risks in LIB modules. By combining the latent heat absorption of PCMs with the continuous heat dissipation of liquid cooling, hybrid systems achieve superior control over both peak temperatures and temperature uniformity. Studies have demonstrated that hybrid configurations can significantly delay TRP events and prevent adjacent cells from reaching critical failure conditions. However, the complexity of integrating active and passive components increases design difficulty, system volume, and cost. Inefficient thermal coupling between PCM layers and liquid cooling structures can limit overall system performance, particularly during rapid TR events. Additionally, PCM saturation over prolonged abuse conditions may reduce cooling effectiveness. Therefore, future research must focus on optimizing composite materials, refining liquid channel geometries, and developing modular, lightweight, hybrid architectures to fully realize the potential of hybrid cooling in next-generation BTMS designs.

(f) In addition, thermal insulation layers have shown substantial potential as a passive mitigation strategy for TRP in LIB systems. By introducing low thermal conductivity materials such as aerogels or glass fiber composites between adjacent cells, insulation layers act as thermal barriers, effectively delaying or even preventing the spread of heat generated during a TR event. Moreover, when combined with active cooling systems such as cold plates or immersion cooling setups, insulation layers allow for the cooling system to have more time to remove heat before it reaches critical levels. Despite their advantages, insulation layers can accumulate heat over prolonged operation and reduce energy density due to added volume, necessitating careful thermal design and material selection. Nonetheless, their simplicity, reliability, and proven efficacy make them a valuable component in next-generation BTMS aimed at enhancing safety and suppressing TRP in high-energy-density LIB packs.

Author Contributions: Conceptualization, L.D.T. and M.-Y.L.; methodology, L.D.T. and M.-Y.L.; formal analysis, L.D.T. and M.-Y.L.; investigation, L.D.T. and M.-Y.L.; resources, L.D.T. and M.-Y.L.; data curation, L.D.T. and M.-Y.L.; writing—original draft preparation, L.D.T.; writing—review and editing, L.D.T. and M.-Y.L.; visualization, L.D.T. and M.-Y.L.; supervision, M.-Y.L.; project administration, M.-Y.L.; funding acquisition, M.-Y.L. All authors have read and agreed to the published version of the manuscript.

Funding: This work was supported by the Dong-A University research fund.

Data Availability Statement: The data presented in this study are available upon request to the corresponding author. The data are not publicly available due to privacy.

Conflicts of Interest: The authors declare no conflicts of interest.

References

1. Febriyanto, R.; Ariyadi, H.M.; Pranoto, I.; Rahman, M.A. Experimental study of serpentine channels immersion cooling for lithium-ion battery thermal management using single-phase dielectric fluid. *J. Energy Storage* **2024**, *97*, 112799. [CrossRef]
2. Berwal, P.; Mehrotra, A.; Lee, Y.; Padhi, U.P.; Yoh, J. Combustion theory based cooling strategies for preventing thermal runaway in modern batteries. *Combust. Flame* **2025**, *274*, 113995. [CrossRef]
3. Hwang, S.-G.; Lee, M.-Y.; Ko, B.-S. Numerical Analysis on Cooling Performances for Connectors Using Immersion Cooling in Ultra-Fast Chargers for Electric Vehicles. *Symmetry* **2025**, *17*, 624. [CrossRef]
4. Chandrasekaran, M.; Jithin, K.V.; Soundarya, T.; Rajesh, P.K. Comprehensive experimental study of battery thermal management using single-phase liquid immersion cooling. *J. Energy Storage* **2025**, *111*, 115445.
5. Kim, S.-H.; Park, S.-Y. Computational Analysis for Efficient Battery Cooling for Eletric Vehicles using Various Cooling Methods. *J. Korea Acad. Ind. Coop. Soc.* **2025**, *26*, 9–20.
6. Song, Y.; Wang, Z.; Jia, D.; Yuan, D.; Chang, Y.; Zheng, X.; Zhang, S. Effect of immersion cooling design optimization on thermal management for lithium battery module. *Appl. Therm. Eng.* **2025**, *272*, 126401. [CrossRef]
7. Wahab, A.; Najmi, A.; Senobar, H.; Amjady, N.; Kemper, H.; Khayyam, H. Immersion cooling innovations and critical hurdles in Li-ion battery cooling for future electric vehicles. *Renew. Sustain. Energy Rev.* **2025**, *211*, 115268. [CrossRef]
8. Baek, S.-H.; Ma, B.-C. Optimization of thermal management for immersion cooled lithium-ion batteries based on experimental design methods. *J. Korea Acad. Ind. Coop. Soc.* **2025**, *26*, 21–28.
9. Tai, L.D.; Garud, K.S.; Lee, M. Experimental Study on Thermal Management of 5S7P Battery Module with Immersion Cooling Under High Charging/Discharging C-Rates. *Batteries* **2025**, *11*, 59. [CrossRef]
10. Tai, L.D.; Garud, K.S.; Hwang, S.; Lee, M. A Review on Advanced Battery Thermal Management Systems for Fast Charging in Electric Vehicles. *Batteries* **2024**, *10*, 372. [CrossRef]
11. Garud, K.S.; Tai, L.D.; Hwang, S.; Nguyen, N.; Lee, M. A Review of Advanced Cooling Strategies for Battery Thermal Management Systems in Electric Vehicles. *Symmetry* **2023**, *15*, 1322. [CrossRef]
12. Hussain, M.; Khan, M.K.; Pathak, M. Thermal management of high-energy lithium titanate oxide batteries using an effective channeled dielectric fluid immersion cooling system. *Energy Convers. Manag.* **2024**, *313*, 118644. [CrossRef]
13. Han, J.; Garud, K.S.; Hwang, S.; Lee, M. Experimental Study on Dielectric Fluid Immersion Cooling for Thermal Management of Lithium-Ion Battery. *Symmetry* **2022**, *14*, 2126. [CrossRef]
14. Han, J.; Garud, K.S.; Kang, E.; Lee, M. Numerical Study on Heat Transfer Characteristics of Dielectric Fluid Immersion Cooling with Fin Structures for Lithium-Ion Batteries. *Symmetry* **2022**, *15*, 92. [CrossRef]
15. Huang, H.; Li, W.; Xiong, S.; Luo, Z.; Ahmed, M. Single-phase static immersion-cooled battery thermal management system with finned heat pipes. *Appl. Therm. Eng.* **2024**, *254*, 123931. [CrossRef]
16. Garud, K.S.; Han, J.; Hwang, S.; Lee, M. Artificial Neural Network Modeling to Predict Thermal and Electrical Performances of Batteries with Direct Oil Cooling. *Batteries* **2023**, *9*, 559. [CrossRef]
17. Moon, J.-H. Air-Cooled Battery Cooling Simulation Techniques Using Simulink and Simscape. *J. Korea Acad. Ind. Coop. Soc.* **2025**, *26*, 35–40. [CrossRef]
18. Feng, X.; Ouyang, M.; Liu, X.; Lu, L.; Xia, Y.; He, X. Thermal runaway mechanism of lithium ion battery for electric vehicles: A review. *Energy Storage Mater.* **2017**, *10*, 246–267. [CrossRef]
19. Shahid, S.; Agelin-Chaab, M. A review of thermal runaway prevention and mitigation strategies for lithium-ion batteries. *Energy Convers. Manag. X* **2022**, *16*, 100310. [CrossRef]
20. Chavan, S.; Venkateswarlu, B.; Prabakaran, R.; Salman, M.; Joo, S.W.; Choi, G.S.; Kim, S.C. Thermal runaway and mitigation strategies for electric vehicle lithium-ion batteries using battery cooling approach: A review of the current status and challenges. *J. Energy Storage* **2023**, *72*, 108569. [CrossRef]

21. Yang, Y.; Wang, R.; Shen, Z.; Yu, Q.; Xiong, R.; Shen, W. Towards a safer lithium-ion batteries: A critical review on cause, characteristics, warning and disposal strategy for thermal runaway. *Adv. Appl. Energy* **2023**, *11*, 100146. [CrossRef]
22. Dhuchakallaya, I.; Saechan, P. Numerical study of a pressurized gas cooling system to suppress thermal runaway propagation in high-energy-density lithium-ion battery packs. *J. Energy Storage* **2024**, *101*, 113916. [CrossRef]
23. Hong, Y.; Wu, H.; Wong, S.K.; Jin, C.; Xu, C.; Wang, H.; Peng, Y.; Zheng, Y.; Feng, X.; Ouyang, M. Dynamic thermophysical modeling and parametric sensitivity analysis of flood cooling suppressing the thermal runaway propagation for electric bicycle battery. *J. Energy Storage* **2024**, *98*, 113084. [CrossRef]
24. Ye, Y.; Mao, Y.; Zhao, L.; Chen, Y.; Chen, M. Experimental investigation of thermal runaway behavior and propagation inhibition of lithium-ion battery by immersion cooling. *Appl. Therm. Eng.* **2024**, *256*, 124093. [CrossRef]
25. Liu, X.; Zhou, Z.; Wu, W.; Wei, L.; Wu, W.; Li, Y.; Gao, L.; Li, Y.; Song, Y. Modelling for the mitigation of lithium ion battery thermal runaway propagation by using phase change material or liquid immersion cooling. *Case Stud. Therm. Eng.* **2023**, *52*, 103749. [CrossRef]
26. Tong, B.; Shi, J.; Cao, M.; Xuan, W.; Chen, J.; Jin, K.; Sun, J.; Wang, Q. Comprehensive comparison study on battery thermal management modules with indirect and direct liquid cooling. *Appl. Therm. Eng.* **2025**, *268*, 125945. [CrossRef]
27. Zhao, L.; Song, G.; Yang, X.; Pan, R. Experiments on the effects of liquid immersion cooling on the thermal runaway (TR) behaviors of 280 Ah lithium-ion batteries subjected to different TR-triggering conditions. *J. Energy Storage* **2025**, *121*, 116358. [CrossRef]
28. Kshetrimayum, K.S.; Yoon, Y.; Gye, H.; Lee, C. Preventing heat propagation and thermal runaway in electric vehicle battery modules using integrated PCM and micro-channel plate cooling system. *Appl. Therm. Eng.* **2019**, *159*, 113797. [CrossRef]
29. Luo, P.; Gao, K.; Hu, L.; Chen, B.; Zhang, Y. Adaptive hybrid cooling strategy to mitigate battery thermal runaway considering natural convection in phase change material. *Appl. Energy* **2024**, *361*, 122920. [CrossRef]
30. Choi, D.-H.; Kim, M.-J.; Yu, J.-S.; Hong, D.-S. Experimental Study on the Heat Charging and Discharging Characteristics of a Heat Absorption Device Applied to the Cooling System of Electric Vehicle Chargers. *J. Korea Acad. Ind. Coop. Soc.* **2024**, *25*, 536–542.
31. Choi, D.-H.; Kim, T.-W.; Kim, M.-J.; Yu, J.-S.; Hong, D.-S. Experimental Study on the Thermal Characteristics and Melting Behavior of Phase Change Materials Applied to the Cooling System of Electric Vehicle Chargers. *J. Korea Acad. Ind. Coop. Soc.* **2024**, *25*, 349–356.
32. Jiang, Z.; Li, H.; Qu, Z.; Zhang, J. Recent progress in lithium-ion battery thermal management for a wide range of temperature and abuse conditions. *Int. J. Hydrogen Energy* **2022**, *47*, 9428–9459. [CrossRef]
33. Zhi, M.; Liu, Q.; Xu, Q.; Pan, Z.; Sun, Q.; Su, B.; Zhao, H.; Cui, H.; He, Y. Review of prevention and mitigation technologies for thermal runaway in lithium-ion batteries. *Aerosp. Traffic Saf.* **2024**, *1*, 55–72. [CrossRef]
34. Yang, L.; Liu, F.; Li, F.; Chen, Z.; Wang, J.; Gao, L.; Xiao, F.; Sun, J.; Romagnoli, A. Thermal runaway prevention and mitigation for lithium-ion battery-powered electric aircraft: Challenges and perspectives. *Aerosp. Traffic Saf.* **2024**, *1*, 103–118. [CrossRef]
35. Leising, R.A.; Palazzo, M.J.; Takeuchi, E.S.; Takeuchi, K.J. Abuse Testing of Lithium-Ion Batteries: Characterization of the Overcharge Reaction of $LiCoO_2$/Graphite Cells. *J. Electrochem. Soc.* **2001**, *148*, 8. [CrossRef]
36. Saito, Y.; Takano, K.; Negishi, A. Thermal behaviors of lithium-ion cells during overcharge. *J. Power Sources* **2001**, *97–98*, 693–696. [CrossRef]
37. Wen, J.; Yu, Y.; Chen, C. A review on lithium-ion batteries safety issues: Existing problems and possible solutions. *Mater. Express* **2012**, *2*, 197–212. [CrossRef]
38. Lin, C.; Ren, Y.; Amine, K.; Qin, Y.; Chen, Z. In situ high-energy X-ray diffraction to study overcharge abuse of 18,650-size lithium-ion battery. *J. Power Sources* **2013**, *230*, 32–37. [CrossRef]
39. Zhang, L.; Ma, Y.; Cheng, X.; Du, C.; Guan, T.; Cui, Y.; Sun, S.; Zuo, P.; Gao, Y.; Yin, G. Capacity fading mechanism during long-term cycling of over-discharged $LiCoO_2$/mesocarbon microbeads battery. *J. Power Sources* **2015**, *293*, 1006–1015. [CrossRef]
40. Guo, R.; Lu, L.; Ouyang, M.; Feng, X. Mechanism of the entire overdischarge process and overdischarge-induced internal short circuit in lithium-ion batteries. *Sci. Rep.* **2016**, *6*, 1–9. [CrossRef]
41. Maleki, H.; Howard, J.N. Effects of overdischarge on performance and thermal stability of a Li-ion cell. *J. Power Sources* **2006**, *160*, 1395–1402. [CrossRef]
42. Beauregard, G.P.; Phoenix, A.Z. *Report of Investigation: Hybrids Plus Plug in Hybrid Electric Vehicle*; Revision 1; US Department of Energy and National Rural Electric Cooperative Association, Inc.: Phoenix, AZ, USA, 2008.
43. Feng, X.; Sun, J.; Ouyang, M.; Wang, F.; He, X.; Lu, L.; Peng, H. Characterization of penetration induced thermal runaway propagation process within a large format lithium ion battery module. *J. Power Sources* **2015**, *275*, 261–273. [CrossRef]
44. Chen, M.; Dongxu, O.; Liu, J.; Wang, J. Investigation on thermal and fire propagation behaviors of multiple lithium-ion batteries within the package. *Appl. Therm. Eng.* **2019**, *157*, 113750. [CrossRef]
45. Li, H.; Duan, Q.; Zhao, C.; Huang, Z.; Wang, Q. Experimental investigation on the thermal runaway and its propagation in the large format battery module with $Li(Ni_{1/3}Co_{1/3}Mn_{1/3})O_2$ as cathode. *J. Hazard. Mater.* **2019**, *375*, 241–254. [CrossRef]

46. Xiao, H.; Jiaqiang, E.; Tian, S.; Huang, Y.; Song, X. Effect of composite cooling strategy including phase change material and liquid cooling on the thermal safety performance of a lithium-ion battery pack under thermal runaway propagation. *Energy* **2024**, *295*, 131093. [CrossRef]
47. Xu, H.; Cheng, L.; Paizulamu, D.; Zheng, H. Safety and Reliability Analysis of Reconfigurable Battery Energy Storage System. *Batteries* **2024**, *11*, 12. [CrossRef]
48. Fang, J.; Yin, X.; Guan, J.; Han, X.; Jia, F.; Chen, B.; Cao, F.; Wang, X. Comparative study on the thermal performance of the battery two-phase direct CO_2 cooling system with parallel and half-series configuration for electric vehicles. *Renew. Energy* **2025**, *242*, 122461. [CrossRef]
49. Min, Z.; Li, M.; Liu, H.; Gan, Y.; Wang, Y.; You, J.; Jiang, Y. Battery temperature performance of external cooling method for electric vehicles under different working conditions. *Appl. Therm. Eng.* **2025**, *269*, 126010. [CrossRef]
50. Wang, C.; Chen, Z.; Shen, Y.; Li, J. Simulation analysis of battery thermal management system for electric vehicles based on direct cooling cycle optimization. *Appl. Therm. Eng.* **2025**, *268*, 125938. [CrossRef]
51. Wankhede, S.; Pingale, A.D.; Kale, A. Experimental investigation on thermal management of lithium-ion battery pack for formula student electric vehicle using air-cooling system. *Energy Storage Sav.* **2025**, *4*, 38–47. [CrossRef]
52. Luo, Y.; Qiu, X.; Wang, S.; Jia, Z. Optimizing a direct flow cooling battery thermal management with bod baffles for electric vehicles: An experimental and simulation study. *J. Energy Storage* **2023**, *74*, 109410. [CrossRef]
53. Larrañaga-Ezeiza, M.; Vertiz, G.; Arroiabe, P.; Martinez-Agirre, M.; Berasategi, J. A novel direct liquid cooling strategy for electric vehicles focused on pouch type battery cells. *Appl. Therm. Eng.* **2022**, *216*, 118869. [CrossRef]
54. Hong, S.H.; Jang, D.S.; Park, S.; Yun, S.; Kim, Y. Thermal performance of direct two-phase refrigerant cooling for lithium-ion batteries in electric vehicles. *Appl. Therm. Eng.* **2020**, *173*, 115213. [CrossRef]
55. Suresh, C.; Awasthi, A.; Kumar, B.; Im, S.; Jeon, Y. Advances in battery thermal management for electric vehicles: A comprehensive review of hybrid PCM-metal foam and immersion cooling technologies. *Renew. Sustain. Energy Rev.* **2025**, *208*, 115021.
56. Sadeh, M.; Tousi, M.; Sarchami, A.; Sanaie, R.; Kiani, M.; Ashjaee, M.; Houshfar, E. A novel hybrid liquid-cooled battery thermal management system for electric vehicles in highway fuel-economy condition. *J. Energy Storage* **2024**, *86*, 111195. [CrossRef]
57. Mohammed, A.G.; Wang, Q.; Elfeky, K.E. Rapid cooling effectiveness of Li-ion battery module with multiple phase change materials for plug-in hybrid electric vehicle. *Int. J. Therm. Sci.* **2023**, *185*, 108040. [CrossRef]
58. Yetik, O. Innovative-serpentine cooling method of batteries: Both thermal and statistical method approach. *Int. J. Therm. Sci.* **2025**, *208*, 109437. [CrossRef]
59. Fu, H.; Wang, J.; Li, L.; Gong, J.; Wang, X. Numerical study of mini-channel liquid cooling for suppressing thermal runaway propagation in a lithium-ion battery pack. *Appl. Therm. Eng.* **2023**, *234*, 121349. [CrossRef]
60. Ke, Q.; Li, X.; Guo, J.; Cao, W.; Wang, Y.; Jiang, F. The retarding effect of liquid-cooling thermal management on thermal runaway propagation in lithium-ion batteries. *J. Energy Storage* **2022**, *48*, 104063. [CrossRef]
61. Rui, X.; Feng, X.; Wang, H.; Yang, H.; Zhang, Y.; Wan, M.; Wei, Y.; Ouyang, M. Synergistic effect of insulation and liquid cooling on mitigating the thermal runaway propagation in lithium-ion battery module. *Appl. Therm. Eng.* **2021**, *199*, 117521. [CrossRef]
62. Liu, X.; Zhou, Z.; Wu, T.; Lv, J.; Hu, C.; Gao, L.; Li, Y.; Li, Y.; Song, Y. Inhibition of Thermal Runaway Propagation in Lithium-Ion Battery Pack by Minichannel Cold Plates and Insulation Layers. *Int. J. Energy Res.* **2022**, *2023*, 6661693. [CrossRef]
63. Yang, S.; Luo, X.; Li, X.; Nian, V.; Liu, S.; Wang, Y.; Li, H. Comparing different battery thermal management systems for suppressing thermal runaway propagation. *J. Energy Storage* **2024**, *101*, 114005. [CrossRef]
64. Du, X.; Wang, Z.; Gao, Q.; Yang, H.; Bao, R.; Xiong, S. Study on novel battery thermal management using triply periodic minimal surface porous structures liquid cooling channel. *Appl. Therm. Eng.* **2024**, *257*, 124384. [CrossRef]
65. Ma, Y.; Zhang, Y.; Chen, N.; Lai, Z.; Zhu, L.; Chen, S.; Teng, Z.; Zhang, N.; Mao, B. Thermal runaway propagation behavior and cooling effect of water mist within a 18650-type $LiFePO_4$ battery module under different conditions. *Process Saf. Environ. Prot.* **2024**, *185*, 1362–1372. [CrossRef]
66. Liu, T.; Huang, J.; Hu, X.; Hu, J.; Wang, X. Experimental study on the cooling effect of fine water mist on the thermal runaway in a single lithium ion battery. *Appl. Therm. Eng.* **2024**, *240*, 122194. [CrossRef]
67. Liu, T.; Tao, C.; Wang, X. Cooling control effect of water mist on thermal runaway propagation in lithium ion battery modules. *Appl. Energy* **2020**, *267*, 115087. [CrossRef]
68. Yao, Y.; Jiang, Y.; Chen, F.; Wang, Y.; Zhou, H.; Zhao, J. Experimental study of the strategy of $C_6F_{12}O$ and water mist intermittent spray to suppress lithium-ion batteries thermal runaway propagation. *J. Loss Prev. Process Ind.* **2024**, *91*, 105407. [CrossRef]
69. Xu, J.; Duan, Q.; Zhang, L.; Liu, Y.; Sun, J.; Wang, Q. The enhanced cooling effect of water mist with additives on inhibiting lithium ion battery thermal runaway. *J. Loss Prev. Process Ind.* **2022**, *77*, 104784. [CrossRef]
70. Sadar, A.; Mohammad, N.; Amir, M.; Haque, A. An experimental study on lithium-ion electric vehicles battery packs behavior under extreme conditions for prevention of thermal runaway. *Process Saf. Environ. Prot.* **2024**, *191*, 1024–1034. [CrossRef]
71. Yang, C.; Liu, Q.; Liu, M.; Shi, Q.; Cao, X.; Pei, J.; Cao, C.; Lei, H.; Ping, X.; Ju, X. Investigation of the immersion cooling system for 280Ah $LiFePO_4$ batteries: Effects of flow layouts and fluid types. *Case Stud. Therm. Eng.* **2024**, *61*, 104922. [CrossRef]

72. Patil, M.S.; Seo, J.; Lee, M. A novel dielectric fluid immersion cooling technology for Li-ion battery thermal management. *Energy Convers. Manag.* **2021**, *229*, 113715. [CrossRef]
73. Zhi, M.; Fan, R.; Zheng, L.; Yue, S.; Pan, Z.; Sun, Q.; Liu, Q. Experimental investigation on hydrated salt phase change material for lithium-ion battery thermal management and thermal runaway mitigation. *Energy* **2024**, *307*, 132685. [CrossRef]
74. Lu, D.; Cui, N.; Zhou, J.; Li, C. Hybrid cooling system with phase change material and liquid microchannels to prevent thermal runaway propagation within lithium-ion battery packs. *Appl. Therm. Eng.* **2024**, *247*, 123118. [CrossRef]
75. Ouyang, T.; Liu, B.; Wang, C.; Ye, J.; Liu, S. Novel hybrid thermal management system for preventing Li-ion battery thermal runaway using nanofluids cooling. *Int. J. Heat Mass Transf.* **2023**, *201*, 123652. [CrossRef]
76. Zhang, W.; Liang, Z.; Yin, X.; Ling, G. Avoiding thermal runaway propagation of lithium-ion battery modules by using hybrid phase change material and liquid cooling. *Appl. Therm. Eng.* **2021**, *184*, 116380. [CrossRef]
77. Wang, Y.; Wang, Y.; He, T.; Mao, N. A numerical study on a hybrid battery thermal management system based on PCM and wavy microchannel liquid cooling. *Renew. Energy* **2024**, *235*, 121273. [CrossRef]
78. Gu, X.; Lei, W.; Xi, J.; Song, M. Investigation and optimization of battery thermal management system based on composite phase change material and variable wall liquid cooling plate. *Int. J. Thermofluids* **2024**, *24*, 100886. [CrossRef]
79. Zonouzi, S.A.; Yousefi, A.; Hosseini, S.H.; Song, M. Improvement of the thermal management of lithium-ion battery with helical tube liquid cooling and phase change material integration. *J. Energy Storage* **2024**, *102*, 114113. [CrossRef]
80. Liu, Y.; Zhou, Z.; Wu, W.; Wei, L.; Lyu, J.; Li, Y.; Liu, X.; Li, Y.; Song, Y. Simulations on hybrid thermal management of mini-channel cold plate and PCM for lithium-ion batteries under discharging and thermal runaway conditions. *Case Stud. Therm. Eng.* **2024**, *60*, 104837. [CrossRef]
81. Ji, W.; Dang, Y.; Yu, Y.; Zhou, X.; Li, L. Combination of Phase Change Composite Material and Liquid-Cooled Plate Prevents Thermal Runaway Propagation of High-Specific-Energy Battery. *Appl. Sci.* **2024**, *15*, 1274. [CrossRef]

batteries

MDPI

Article

Characteristic Differences of Thermal Runaway Triggered by Overheating and Overcharging in Lithium-Ion Batteries and Multi-Dimensional Safety Protection Strategies

Yao Yao [1], Lu Liu [1], Juan Gu [1], Haozhe Xing [1], Huachao Liu [1], Yihao Cheng [1], Youning Wang [1], Songlin Yue [1], Yanyu Qiu [1] and Zhi Zhang [2,*]

1 State Key Laboratory for Disaster Prevention & Mitigation of Explosion & Impact, Army Engineering University of PLA, Nanjing 210007, China; gujuan0617@126.com (J.G.); haozhexing@hotmail.com (H.X.); wangyouning9856@163.com (Y.W.); yslseu@hotmail.com (S.Y.); qiuyanyu78@sina.com (Y.Q.)

2 Position Engineering Research Office, Army Engineering University of PLA, Nanjing 210007, China

* Correspondence: zhangnjjn@outlook.com

Abstract: Overheating and overcharging are the core triggering conditions for the thermal runaway of lithium-ion batteries. Studying the behavioral differences of thermal runaway of lithium-ion batteries under these two conditions is crucial for the safety design and protection of lithium-ion batteries. In this study, we investigated the temperature, pressure, gas generation, and heat generation characteristics of lithium batteries under these two conditions. Under overheating conditions, the release of lattice oxygen in the cathode and the decomposition of the electrolyte trigger a self-catalytic reaction, generating CO_2 (54.7%) and H_2 (29.7%), with a total heat release of 17.6 kJ and a heat accumulation rate of 24.3 W, forming a local high-temperature core area. Under overcharging conditions, the voltage drop, capacity attenuation of 21.1% (2230→1762 mAh), and internal resistance surge (6→21 mΩ) reflect severe damage to the electrode. Accompanied by the oxygenation of the EC electrolyte (CO_3^{2-} + C_2H_4↑), the gas production rate is faster. The middle pressure was 0.601 MPa, and the proportion of CO_2 was 67.4%. However, the triggering of thermal runaway relies on the synergistic effect of internal electrochemical reactions and ohmic heat accumulation, resulting in a relatively low rate of energy accumulation.

Keywords: lithium-ion battery; thermal runaway; overheating; overcharging

1. Introduction

Lithium-ion batteries, renowned for their high energy density, long cycle life, and environmental friendliness, have become the core energy carriers for electric vehicles (EVs), energy storage systems, and portable electronics [1–4]. However, with the pursuit of higher energy density (e.g., the widespread adoption of ternary cathode materials like $Li(NiCoMn)O_2$), the risk of thermal runaway (TR) has significantly increased, potentially leading to fire and explosion hazards [5–10].

Thermal runaway is typically triggered by mechanical, electrical, or thermal abuses, with overheating-induced TR (temperature > 120 °C) and overcharging-induced TR (voltage > 4.2 V) being two typical modes [11,12]. The mechanism of overheating TR involves a chain reaction: when the temperature exceeds the decomposition threshold of the solid electrolyte interphase (SEI, ~120 °C), SEI breakdown exposes the anode to electrolyte, triggering exothermic reactions (ΔH > 2000 kJ/kg). Subsequent phase transitions in cathode materials (e.g., NCM ternary cathodes shifting from layered to spinel structures) release oxygen, accelerating electrolyte oxidation and culminating in thermal runaway [13–15]. In contrast, overcharging TR arises from electrochemical imbalance: excessive charging causes lithium plating on the anode, forming dendrites that pierce the separator (<20 μm thickness exacerbates risks), inducing internal short circuits. Concurrently, delithiated cathodes collapse, releasing oxygen that reacts violently with the electrolyte [16–18].

In terms of failure manifestations, overheating TR often involves jet flames and explosions. For instance, the 2023 Beijing residential battery explosion (three fatalities) released TR gases dominated by CO (18%) and H_2 (32%), igniting surrounding combustibles. Overcharging TR, however, typically manifests as smoldering, as seen in the 2024 Nanjing e-bike fire, where prolonged electrolyte decomposition sustained flame propagation.

Current research on TR mechanisms has advanced [6,8,19]. At the Institute of Energy Storage, Qingdao Energy Research Institute, in situ analysis revealed that H^- ions at the anode of NCM cathode batteries exhibit poor thermal compatibility with electrolytes, acting as a primary trigger for chain reactions [20]. Experimental methods like adiabatic accelerating rate calorimetry (ARC) and differential scanning calorimetry (DSC) are widely used to determine TR critical temperatures but lack the capacity to analyze multi-factor coupled TR behaviors [21]. The state of charge (SOC) significantly impacts TR severity: high-SOC batteries exhibit more violent reactions due to higher active lithium content [22–24]. For example, Zeng et al. observed complete collapse of $LiCoO_2$ cathodes at critical overcharge states (x = 0.16), releasing substantial heat [25]. For overheating TR mitigation, strategies include developing high-temperature-resistant separators (e.g., Al_2O_3 coatings increasing TR onset temperature to 180 °C at 5 μm thickness) and optimizing thermal management (liquid cooling systems reducing temperature rise rates by 60%) [26–29]. For overcharging TR, BMS voltage thresholds (<4.3 V) and electrolyte additives (e.g., LiBOB for aluminum current collector corrosion inhibition) are mainstream solutions [30–32]. However, existing studies focus on single-condition TR mechanisms, lacking systematic analysis of differences between overheating and overcharging TR or multi-scale suppression strategies.

This study investigates multi-mode TR triggers in 18650 ternary lithium-ion batteries ($Li(NiCoMn)O_2$ cathode), including overheating TR (external heating) and overcharging TR (voltage triggering). The goal is to elucidate temperature/pressure evolution, gas generation characteristics, jet flame behavior, and the influence of trigger mechanisms on TR dynamics.

2. Materials and Methods

2.1. Experimental Materials and Sample Preparation

The battery adopts a commercial 18650 lithium-ion cell (nominal capacity: 2.5 Ah, voltage: 3.6 V). The cathode material is $LiNi_{0.6}Co_{0.2}Mn_{0.2}O_2$ (NCM622), the anode is graphite, and the electrolyte is 1 M $LiPF_6$. The separator is a PE-based ceramic-coated membrane (thickness: 20 μm, porosity: 45%). Its characteristics are as follows: dimensions: 18.2 mm (D) × 65.0 mm (H); charging cutoff voltage: 4.2 ± 0.05 V; nominal voltage: 3.7 V; rated capacity: 2450 mAh; storage temperature range: −20 °C to 40 °C; and cell weight: 44 ± 0.5 g.

2.2. Experimental Procedures

The schematic diagram of the experimental procedures is shown in Figure 1a. The battery was placed in a 25 °C constant-temperature chamber and left undisturbed for 24 h to ensure uniform state of charge (SOC) distribution. A high-precision Type K thermocouple array (accuracy ± 0.5 °C) was deployed to monitor temperatures at the cathode, anode, and sidewall center. The battery was secured in a thermal runaway test chamber, with ceramic-coated heating plates (power density: 0.5 W/cm^2, coverage ≥ 90%) clamped on both sides. The assembly was heated at 5 °C/min until it reached the target temperature. When the thermocouples detected the onset temperature (T_0) of self-heating reactions, external heating was halted, and a multichannel data acquisition system (sampling rate: 10 Hz) was activated to continuously record temperature, pressure, and visual dynamics. For overcharge tests, the battery was charged via a programmable charger (Neware BTS-4000, Neware Technology Limited, Dongguan, China) with a constant current of 0.5 C–4.2 V, followed by constant-voltage charging (positive electrode vent valve deactivated). Temperature evolution at the cathode, mid-layer, and anode (burst core distance: 0.15 m) was monitored in real time, while pressure data and released gases were collected.

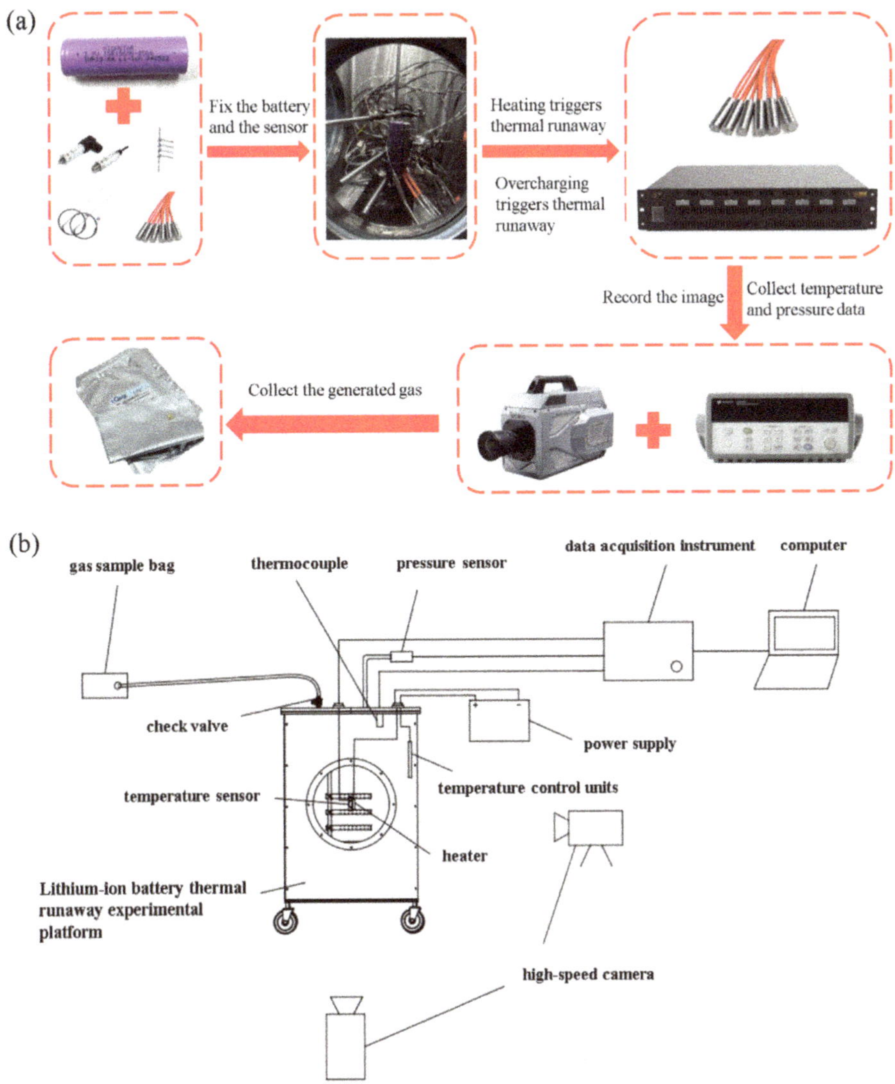

Figure 1. (**a**) Schematic diagram of the experimental procedures; (**b**) experimental platform for thermal runaway of lithium-ion batteries.

2.3. Experimental Platform

The experimental platform employs an in-house developed 18650 lithium-ion battery thermal runaway test system, as shown in Figure 1b. Thermodynamic data were collected using temperature sensors and a K-type thermocouple array. Pressure data were acquired via pressure sensors. Electrochemical parameters of the battery, including voltage precision (±0.1 mV), were recorded using a battery tester (Neware BTS-4000, Neware Technology Limited, Dongguan, China). Gas evolution characteristics during thermal runaway were analyzed using a gas chromatography–mass spectrometry (GC-MS) system (Agilent 7890B-5977A, Agilent Technologies, Palo Alto, CA, USA). Flame characteristics during thermal runaway were captured using a high-speed camera (Phantom V2512, Phantom V2512,Vision Research, Wayne, NJ, USA, resolution: 1280 × 800, frame rate: 10,000 fps).

3. Results

3.1. Temperature Characteristics of Lithium-Ion Batteries Under Different Thermal Runaway Conditions

The video images recorded through the observation window under overheating-induced thermal runaway conditions are shown in Figure 2. Initially, no significant changes were observed in the battery during heating. As the temperature rose, internal thermal runaway reactions began. At 195 s, white smoke started emitting from the battery, followed by jet flames erupting from the cathode. Approximately 5 s later, the battery exploded, with the cathode rupturing and the cell ejecting. Post-test battery images are shown in Figure 2e,f. Under overheating conditions, the temperature evolution at the cathode, mid-layer, and anode (Figure 3a) revealed distinct phases: From 0–100 s, minimal temperature variation at the cathode indicated slow Joule heat accumulation from external heat sources or localized internal short circuits, with temperatures remaining below the separator pore-closing threshold (~80 °C). During this period, only minor SEI (solid electrolyte interphase) film decomposition occurred, generating limited heat. Once the local temperature exceeded 80 °C, PE/PP-based separators underwent pore closure, blocking lithium-ion transport and sharply increasing internal resistance. This accelerated Joule heat generation, driving a gradual cathode temperature rise from 100 to 150 s [8,33]. Between 150 and 195 s, the cathode temperature surged from 80 °C to 516 °C due to chain exothermic reactions (e.g., electrolyte decomposition: $LiPF_6 \rightarrow LiF + PF_5\uparrow$; $EC \rightarrow CO\uparrow + C_2H_4O\uparrow$) at temperatures above 130 °C, which released flammable gases (CO, CH_4, etc.) and approximately 500 J/g of heat. Concurrently, lattice oxygen release from NCM622 cathodes (>200 °C) further intensified heat generation, causing a 436 °C temperature spike within 35 s [34].

The mid-layer temperature rose from ambient to 300 °C between 50 and 180 s, followed by a sharp increase to 460 °C at 185–195 s, before cooling to ambient temperature post-test. The anode temperature began rising at 100 s, reaching 50 °C at the bottom by 190 s. A rapid temperature surge to 425 °C (peak) occurred between 190 and 220 s, driven by internal short circuits and exothermic reactions, before gradual cooling to ambient temperature.

The peak temperatures and corresponding timestamps at the cathode, mid-layer, and anode of lithium-ion batteries under overheating-induced thermal runaway conditions are summarized in Table S1. The highest peak temperature was observed at the cathode (516 °C), primarily attributed to high-temperature oxygen evolution reactions and their self-catalytic nature. The cathode active material ($LiNiCoMnO_2$) becomes structurally unstable above 200 °C, releasing lattice oxygen ($LiNiCoMnO_2 \rightarrow NiO + CoO + MnO + O_2\uparrow$) accompanied by significant heat generation [35,36]. The released O_2 reacts violently with gases (e.g., CO and CH_4) generated from electrolyte decomposition, further amplifying heat release. Due to the spatial position of the internal short-circuit point and the influence of the thermal diffusion path, the mid-layer reaches the peak (460 °C) first. Its peak temperature is lower than that of the positive electrode but higher than that of the negative electrode. The mid-layer exhibited the earliest temperature rise (starting at 50 s), indicating that the internal short-circuit point proximity to the geometric center of the battery. During initial short-circuiting, Joule heat diffused radially via current collectors, prioritizing mid-layer heating. However, despite initiating the shortest path, the mid-layer's lower exothermic capacity compared to cathode active materials and faster heat dissipation to the outer casing resulted in a lower final peak temperature than the cathode. The anode exhibited the latest peak timing, and its peak temperature (425 °C) is significantly lower than both the cathode and mid-layer. This is due to the delayed decomposition of the SEI film on graphite anodes (>250 °C, $Li_2CO_3 \rightarrow Li_2O + CO_2\uparrow$) with lower heat output [37]. Additionally, the high activation energy of graphite oxidation ($C + O_2 \rightarrow CO_2\uparrow$) slows reaction kinetics compared to cathode oxygen release processes [38].

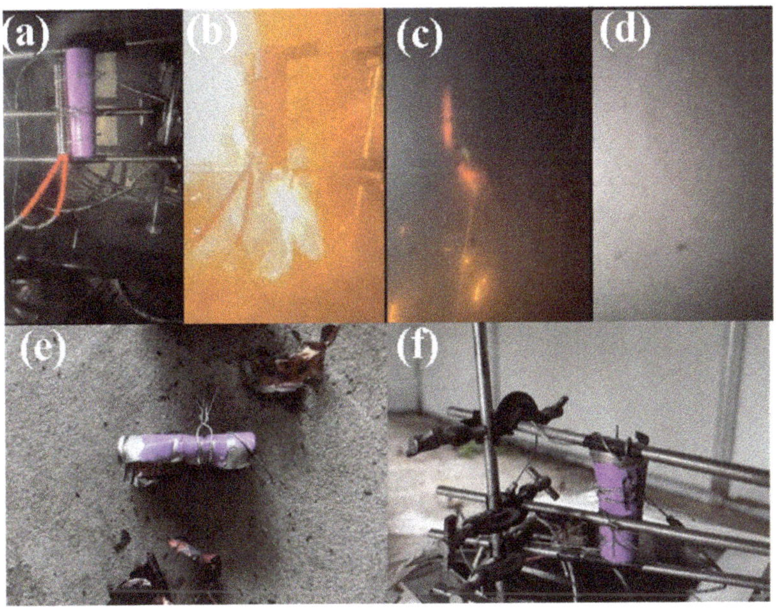

Figure 2. Photos of the experimental process of thermal runaway of lithium-ion batteries under overheating conditions: (**a**) initial state; (**b**) jet flame ejection; (**c**) the flame weakens; (**d**) generation of a large amount of smoke; (**e**,**f**) the battery after thermal runaway.

Under overcharging-induced thermal runaway conditions, the video images recorded through the observation window are shown in Figure 4. As the lithium-ion battery was continuously overcharged, slight swelling occurred in the mid-layer of the battery without any white smoke emission. Subsequently, jet flames erupted from the cathode, accompanied by massive white smoke release, followed by an immediate explosion.

The temperature evolution at the cathode, mid-layer, and anode under overcharging conditions is illustrated in Figure 3b. During overcharging, the cathode exhibited a slow linear temperature rise from 50 to 800 s, attributed to localized internal short circuits caused by lithium dendrite growth and SEI (solid electrolyte interphase) film decomposition. After 800 s, the cathode temperature sharply increased to a peak of 411 °C, driven by the combined effects of large-scale internal short circuits induced by lithium dendrites and structural collapse of the cathode material [39]. The cathode temperature rise rate was significantly higher than those of the mid-layer and anode, indicating that cathode oxidation reactions dominated the runaway process. When the voltage exceeded 4.5 V, electrolyte oxidation decomposition occurred on the cathode surface (reaction: EC $\rightarrow$ CO_3^{2-} + $C_2H_4\uparrow$), releasing CO_2 and O_2, which further accelerated thermal runaway.

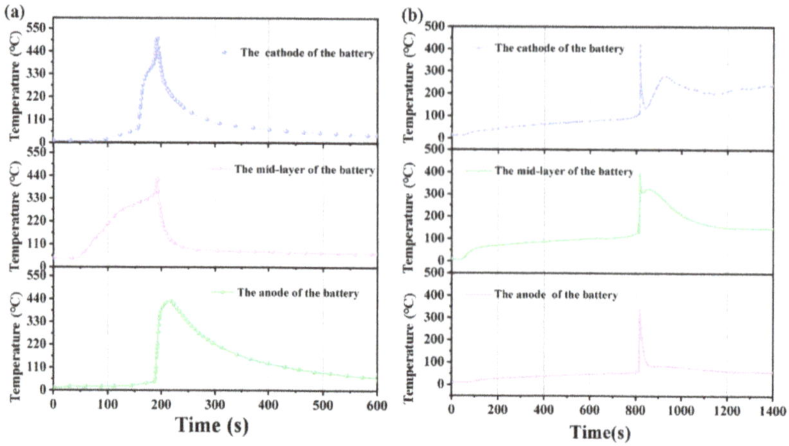

Figure 3. (**a**) The temperature variation curves of the cathode, mid-layer, and anode of lithium-ion batteries during thermal runaway under overheating conditions; (**b**) The temperature variation curves of the cathode, mid-layer, and anode of lithium-ion batteries during thermal runaway under overheating conditions.

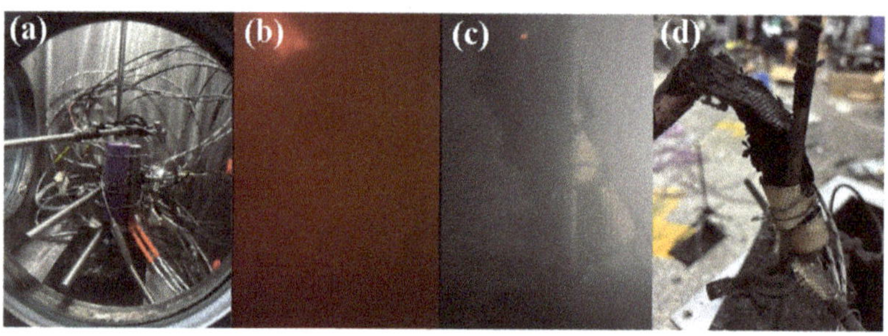

Figure 4. Photos of the experimental process of thermal runaway of lithium-ion batteries under overcharging conditions: (**a**) initial state; (**b**) jet flame ejection; (**c**) generation of a large amount of smoke; (**d**) the battery after thermal runaway.

Notably, the cathode temperature remained higher than the mid-layer (400 °C) and anode throughout the process. The 11 °C temperature difference between the cathode (411 °C) and mid-layer (400 °C) is attributed to uneven current density distribution caused by tortuous conductive paths formed by lithium dendrites penetrating the separator, leading to delayed mid-layer heating. A secondary temperature peak in the cathode between 800 and 900 s may arise from the disproportionation reaction of residual NiO and secondary reactions of $LiPF_6$ decomposition products.

The peak temperatures and corresponding timestamps at the cathode, mid-layer, and anode under overcharging conditions are summarized in Table S2. The cathode exhibited the highest peak temperature, followed by the mid-layer and anode. This aligns with the flame propagation pattern (jet flames primarily along the cathode direction) observed under overcharging conditions, consistent with the temperature and flame dynamics during overheating-induced thermal runaway. However, the temperatures under overcharging were slightly lower than those under overheating conditions. This discrepancy may stem from differences in thermal measurement methods: thermocouple readings during overcharging reflect ambient temperatures, whereas overheating experiments involved localized heating rods (>700 °C), potentially elevating surface temperatures beyond actual battery surface values.

3.2. Pressure Characteristics of Lithium-Ion Batteries Under Different Thermal Runaway Conditions

The pressure evolution profiles at the cathode, mid-layer, and anode under overheating-induced thermal runaway conditions are shown in Figure 5a. All three regions exhibited instantaneous large-pressure spikes, with the cathode showing the highest peak pressure (0.537 MPa), followed by the mid-layer (0.392 MPa) and anode (0.310 MPa). This gradient arises because the cathode's layered structure collapses during thermal runaway, releasing lattice oxygen ($LiNiCoMnO_2 \rightarrow NiO + CoO + MnO + O_2\uparrow$) that reacts violently with electrolytes (e.g., $EC \rightarrow CO_2\uparrow + H_2O$ vapor), generating high gas volumes (CO_2, H_2O, etc.). Meanwhile, gas accumulation in the mid-layer is constrained by the separator, leading to slower volume expansion compared to the cathode, resulting in a slightly lower peak pressure. The anode's minimal pressure increase (0.310 MPa) stems from slow SEI (solid electrolyte interphase) film decomposition (e.g., $Li_2CO_3 \rightarrow Li_2O + CO_2\uparrow$), which produces gases at a slower rate.

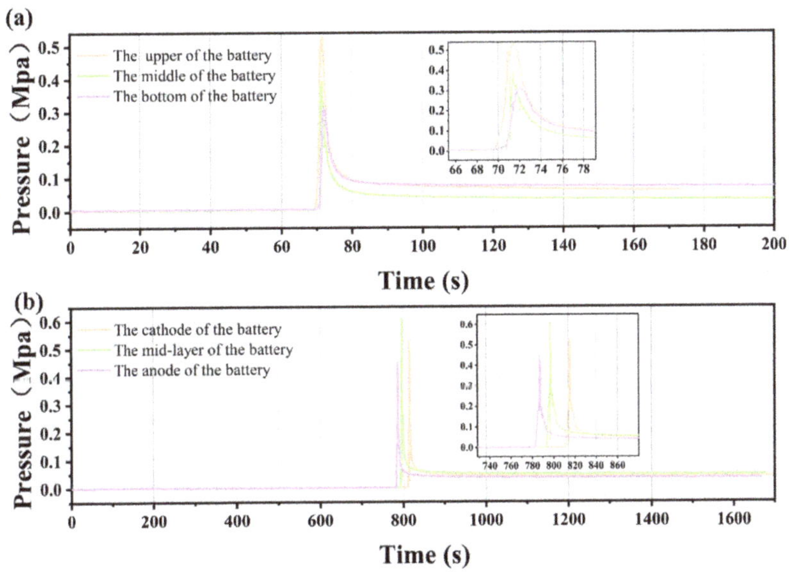

Figure 5. (**a**) The pressure variation curves of the cathode, mid-layer, and anode of lithium-ion batteries during thermal runaway under overheating conditions; (**b**) the pressure variation curves of the cathode, mid-layer, and anode of lithium-ion batteries during thermal runaway under overcharging conditions.

Under overcharging-induced thermal runaway (Figure 5b), the pressure peaks were 0.531 MPa (cathode), 0.601 MPa (mid-layer), and 0.452 MPa (anode). The mid-layer's higher pressure reflects intense electrolyte decomposition (e.g., $LiPF_6 \rightarrow LiF + PF_5\uparrow$) combined with restricted gas diffusion, while the cathode's slightly lower peak is due to rapid oxygen release constrained by structural integrity. The anode's minimal pressure increase is attributed to slow gas generation (e.g., $C + O_2 \rightarrow CO_2\uparrow$) and efficient venting paths. During the overcharging thermal runaway process of lithium-ion batteries, there are significant differences in the time at which the pressure peaks in the cathode, mid-layer, and anode reach their peak values. This temporal characteristic is closely related to the reaction kinetics and mass transfer processes in different regions. The earliest peak of the anode electrode is due to the SEI film decomposition and lithium solvent reaction triggered by lithium deposition on the negative electrode, which are activated in the early stage of overcharging. The delay in peak pressure in the mid-layer is due to the combined effect of electrolyte decomposition (such as EC $\rightarrow$ CO_2) and gas diffusion obstruction, resulting in a rapid increase in pressure during the overcharge phase. The pressure peak in the cathode region appeared latest, primarily due to the structural collapse of the positive electrode material (such as NCM layered decomposition) and the need for higher temperatures (>700 °C) for oxygen release. Therefore, its pressure peak usually appears in the later stage of thermal runaway. This temporal difference reflects the dynamic competitive relationship between material degradation mechanisms (such as SEI rupture and positive electrode phase transition) and thermal electrochemical coupling processes.

3.3. Gas Production Characteristics of Lithium-Ion Batteries Under Different Thermal Runaway Conditions

The gas components generated by lithium-ion batteries after overheating and losing control include a large number of organic and flammable gases such as CH_4, C_2H_6, C_2H_4, C_3H_8, C_4H_8, C_5H_{12}, CO_2, CO, and H_2. The normalized gas composition is shown in Figure 6a. Among them, CO_2 and H_2 have the highest proportions, accounting for 54.7% and 29.7%, respectively. CO_2 may originate from the thermal decomposition of SEI film, lithium carbonate, electrolyte lithium salt LiPF6, and the combustion of electrolyte. H_2 may originate from electrode adhesives such as polyvinylidene fluoride (PVDF) and sodium carboxymethyl cellulose (CMC) reacting with Li. CO accounts for 7.5%, being mainly from the reaction between Li embedded on the negative electrode and CO_2, as well as the reaction between Li embedded on the negative electrode and electrolyte [40]. The total proportion of hydrocarbon gases such as methane and ethylene is less than 10%, being mainly derived from the reaction between electrolyte solvents and Li. The main components of the gas generated by overcharging runaway in lithium-ion batteries are similar to those under overheating runaway conditions, still dominated by CO_2, H_2, CO, and olefin gases. The normalized gas composition is shown in Figure 6b. However, it can be observed that compared to the condition of overheating out of control, the proportion of hydrogen gas decreases under overcharge out-of-control conditions, indicating that overcharge conditions inhibit hydrogen evolution reactions. Under overcharge conditions, the high voltage preferentially triggers electrolyte oxidation rather than lithium deposition. At this time, H_2 mainly comes from the secondary reaction after the release of positive electrode oxygen, and the rate of H_2 production decreases. On the other hand, under overcharge conditions, the proportion of CO_2 increases, indicating that a high voltage accelerates the decomposition of positive electrode materials. Overall, the gas produced by overheating and loss of control is mainly CO_2 and H_2, reflecting the dominant decomposition of positive electrode materials; the proportion of H_2 decreases due to overcharging and thermal runaway, but the proportion of CO and hydrocarbon gases increases, indicating an intensification of electrolyte oxidation and electrochemical reactions. There are essential differences between the two in terms of thermal runaway pathways, gas toxicity, and safety risks, and targeted thermal management and monitoring strategies need to be designed.

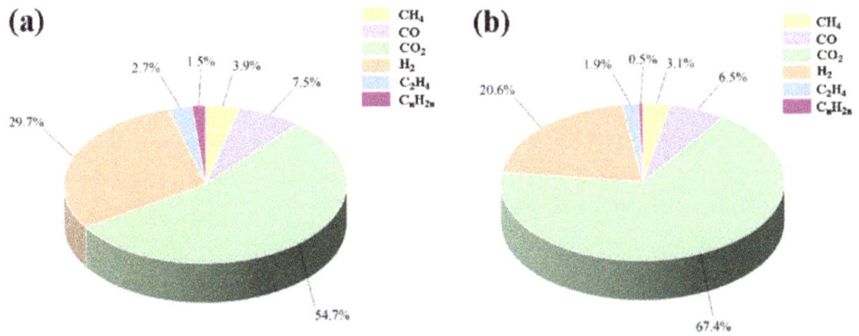

Figure 6. (**a**) The main types and volume fractions of gases released during thermal runaway of lithium-ion batteries under overheating conditions; (**b**) the main types and volume fractions of gases released during thermal runaway of lithium-ion batteries under overcharging conditions.

4. Discussion

4.1. Heat Release During the Thermal Runaway Process

The heat accumulation rate of the battery before thermal runaway can be calculated using Equation (1).

$$P_b = c_b m_b \frac{dT}{dt} \quad (1)$$

In the formula, $c_b = 1.0$ kJ/(kg·°C) represents the heat capacity of the battery; m_b represents the mass of the battery after the test; and $\frac{dT}{dt}$ is the first-order derivative of temperature varying with time.

As can be seen in Figure 3b, when thermal runaway occurs under overheating conditions, the temperature distribution in the cathode, mid-layer, and anode of the lithium battery is uneven, and the changes are also different. Generally, they can all be divided into two stages, as shown in Table S3. The average heat accumulation power can be calculated using the average value of each stage. After the test, the mass of the battery was measured (m_b = 38.7 g), and the average heat accumulation rate was calculated to be 24.3 W.

$$\overline{P_{bh}} = \frac{c_b m_b \frac{dT_1}{dt_1} + c_b m_b \frac{dT_2}{dt_2} + c_b m_b \frac{dT_3}{dt_3} + c_b m_b \frac{dT_4}{dt_4} + c_b m_b \frac{dT_5}{dt_5} + c_b m_b \frac{dT_6}{dt_6}}{6} = 24.3\text{W}$$

As can be seen in Figure 3b, when thermal runaway occurs under overcharging conditions, the temperature distribution of the battery is uneven. Therefore, the average temperature change rates of the positive electrode, negative electrode, and the middle region are used for calculation. The rates of temperature changes of the cathode, mid-layer, and anode are shown in Table S4. After the test, the mass of the battery is 37.9 g, and the average heat accumulation rate calculation is 4W. The average heat accumulation rate under overheating conditions is much higher than that under overcharging conditions. This is because the heat lost control due to overheating mainly comes from the direct input of the external heating device, while the heat lost control due to overcharging is only the internal electrochemical reaction and ohmic heat, which accumulates with the charging time and takes a relatively long time to reach the critical temperature.

The total heat release can be calculated using Equation (2).

$$H_t = c_b m_b (T_t - T_0) \quad (2)$$

In the formula, H_t represents the total heat release of the battery; c_b represents the heat capacity of the battery; m_b represents the quality of the battery after testing; T_t represents the peak temperature; and T_0 represents the ambient temperature.

As can be seen in Figure 3, the heat distribution of lithium batteries during thermal runaway is uneven. Further considering the mass distribution of the cathode, mid-layer, and anode of the battery, when calculating the total heat, the mass fractions of the cathode, mid-layer, and anode (25%, 50%, and 25%) can be multiplied by the heat released.

Firstly, the temperature variation under overheating conditions was analyzed. The initial test temperature was 10 °C, the peak temperature of the positive electrode was 516 °C, the peak temperature of the middle region was 460 °C, and the peak temperature of the negative electrode region was 425 °C. Therefore, the ΔT ($T_t - T_0$) of the cathode is 506 K, the ΔT of the mid-layer is 450 K, and the ΔT of the anode is 415 K.

Based on this calculation, it can be known that the total heat release of thermal runaway of lithium-ion batteries under overheating conditions is 17.6 KJ. However, since the temperature sensor collects the field temperature, during the process of overheating and runaway, which is affected by the high temperature of the heating rod, the temperature data collected by the sensor may be higher than the surface temperature of the lithium-ion battery. Therefore, the total heat release during the thermal runaway of the lithium-ion battery will be slightly lower than 17.6 KJ. The total heat release from thermal runaway of lithium-ion batteries under overcharge runaway conditions is estimated to be approximately 14.2 KJ by a similar method. It should also be noted that these data are an approximate estimation of the heat release during the thermal runaway process of lithium-ion batteries based on temperature data and under ideal adiabatic conditions.

During the thermal runaway process of lithium-ion batteries, the internal energy of the battery decreases and is converted into thermal energy for release. Under near-adiabatic conditions, all the heat released leads to an increase in temperature. Therefore, the change in internal energy during the thermal runaway process of lithium-ion batteries can be inferred through the variation of temperature.

The relationship between internal energy and temperature can be expressed as Equation (3).

$$Q = c_b m_b \Delta T \tag{3}$$

Thermal runaway under overheating conditions mainly has two stages. In the first stage, when the temperature is less than 80 °C, only a slight decomposition of the SEI film occurs, with a relatively low heat generation and an internal energy of 3.15 KJ. In the second stage, the battery undergoes a chain exothermic reaction, with the temperature sharply rising to 516 °C, releasing a large amount of heat. At this point, the internal energy is 19.6 KJ.

Under overcharging conditions, the initial temperature rose slowly and linearly to 100 °C. This was due to the local internal short circuit caused by the growth of lithium dendrites and the decomposition of the SEI film. The internal energy was calculated to be 3.4 KJ. After 800 s, the temperature rose sharply to the peak of 411 °C, and the internal energy was calculated to be 15.2 KJ. This heat is derived from the coupling effect of large-area internal short circuits caused by lithium dendrites and the structural collapse of the cathode material.

4.2. Security Protection Strategy

Figure S1 shows the charge–discharge curves of the lithium battery at a rate of 0.5 C: in (a), the cut-off voltage is 4.2 V, and in (b), the cut-off voltage is 4.6V. Figure S1b shows that the voltage curve becomes steep and the platform disappears after the battery voltage exceeds 4.2 V, indicating that polarization intensifies at this time. After the battery is overcharged to 4.5 V, the voltage drops sharply, indicating that the electrolyte undergoes oxidation and decomposition, generating a large amount of gas and causing a sudden increase in the internal pressure of the battery. The accumulation of gas causes the electrode spacing to increase, the internal resistance rises sharply, and the current is forced to be interrupted. Compared with the normal charge and discharge capacity of 2230 mAh, the capacity of lithium-ion batteries decreases to 1762 mAh after overcharging, which is closely related to the damage of the SEI film and the lithium plating phenomenon caused by overcharging. The internal resistance of the lithium battery after overcharge was measured by the four-wire method of the resistance meter to be 21 mΩ. Compared with the internal resistance of 6 mΩ of the battery at normal charge and discharge, it has increased significantly, indicating that overcharging has led to the damage of the electrode structure.

Overheating-driven gas profiles reflect cathode material decomposition (e.g., NCM622 $\rightarrow$ NiO + CoO + MnO + $O_2\uparrow$) and self-catalytic electrolyte combustion ($LiPF_6 \rightarrow LiF + PF_5\uparrow$; EC $\rightarrow$ CO$\uparrow$ + $C_2H_4O\uparrow$), producing rapid O_2 release and high-pressure peaks at the cathode (0.537 MPa). In contrast, overcharging triggers lithium dendrite-induced internal short circuits (post-800 s), combining cathode collapse and electrolyte oxidation (EC $\rightarrow$ CO_3^{2-} + $C_2H_4\uparrow$) to generate faster gas release with mid-layer pressure spikes (0.601 MPa) due to restricted gas diffusion. The distinct gas compositions—CO_2/H_2 dominance in overheating versus elevated CO_2 and suppressed H_2 in overcharging—highlight divergent failure mechanisms: cathode-driven thermal runaway versus electrolyte oxidation and electrochemical intensification under high voltage.

Therefore, safety strategies must prioritize material-specific and system-level controls. For overheating mitigation, Al^{3+} doping or Al_2O_3 coating suppresses cathode oxygen release, while PTC materials integrated into electrodes enable current self-limiting during thermal runaway. Graded pressure valves (0.4 MPa threshold) combined with N_2 inerting mitigate re-ignition risks. For overcharging protection, BMS redundancy (primary 4.2 V + redundant 4.3 V thresholds) triggers charging termination and balancing circuits, supplemented by FEC (fluoroethylene carbonate) additives to stabilize SEI and TPP (triphenyl phosphate) to suppress electrolyte oxidation. High-temperature ceramic-coated PP separators delay gas diffusion and reduce mid-layer pressure peaks.

These findings underscore the necessity for tailored thermal management and multi-layered safety protocols addressing the unique gas dynamics, reaction pathways, and failure modes of overheating versus overcharging scenarios in high-energy-density lithium-ion batteries.

5. Conclusions

This study systematically contrasts the thermal runaway behaviors and safety risks of lithium-ion batteries under overheating-induced and overcharging-induced thermal runaway conditions. Through a customized experimental platform, the dynamic evolution mechanisms and safety protection requirements of thermal runaway in both abuse scenarios were revealed. Based on key experimental differences, differentiated safety strategies were proposed, providing theoretical and technical pathways for designing and protecting high-safety lithium-ion batteries. In overheating-induced thermal runaway, the cathode material releases lattice oxygen above 200 °C (NCM622 $\rightarrow$ NiO + CoO + MnO + $O_2\uparrow$), initiating a self-catalytic chain reaction with electrolyte decomposition ($LiPF_6 \rightarrow LiF + PF_5\uparrow$; EC $\rightarrow$ CO$\uparrow$ + $C_2H_4O\uparrow$), causing the cathode temperature to surge from 80 °C to 516 °C within 195 s, forming the core heat generation zone with an instantaneous pressure peak of 0.537 MPa. Gas composition is dominated by CO_2 (54.7%) and H_2 (29.7%). Mitigation strategies include Al^{3+} doping to suppress lattice oxygen release and integrating PTC materials for current self-limiting to block thermal propagation. In overcharging-induced thermal runaway, lithium dendrites piercing the separator after 800 s trigger internal short circuits combined with cathode structural collapse and electrolyte oxidation (EC → CO_3^{2-} + C_2H_4↑), resulting in a lower cathode peak temperature (411 °C) but faster gas generation. Restricted gas diffusion in the mid-layer creates a pressure peak of 0.601 MPa, with CO_2 dominating (67.4%) and H_2 reduced to 20.6%, indicating intensified electrolyte oxidation. Protective measures involve optimizing BMS thresholds (4.2 V primary + 4.3 V redundant) and incorporating FEC/TPP additives to stabilize SEI and suppress gas evolution. Comparative analysis highlights distinct thermal runaway pathways: overheating is driven by cathode oxygen release and self-catalytic electrolyte combustion, while overcharging involves electrolyte oxidation and dendrite-induced short circuits. Safety strategies must therefore prioritize material-specific modifications (e.g., Al^{3+} doping and SEI-stabilizing additives) and system-level controls (e.g., redundant BMS thresholds and high-temperature separators) to address the unique risks of each abuse scenario, ensuring comprehensive protection against complex thermal runaway triggers in high-energy-density batteries.

Supplementary Materials: The following supporting information can be downloaded at: https://www.mdpi.com/article/10.3390/batteries11070242/s1. Figure S1: (a) The charge–discharge curve of lithium-ion batteries at a rate of 0.5C; (b) the charge–discharge curve of lithium-ion batteries under overcharge conditions at a rate of 0.5C; Table S1. Peak temperatures and times of the cathode, mid-layer, and anode of lithium-ion batteries under overheating conditions; Table S2. Peak temperatures and times of the cathode, mid-layer, and anode of lithium-ion batteries under overcharging conditions; Table S3: The rate of temperature change of lithium batteries before thermal runaway under overheating conditions; Table S4: The rate of temperature change of lithium batteries before thermal runaway under overcharging conditions; Figure S2: Comparison of the representative gas (CO_2, H_2, and CO) production volumes generated by thermal runaway of lithium batteries under overheating and overcharging conditions.

Author Contributions: Conceptualization, Y.Y.; Data curation, H.L.; Investigation, L.L., H.X., Y.C., Y.W. and Z.Z.; Methodology, Y.Y., J.G. and Z.Z.; Writing—original draft, Y.Y.; Writing—review and editing, Z.Z. Resources, Y.Q. and S.Y. All authors have read and agreed to the published version of the manuscript.

Funding: This work was supported by the State Key Laboratory of Disaster Prevention & Mitigation of Explosion & Impact (No. LGD-SKL-202203), the Nature Science Foundation of Jiangsu Province (No. BK20241602), and the Army Engineering University Youth Independent Innovation Fund (No. KYFYJKQTZQ23006).

Data Availability Statement: The data that support the findings of this study are available from the corresponding author upon reasonable request.

Conflicts of Interest: The authors declare no conflicts of interest.

References

1. Kim, T.; Song, W.; Son, D.-Y.; Ono, L.K.; Qi, Y. Lithium-ion batteries: Outlook on present, future, and hybridized technologies. *J. Mater. Chem. A* **2019**, *7*, 2942–2964. [CrossRef]
2. Li, M.; Lu, J.; Chen, Z.; Amine, K. 30 years of lithium-ion batteries. *Adv. Mater.* **2018**, *30*, 1800935–1809648. [CrossRef] [PubMed]
3. Manthiram, A. An outlook on lithium ion battery technology. *ACS Cent. Sci.* **2017**, *3*, 1063–1069. [CrossRef]
4. Xie, J.; Lu, Y.-C. A retrospective on lithium-ion batteries. *Nat. Commun.* **2020**, *11*, 2499. [CrossRef] [PubMed]
5. Feng, X.; Ren, D.; He, X.; Ouyang, M. Mitigating thermal runaway of lithium-ion batteries. *Joule* **2020**, *4*, 743–770. [CrossRef]
6. Wang, Q.; Ping, P.; Zhao, X.; Chu, G.; Sun, J.; Chen, C. Thermal runaway caused fire and explosion of lithium ion battery. *J. Power Sources* **2012**, *208*, 210–224. [CrossRef]
7. Kong, D.; Lv, H.; Ping, P.; Wang, G. A review of early warning methods of thermal runaway of lithium ion batteries. *J. Energy Storage* **2023**, *64*, 107073. [CrossRef]
8. Feng, X.; Ouyang, M.; Liu, X.; Lu, L.; Xia, Y.; He, X. Thermal runaway mechanism of lithium ion battery for electric vehicles: A review. *Energy Storage Mater.* **2018**, *10*, 246–267. [CrossRef]

9. Mallick, S.; Gayen, D. Thermal behaviour and thermal runaway propagation in lithium-ion battery systems—A critical review. *J. Energy Storage* **2023**, *62*, 106894. [CrossRef]
10. Hu, X.; Gao, F.; Xiao, Y.; Wang, D.; Gao, Z.; Huang, Z.; Ren, S.; Jiang, N.; Wu, S. Advancements in the safety of Lithium-Ion Battery: The Trigger, consequence and mitigation method of thermal runaway. *Chem. Eng. J.* **2024**, *481*, 148450. [CrossRef]
11. Kong, L.; Li, Y.; Feng, W. Strategies to solve lithium battery thermal runaway: From mechanism to modification. *Electrochem. Energy Rev.* **2021**, *4*, 633–679. [CrossRef]
12. Galushkin, N.E.; Yazvinskaya, N.N.; Galushkin, D.N. Causes and mechanism of thermal runaway in lithium-ion batteries, contradictions in the generally accepted mechanism. *J. Energy Storage* **2024**, *86*, 111372. [CrossRef]
13. Yang, Y.; Wang, R.; Shen, Z.; Yu, Q.; Xiong, R.; Shen, W. Towards a safer lithium-ion batteries: A critical review on cause, characteristics, warning and disposal strategy for thermal runaway. *Adv. Appl. Energy* **2023**, *11*, 100146. [CrossRef]
14. He, D.; Wang, J.; Peng, Y.; Li, B.; Feng, C.; Shen, L.; Ma, S. Research advances on thermal runaway mechanism of lithium-ion batteries and safety improvement. *Sustain. Mater. Technol.* **2024**, *41*, e01017. [CrossRef]
15. Feng, X.; Zheng, S.; Ren, D.; He, X.; Wang, L.; Cui, H.; Liu, X.; Jin, C.; Zhang, F.; Xu, C. Investigating the thermal runaway mechanisms of lithium-ion batteries based on thermal analysis database. *Appl. Energy* **2019**, *246*, 53–64. [CrossRef]
16. Yan, W.; Huang, W.; Yang, Y.; Wei, Z.; Zhen, H.; Lin, Y. Research on overcharge mitigations and thermal runaway risk of 18650 lithium-ion batteries. *J. Energy Storage* **2025**, *120*, 116372. [CrossRef]
17. Liu, Z.; Guo, X.; Meng, N.; Yu, Z.; Yang, H. Study of thermal runaway and the combustion behavior of lithium-ion batteries overcharged with high current rates. *Thermochim. Acta* **2022**, *715*, 179276. [CrossRef]
18. Zhou, G.; Lu, H.; Zhang, Q.; Yang, S.; Liu, Y.; Niu, C.; Kong, Y.; Huang, Q.; Wei, Z. Experimental study on thermal runaway and flame eruption characteristics of NCM523 lithium-ion battery induced by the coupling stimulations of overcharge-penetration. *Process Saf. Environ. Prot.* **2024**, *191*, 131–145. [CrossRef]
19. Li, Y.; Liu, X.; Wang, L.; Feng, X.; Ren, D.; Wu, Y.; Xu, G.; Lu, L.; Hou, J.; Zhang, W. Thermal runaway mechanism of lithium-ion battery with $LiNi_{0.8}Mn_{0.1}Co_{0.1}O_2$ cathode materials. *Nano Energy* **2021**, *85*, 105878. [CrossRef]
20. Huang, L.; Xu, G.; Du, X.; Li, J.; Xie, B.; Liu, H.; Han, P.; Dong, S.; Cui, G.; Chen, L. Uncovering LiH triggered thermal runaway mechanism of a high-energy $LiNi_{0.5}Co_{0.2}Mn_{0.3}O_2$/graphite pouch cell. *Adv. Sci.* **2021**, *8*, 2100676. [CrossRef]
21. Yue, Y.; Jia, Z.; Li, Y.; Wen, Y.; Lei, Q.; Duan, Q.; Sun, J.; Wang, Q. Thermal runaway hazards comparison between sodium-ion and lithium-ion batteries using accelerating rate calorimetry. *Process Saf. Environ. Prot.* **2024**, *189*, 61–70. [CrossRef]
22. Qi, C.; Liu, Z.; Lin, C.; Liu, X.; Liu, D.; Li, Z.; Yi, A. The gas production characteristics and catastrophic hazards evaluation of thermal runaway for $LiNi_{0.5}Co_{0.2}Mn_{0.3}O_2$ lithium-ion batteries under different SOCs. *J. Energy Storage* **2024**, *88*, 111678. [CrossRef]
23. Liao, Z.; Zhang, S.; Li, K.; Zhao, M.; Qiu, Z.; Han, D.; Zhang, G.; Habetler, T.G. Hazard analysis of thermally abused lithium-ion batteries at different state of charges. *J. Energy Storage* **2020**, *27*, 101065. [CrossRef]
24. Zhang, Q.; Niu, J.; Zhao, Z.; Wang, Q. Research on the effect of thermal runaway gas components and explosion limits of lithium-ion batteries under different charge states. *J. Energy Storage* **2022**, *45*, 103759. [CrossRef]
25. Zeng, Y.; Wu, K.; Wang, D.; Wang, Z.; Chen, L. Overcharge investigation of lithium-ion polymer batteries. *J. Power Sources* **2006**, *160*, 1302–1307. [CrossRef]
26. Song, Y.; Liu, X.; Ren, D.; Liang, H.; Wang, L.; Hu, Q.; Cui, H.; Xu, H.; Wang, J.; Zhao, C. Simultaneously blocking chemical crosstalk and internal short circuit via gel-stretching derived nanoporous non-shrinkage separator for safe lithium-ion batteries. *Adv. Mater.* **2022**, *34*, 2106335. [CrossRef]
27. Liu, H.; Liu, C.; Zhou, Y.; Zhang, Y.; Deng, W.; Zou, G.; Hou, H.; Ji, X. The application of Al_2O_3 in separators and solid electrolytes of lithium-ion battery: A review. *Energy Storage Mater.* **2024**, *71*, 103575. [CrossRef]
28. Shang, Z.; Qi, H.; Liu, X.; Ouyang, C.; Wang, Y. Structural optimization of lithium-ion battery for improving thermal performance based on a liquid cooling system. *Int. J. Heat Mass Transf.* **2019**, *130*, 33–41. [CrossRef]
29. Akbarzadeh, M.; Kalogiannis, T.; Jaguemont, J.; Jin, L.; Behi, H.; Karimi, D.; Beheshti, H.; Van Mierlo, J.; Berecibar, M. A comparative study between air cooling and liquid cooling thermal management systems for a high-energy lithium-ion battery module. *Appl. Therm. Eng.* **2021**, *198*, 117503. [CrossRef]
30. Gabbar, H.A.; Othman, A.M.; Abdussami, M.R. Review of battery management systems (BMS) development and industrial standards. *Technologies* **2021**, *9*, 28. [CrossRef]
31. Li, J.; Yang, J.; Ji, Z.; Su, M.; Li, H.; Wu, Y.; Su, X.; Zhang, Z. Prospective Application, Mechanism, and Deficiency of Lithium Bis (oxalate) Borate as the Electrolyte Additive for Lithium-Batteries. *Adv. Energy Mater.* **2023**, *13*, 2301422. [CrossRef]
32. Haregewoin, A.M.; Wotango, A.S.; Hwang, B.-J. Electrolyte additives for lithium ion battery electrodes: Progress and perspectives. *Energy Environ. Sci.* **2016**, *9*, 1955–1988. [CrossRef]
33. Peng, S.; Li, K.; Liu, P.; Zhang, W.; Wu, W.; Han, X.; Cao, Y.-c.; Wen, J.; Cheng, S. Dynamics of multidimensional signals in lithium-ion battery during thermal runaway under various oven temperatures. *J. Energy Storage* **2025**, *108*, 115071. [CrossRef]

34. Song, Y.; Cui, Y.; Li, B.; Geng, L.; Yan, J.; Zhu, D.; Zhou, P.; Zhou, J.; Yan, Z.; Xue, Q. Revealing the origin of high-thermal-stability of single-crystal Ni-rich cathodes toward higher-safety batteries. *Nano Energy* **2023**, *116*, 108846. [CrossRef]
35. Jia, Y.; Hou, X.; Li, K.; Wang, L.; Zhang, M.; Li, Z.; Xu, X.; Zheng, J. Unraveling the oxygen evolution in layered $LiNiO_2$ with the role of Li/Ni disordering. *Energy Storage Mater.* **2024**, *71*, 103632. [CrossRef]
36. House, R.A.; Marie, J.-J.; Pérez-Osorio, M.A.; Rees, G.J.; Boivin, E.; Bruce, P.G. The role of O_2 in O-redox cathodes for Li-ion batteries. *Nat. Energy* **2021**, *6*, 781–789. [CrossRef]
37. Liu, X.; Yin, L.; Ren, D.; Wang, L.; Ren, Y.; Xu, W.; Lapidus, S.; Wang, H.; He, X.; Chen, Z. In situ observation of thermal-driven degradation and safety concerns of lithiated graphite anode. *Nat. Commun.* **2021**, *12*, 4235. [CrossRef]
38. Shalnev, S.; Krzack, S.; Küster, F.; Friedrich, H.-J.; Meyer, B.; Gräbner, M. An experimental investigation into the conversion of nuclear graphite by oxygen and carbon dioxide: Kinetics and change in pore structure. *Fuel* **2024**, *357*, 129756. [CrossRef]
39. Yu, R.; Zeng, W.; Zhou, L.; Van Tendeloo, G.; Mai, L.; Yao, Z.; Wu, J. Layer-by-layer delithiation during lattice collapse as the origin of planar gliding and microcracking in Ni-rich cathodes. *Cell Rep. Phys. Sci.* **2023**, *4*, 101480. [CrossRef]
40. Rowden, B.; Garcia-Araez, N. A review of gas evolution in lithium ion batteries. *Energy Rep.* **2020**, *6*, 10–18. [CrossRef]

MDPI AG
Grosspeteranlage 5
4052 Basel
Switzerland
Tel.: +41 61 683 77 34

Batteries Editorial Office
E-mail: batteries@mdpi.com
www.mdpi.com/journal/batteries

www.ingramcontent.com/pod-product-compliance
Lightning Source LLC
Chambersburg PA
CBHW040823250926
51850CB00007BA/1248
9783725856930